Lecture Notes in Computer Science 2410

Edited by G. Goos, J. Hartmanis, and J. van Leeuwen

Springer
Berlin
Heidelberg
New York
Barcelona
Hong Kong
London
Milan
Paris
Tokyo

Victor A. Carreño
César A. Muñoz Sofiène Tahar (Eds.)

Theorem Proving in Higher Order Logics

15th International Conference, TPHOLs 2002
Hampton, VA, USA, August 20-23, 2002
Proceedings

Series Editors

Gerhard Goos, Karlsruhe University, Germany
Juris Hartmanis, Cornell University, NY, USA
Jan van Leeuwen, Utrecht University, The Netherlands

Volume Editors

Victor A. Carreño
NASA Langley Research Center
MS 130, Hampton, VA 23681, USA
E-mail: v.a.carreno@larc.nasa.gov

César A. Muñoz
ICASE-Langley Research Center
MS 132C, Hampton, VA 23681, USA
E-mail: munoz@icase.edu

Sofiène Tahar
Concordia University, Electrical and Computer Engineering
1455 de Maisonneuve Blvd. W.
Montréal, Québec H3G 1M8, Canada
E-Mail: tahar@ece.concordia.ca

Cataloging-in-Publication Data applied for

Die Deutsche Bibliothek - CIP-Einheitsaufnahme

Theorem proving in higher order logics : 15th international conference ; proceedings / TPHOLs 2002, Hampton, VA, USA, August 20 - 23, 2002. Victor A. Carreno (ed.). - Berlin ; Heidelberg ; New York ; Barcelona ; Hong Kong ; London ; Milan ; Paris ; Tokyo : Springer, 2002
(Lecture notes in computer science ; Vol. 2410)
ISBN 3-540-44039-9

CR Subject Classification (1998): F.4.1, I.2.3, F.3.1, D.2.4, B.6.3

ISSN 0302-9743
ISBN 3-540-44039-9 Springer-Verlag Berlin Heidelberg New York

Springer-Verlag Berlin Heidelberg New York,
a member of BertelsmannSpringer Science+Business Media GmbH

http://www.springer.de

Typesetting: Camera-ready by author, data conversion by Christian Grosche, Hamburg
Printed on acid-free paper SPIN: 10873706 06/3142 5 4 3 2 1 0

Preface

This volume contains the proceedings of the *15th International Conference on Theorem Proving in Higher Order Logics* (TPHOLs 2002) held on 20–23 August 2002 in Hampton, Virginia, USA. The conference serves as a venue for the presentation of work in theorem proving in higher-order logics, and related areas in deduction, formal specification, software and hardware verification, and other applications.

Each of the 34 papers submitted in the full research category was refereed by at least three reviewers from the program committee or by a reviewer appointed by the program committee. Of these submissions, 20 papers were accepted for presentation at the conference and publication in this volume.

Following a well-established tradition in this conference series, TPHOLs 2002 also offered a venue for the presentation of work in progress. For the work in progress track, short introductory talks were given by researchers, followed by an open poster session for further discussion. Papers accepted for presentation in this track have been published as Conference Proceedings CP NASA-2002-211736.

The organizers would like to thank Ricky Butler and Gérard Huet for gracefully accepting our invitation to give talks at TPHOLs 2002. Ricky Butler was instrumental in the formation of the Formal Methods program at the NASA Langley Research Center and has led the group since its beginnings. The NASA Langley Formal Methods group, under Ricky Butler's guidance, has funded, been involved in, or influenced many formal verification projects in the US over more than two decades. In 1998 Gérard Huet received the prestigious Herbrand Award for his fundamental contributions to term rewriting and theorem proving in higher-order logic, as well as many other key contributions to the field of automated reasoning. He is the originator of the Coq System, under development at INRIA-Rocquencourt. Dr. Huet's current main interest is computational linguistics, however his work continues to influence researchers around the world in a wide spectrum of areas in theoretical computer science, formal methods, and software engineering.

The venue of the TPHOLs conference traditionally changes continent each year in order to maximize the likelihood that researchers from all over the world will attend. Starting in 1993, the proceedings of TPHOLs and its predecessor workshops have been published in the following volumes of the Springer-Verlag *Lecture Notes in Computer Science* series:

1993 (Canada)	780	1998 (Australia)	1479
1994 (Malta)	859	1999 (France)	1690
1995 (USA)	971	2000 (USA)	1869
1996 (Finland)	1125	2001 (UK)	2152
1997 (USA)	1275		

The 2002 conference was organized by a team from NASA Langley Research Center, the ICASE Institute at Langley Research Center, and Concordia University. Financial support came from Intel Corporation. The support of all these organizations is gratefully acknowledged.

August 2002

Víctor A. Carreño
César A. Muñoz

Organization

TPHOLs 2002 is organized by NASA Langley and ICASE in cooperation with Concordia University.

Organizing Committee

Conference Chair: Víctor A. Carreño (NASA Langley)
Program Chair: César A. Muñoz (ICASE, NASA LaRC)
Sofiène Tahar (Concordia University)

Program Committee

Mark Aagaard (Waterloo)
David Basin (Freiburg)
Víctor Carreño (NASA Langley)
Shiu-Kai Chin (Syracuse)
Paul Curzon (Middlesex)
Gilles Dowek (INRIA)
Harald Ganzinger (MPI Saarbrücken)
Ganesh Gopalakrishnan (Utah)
Jim Grundy (Intel)
Elsa Gunter (NJIT)
John Harrison (Intel)
Doug Howe (Carleton)
Bart Jacobs (Nijmegen)
Paul Jackson (Edinburgh)
Sara Kalvala (Warwick)
Michael Kohlhase (CMU & Saarland)
Thomas Kropf (Bosch)
Tom Melham (Glasgow)
J Strother Moore (Texas, Austin)
César Muñoz (ICASE, NASA LaRC)
Sam Owre (SRI)
Christine Paulin-Mohring (INRIA)
Lawrence Paulson (Cambridge)
Frank Pfenning (CMU)
Klaus Schneider (Karlsruhe)
Henny Sipma (Stanford)
Konrad Slind (Utah)
Don Syme (Microsoft)
Sofiène Tahar (Concordia)
Wai Wong (Hong Kong Baptist)

Additional Reviewers

Otmane Ait-Mohamed
Behzad Akbarpour
Nancy Day
Ben Di Vito
Jean-Christophe Filliâtre
Alfons Geser
Hanne Gottliebsen
Mike Kishinevsky
Hans de Nivelle
Andrew Pitts
Harald Rueß
Leon van der Torre
Tomas Uribe

Invited Speakers

Ricky Butler (NASA Langley)
Gérard Huet (INRIA)

Sponsoring Institutions

NASA Langley
ICASE
Concordia University
INTEL

Table of Contents

Invited Talks

Regular Papers

Formal Methods at NASA Langley

Ricky Butler

Assessment Technology Branch
Mail Stop 130, NASA Langley Research Center
Hampton, VA 23681-2199
r.w.butler@larc.nasa.gov

Extended Abstract

In this talk, a short history of NASA Langley's research in formal methods is presented. The talk begins with an examination of the attempted formal verification of the SIFT (Software Implemented Fault Tolerance) operating system in the late 1970s. The primary goal of the SIFT verification project was to verify an operating system for a fault-tolerant, distributed, real-time, avionics computing platform. The SIFT project was deemed a failure because it did not meet its primary objective. However, important results in the field of computer science were obtained from the SIFT project including fault tolerant clock synchronization, Byzantine agreement (interactive consistency), and others.

After the SIFT project, formal methods at NASA Langley went through a period of hibernation and would have probably died except for the providential visit of John Cullyer to NASA Langley in 1988. Soon after this, a Memorandum of Understanding (MOU) between NASA Langley and The Royal Signal and Radar Establishment (RSRE) was initiated. NASA Langley was asked to perform a critical assessment of the VIPER microprocessor, which had been designed at the RSRE research laboratory. With the help of Computational Logic Inc. (CLI) and the Boeing Commercial Airplane Company, the usefulness of the VIPER processor for aerospace application was investigated. Computational Logic wrote a critical review emphasizing the incompleteness of the VIPER mechanical proofs. Although an unprecedented level of analysis had been performed on the VIPER, the CLI team did not consider hand-proofs to be adequate for something like the VIPER. Once again Langley found itself in the midst of a formal methods controversy. At the heart of this controversy was the nature of proof itself. How rigorous does a proof have to be before one can claim that a *mathematical proof* has been accomplished?

Next, NASA Langley began to fund the direct application of existing formal methods to real problems in the aerospace industry. A competitive procurement resulted in three contracts to Computational Logic Inc, SRI International, and Odyssey Research Associates. During this period NASA Langley had its first recognized successes in formal methods: the formal verification of the Rockwell Collins AAMP5 microprocessor and the formal verification of the Draper FTPP scoreboard circuit. Most of the work centered on fault-tolerance and hardware

V.A. Carreño, C. Muñoz, S. Tahar (Eds.): TPHOLs 2002, LNCS 2410, pp. 1–2, 2002.

verification at this time, but some initial work in modeling software was also initiated. The Jet Select subsystem of the Space Shuttle was modeled and later the GPS upgrade and 3-Engine-Out upgrade was formally analyzed. Both theorem proving and model checking approaches were pursued.

Towards the end of the 1990s the Langley team began to look into ways of using formal methods to improve critical flight deck software. After the Pentium bug mobilized the commercial silicon manufacturers to pursue formal verification technology, the Langley team decided to focus its attention on software and systems. Cooperative work with industry to model flight guidance systems, Ada run-time systems, operating systems that support integrated modular avionics, and other safety-critical applications were begun. In FY2000, the Langley formal methods team began to work for the Aviation Safety Program. The Aviation Safety Program is a NASA focused program with a goal of reducing the fatal aviation accident rate by 80% in ten years and 90% in twenty years. Four new contracts were started with a goal of using formal methods to make flight systems safer: (1) Rockwell Collins Advanced Technology Center to develop formal requirements modeling and analysis techniques to detect and remove flight deck mode confusion, (2) Honeywell Engines and Systems with SRI International to adapt the Time Triggered Architecture (TTA) to a Full Authority Digital Engine Control (FADEC), (3) Barron Associates/Goodrich to develop formal verification methods that can serve as a basis for certifying non-adaptive neural nets, and (4) University of Virginia to develop formal specification tools that integrate formal and informal languages enabling the analysis of the combination. In FY2001, the Langley formal methods team was asked to apply formal methods technology to the Air Traffic Management problem. Current work is looking at conflict detection and resolution algorithms (CD&R) needed in future free flight concepts and at merging and self-spacing algorithms in the terminal area. These later algorithms are being developed in support of NASA Langley's Small Aircraft Transportation Systems (SATS) program. The Langley team has had some early successes with the verification of Air Traffic Management concepts and the outlook is quite promising for this area.

Future goals for the Langley formal methods program include (1) facilitating the transfer of successful formal approaches into the product lines of aerospace companies by developing cost-effective tools that integrate well with existing industrial life-cycle processes, (2) working with the FAA to develop new certification processes that take advantage of the capabilities of formal methods, (3) performing fundamental research in new challenging areas such as systems where there are complex interactions between discrete and continuous components, and (4) integrating formal methods technology into existing engineering tools such as Mathworks.

Higher Order Unification 30 Years Later
(Extended Abstract)

Gérard Huet

INRIA-Rocquencourt
BP 105, F-78153 Le Chesnay Cedex
Gerard.Huet@inria.fr

Abstract. The talk will present a survey of higher order unification, covering an outline of its historical development, a summary of its applications to three fields: automated theorem proving, and more generally engineering of proof assistants, programming environments and software engineering, and finally computational linguistics. It concludes by a presentation of open problems, and a few prospective remarks on promising future directions. This presentation assumes as background the survey by Gilles Dowek in the Handbook of automated theorem proving [28].

1 Theory

1.1 Early History

The problem of automating higher order logic was first posed by J. A. Robinson in [100,101]. Peter Andrews formulated a version of the resolution principle for accommodating Church's Simple Theory of Types [16,4], and as early as 1964 Jim Guard's team at Applied Logic Corporation had been independently investigating higher order logic versions of unification [42,41], but the complexity of the problem appeared when higher order unification was shown to be undecidable in 1972, independently by G. Huet [46] and C. Lucchesi [61], a result to be sharpened by W. Goldfarb [39]. Jensen and Pietrzykowky [89,50], at University of Waterloo, gave a complete recursive enumeration of unifiers, but the process was over-generative in an untractable fashion, since computational quagmires happened where at every step of the process all previous solutions were subsumed by the next one. Huet proposed a more tractable enumeration of preunifiers for better conditioned problems, where rigidity of one of the problem terms led to sequential behavior. By delaying ill-conditioned problems in the form of constraints, it became possible to implement a complete version of the resolution principle for Church's Type Theory [49]. An extensive theory of unification in various algebraic settings, building on early work of G. Plotkin [90] was developed by G. Huet in his Thèse d'Etat [47], while P. Andrews and his collaborators at CMU went on implementing constrained resolution in the proof assistant TPS [5].

Thus the original motivation for higher-order unification was its use as a fundamental algorithm in automated theorem-provers. We shall see below that

V.A. Carreño, C. Muñoz, S. Tahar (Eds.): TPHOLs 2002, LNCS 2410, pp. 3–12, 2002.

the design and implementation of proof assistants is still its main application area, not just in the setting of classical logic with versions of the resolution principle, but in more general meta-theoretic frameworks, where it is the work horse of pattern-matching and more generally proof search. For instance, extensions of the term-rewriting paradigm of equational logic, building on higher-order unification, were proposed by Nipkow, Jouannaud and Okada, Breazu-Tannen, Blanqui, and many others [76,10], while C. Benzmüller investigated the addition of equality paramodulation to higher-order resolution [8,9].

1.2 λ Prolog

It was soon realized that resolution was better behaved in the restricted context of Horn sentences, and Colmerauer and Kowalski took advantage of this quality in the framework of first-order logic with the design of PROLOG, which spurred the whole new field of Logic Programming. The generalization of this promising technology to higher-order logic was successfully accomplished by D. Miller with his λ-PROLOG system [65,73]. G. Nadathur further investigated the λ-PROLOG methodology and its efficient implementation [71,72,70,73,74]. In the Mali team in Rennes, Yves Bekkers and colleagues implemented a λ-PROLOG machine with specific dynamic memory management [7,14]. The most impressive application of the λ-PROLOG formalism was accomplished by F. Pfenning in his ELF system, a logic programming environment for Edinburgh's Logical Framework [85,88].

The theory of higher-order unification in more powerful higher order frameworks, encompassing dependent types, polymorphism, or inductive types, was investigated by C. Elliott[31,32], G. Pym [96], G. Dowek [20,26], F. Pfenning [86,87] etc. in a series of publications in the 80's and 90's. We refer the interested reader to the extensive presentation in G. Dowek's survey [28].

1.3 Matching

An important special case of unification (i.e. solving equations in term structures) is when one of the terms is closed. Unification in this special case is pattern matching, a specially important algorithm in symbolic computation. Huet's semi-decision algorithm turned into an algorithm generating a finite set of solutions in the case of second-order terms [47,48,22]. Furthermore, it was possible to combine pattern-matching for linear terms with first-order unification, as Miller demonstrated with his higher-order patterns [66]. However, the general problem in the full higher-order hierarchy remained open for a long time. G. Dowek showed that the third-order problem was decidable [27], while extensions in various other directions led to undecidability [21,24]. Padovani showed that the fourth-order case was decidable [80,81,82]. Wolfram gave an algorithm which always terminate [115], but whose completeness is dubious. More recently, Loader announced the undecidability of higher-order matching in [59,60], but the argument does not carry over if the η rule of extensionality is added to β computation rule of λ calculus. This vexing but important problem is thus still open after 30 years of intense investigation.

2 Applications

The talk will survey applications in three domains.

2.1 Proof Technology

Early experiments with resolution systems as proof assistants were extremely naive, since full automation was aimed at, for mathematical theories which are undecidable or at least badly untractable. It is interesting that to this day research teams keep the pretense of solving deep mathematical problems in a fully automated fashion, despite all evidence to the contrary.

Hopefully better methodologies exist than attempting to solve mathematics with combinatorial blackboxes. Proof Assistants, following the early models of Automath, LCF and NuPRL, have been designed as interactive systems, where the mathematically sophisticated user drives a proof search process with the help of specialized, programmable tactics implementing decision or semi-decision procedures, which operate on formulas kept in user-understandable terms. Two paradigms are of paramount importance in this newer vision of proof search: the use of general meta-level formalisms (the so-called logical frameworks, where generic pattern-matching and unification algorithms are put to use), and the recognition of proof structures as higher-order objects themselves (through suitable instances of the Curry-Howard correspondence). This gave a second life to higher-order unification, at the level of proof search this time (and not just at the level of terms search). This methodology was clearly articulated by G. Dowek in his thesis [20,23] and by by D. Miller and A. Felty [2,35]. Typical of this approach are the proof assistants Isabelle [83], the Edinburgh Logical Framework and its ELF Logic Programming engine [85,88], and the various versions of the Calculus of Constructions or Intuitionistic Type Theory implementations (Coq, Lego, Alf).

2.2 Program Schemas Manipulation and Software Engineering

The second-order matching algorithm presented in the 70's by Huet and Lang [48] was motivated by an application to program schemas recognition, in a software engineering paradigm of program transformations originally proposed by Burstall and Darlington. Indeed, all the Burstall and Darlington rewrite rules were easily encompasssed by this technology, but it seemed difficult to extrapolate this simple methodology to software engineering in the large. Still, grandiose projects of Inferential Programming based on Semantically Based Programming Tools were lavishly financed in the 80's on such principles [55]. What finally emerged when the dust settled was the notion of Higher-order Abstract

Syntax as a fundamental structure for logic-based programming environments [67,45,84,3,51,64].

2.3 Anaphora Resolution and Other Computational Linguistics Applications

Meanwhile, researchers in computational linguistics working at the frontier between syntax and semantics, and notably within Montague Semantics, had been investigating the resolution of anaphora constructions (pronouns, focus, ellipsis, etc.) through the higher order unification algorithm. Such experiments were facilitated by the widespread adoption of PROLOG as a computation paradigm in this community. Typical representatives of this approach were Dale Miller and Gopalan Nadathur at CMU [68], S. Shieber and colleagues in Harvard [62,114], S. Pulman in SRI/Cambridge [95], and C. Gardent and colleagues in Saarbrücken [15,37,38,36].

3 Prospective

The talk will end with some prospective remarks.

References

1. G. Amiot. The undecidability of the second order predicate unification problem. *Archive for mathematical logic*, 30:193–199, 1990.
2. Dale Miller Amy Felty. Specifying theorem provers in a higher-order logic programming language. In *Proceedings of the 9th Int. Conference on Automated Deduction (CADE-9)*, volume 310 of *Lecture Notes in Computer Science*, pages 61–80. Springer-Verlag, Argonne, Illinois, 1988.
3. Penny Anderson. Represnting proof transformations for program optimization. In *Proceedings of the 12th Int. Conference on Automated Deduction (CADE-12)*, volume 814 of *Lecture Notes in Artificial Intelligence*, pages 575–589, Nancy, France, 1994. Springer-Verlag.
4. P.B. Andrews. Resolution in type theory. *The Journal of Symbolic Logic*, 36(3):414–432, 1971.
5. P.B. Andrews, M. Bishop, S. Issar, D. Nesmith, F. Pfenning, and H. Xi. Tps: a theorem proving system for classical type theory. *J. of Automated Reasoning*, 16:321–353, 1996.
6. J. Avenhaus and C.A. Loría-Sáenz. Higher-order conditional rewriting and narrowing. In J.-P. Jouannaud, editor, *International Conference on Constaints in Computational Logic*, volume 845 of *Lecture Notes in Computer Science*, pages 269–284. Springer-Verlag, 1994.
7. Yves Bekkers, Olivier Ridoux, and Lucien Ungaro. Dynamic memory management for sequential logic programming languages. In *IWMM*, pages 82–102, 1992.
8. C. Benzmüller. Extensional higher-order resolution. In *Proceedings of the 15th Int. Conference on Automated Deduction (CADE-15)*, volume 1421 of *Lecture Notes in Artificial Intelligence*, pages 56–71. Springer-Verlag, Lindau, Germany, 1998.

9. C. Benzmüller. Equality and extensionality in automated higher-order theorem proving. Doctoral Dissertation, Universität des Saarlandes, Saarbrücken, 1999.
10. Frédéric Blanqui. Théorie des types et réécriture. Thèse de Doctorat, Université Paris-Sud, 2001.
11. P. Borovanský. Implementation of higher-order unification based on calculus of explicit substitution. In M. Bartošek, J. Staudek, and J. Wiedermann, editors, *SOFSEM: Theory and Practice of Informatics*, volume 1012 in Lecture Notes in Computer Science, pages 363–368. Springer-Verlag, 1995.
12. A. Boudet and E. Contejean. AC-unification of higher-order patterns. In G. Smolka, editor, *Principles and Practice of Constraint Programming*, volume 1330 of *Lecture Notes in Computer Science*, pages 267–281. Springer-Verlag, 1997.
13. D. Briaud. Higher order unification as typed narrowing. Technical Report 96-R-112, Centre de recherche en informatique de Nancy, 1996.
14. Pascal Brisset and Olivier Ridoux. The architecture of an implementation of lambdaprolog: Prolog/mali. Technical Report RR-2392, INRIA, Oct. 1994.
15. M. Kohlhase, C. Gardent, and K. Konrad. Higher-order colored unification: A linguistic application. *J. of Logic, Language and Information*, 9:313–338, 2001.
16. A. Church. A formulation of the simple theory of types. *The Journal of Symbolic Logic*, 5(1):56–68, 1940.
17. H. Comon and Y. Jurski. Higher-order matching and tree automata. In M. Nielsen and W. Thomas, editors, *Conference on Computer Science Logic*, volume 1414 of *Lecture Notes in Computer Science*, pages 157–176. Springer-Verlag, 1997.
18. D.J. Dougherty. Higher-order unification via combinators. *Theoretical Computer Science*, 114:273–298, 1993.
19. D.J. Dougherty and P. Johann. A combinatory logic approach to higher-order E-unification. In D. Kapur, editor, *Conference on Automated Deduction*, volume 607 of *Lecture Notes in Artificial Intelligence*, pages 79–93. Springer-Verlag, 1992.
20. G. Dowek. Démonstration automatique dans le calcul des constructions. Thèse de Doctorat, Université de Paris VII, 1991.
21. G. Dowek. L'indécidabilité du filtrage du troisième ordre dans les calculs avec types dépendants ou constructeurs de types (the undecidability of third order pattern matching in calculi with dependent types or type constructors). *Comptes Rendus à l'Académie des Sciences*, I, 312(12):951–956, 1991. Erratum, ibid. I, 318, 1994, p. 873.
22. G. Dowek. A second-order pattern matching algorithm in the cube of typed λ-calculi. In A. Tarlecki, editor, *Mathematical Foundation of Computer Science*, volume 520 of *Lecture notes in computer science*, pages 151–160. Springer-Verlag, 1991.
23. G. Dowek. A complete proof synthesis method for the cube of type systems. *Journal of Logic and Computation*, 3(3):287–315, 1993.
24. G. Dowek. The undecidability of pattern matching in calculi where primitive recursive functions are representable. *Theoretical Computer Science*, 107:349–356, 1993.
25. G. Dowek. The undecidability of typability in the lambda-pi-calculus. In M. Bezem and J.F. Groote, editors, *Typed Lambda Calculi and Applications*, volume 664 in Lecture Notes in Computer Science, pages 139–145. Springer-Verlag, 1993.
26. G. Dowek. A unification algorithm for second-order linear terms. Manuscript, 1993.
27. G. Dowek. Third order matching is decidable. *Annals of Pure and Applied Logic*, 69:135–155, 1994.

28. G. Dowek. *Higher-Order Unification and Matching*, chapter 16, pages 1009–1062. Elsevier, 2001.
29. G. Dowek, T. Hardin, and C. Kirchner. Higher-order unification via explicit substitutions. *Information and Computation*, 157:183–235, 2000.
30. G. Dowek, T. Hardin, C. Kirchner, and F. Pfenning. Unification via explicit substitutions: the case of higher-order patterns. In M. Maher, editor, *Joint International Conference and Symposium on Logic Programming*, pages 259–273. The MIT Press, 1996.
31. C.M. Elliott. Higher-order unification with dependent function types. In N. Dershowitz, editor, *Internatinal Conference on Rewriting Techniques and Applications*, volume 355 of *Lecture Notes in Computer Science*, pages 121–136. Springer-Verlag, 1989.
32. C.M. Elliott. *Extensions and applications of higher-order unification.* PhD thesis, Carnegie Mellon University, 1990.
33. W.M. Farmer. A unification algorithm for second-order monadic terms. *Annals of Pure and applied Logic*, 39:131–174, 1988.
34. W.M. Farmer. Simple second order languages for which unification is undecidable. *Theoretical Computer Science*, 87:25–41, 1991.
35. Amy Felty. Implementing tactics and tacticals in a higher-order logic programming language. *Journal of Automated Reasoning*, 11:43–81, 1993.
36. C. Gardent. Deaccenting and higher-order unification. *J. of Logic, Language and Information*, 9:313–338, 2000.
37. C. Gardent and M. Kohlhase. Focus and higher-order unification. In *16th International Conference on Computational Linguistics, Copenhagen*, 1996.
38. C. Gardent and M. Kohlhase. Higher-order coloured unification and natural language semantics. In ACL, editor, *34th Annual Meeting of the Association for Computational Linguistics, Santa Cruz*, 1996.
39. W.D. Goldfarb. The undecidability of the second-order unification problem. *Theoretical Computer Science*, 13:225–230, 1981.
40. J. Goubault. Higher-order rigid E-unification. In F. Pfenning, editor, *5th International Conference on Logic Programming and Automated Reasoning*, volume 822 in Lecture Notes in Artificial Intelligence, pages 129–143. Springer-Verlag, 1994.
41. W.E. Gould. A matching procedure for ω-order logic. Scientific report 4, AFCRL 66-781 (contract AF 19 (628)-3250) AD 646 560.
42. J.R. Guard. Automated logic for semi-automated mathematics. Scientific report 1, AFCRL 64-411 (contract AF 19 (628)-3250) AS 602 710, 1964.
43. M. Hagiya. Programming by example and proving by example using higher-order unification. In M.E. Stickel, editor, *Conference on Automated Deduction*, volume 449 in Lecture Notes in Computer Science, pages 588–602. Springer-Verlag, 1990.
44. M. Hagiya. Higher-order unification as a theorem proving procedure. In K. Furukawa, editor, *International Conference on Logic Programming*, pages 270–284. MIT Press, 1991.
45. John Hannan and Dale Miller. Uses of higher-order unification for implementing programs transformers. In R.A. Kowalski an K.A. Bowen, editor, *International Conference and Symposium on Logic Programming*, pages 942–959, 1988.
46. G. Huet. The undecidability of unification in third order logic. *Information and Control*, 22:257–267, 1973.
47. G. Huet. Résolution d'équations dans les langages d'ordre 1,2, ..., ω. Thèse d'État, Université de Paris VII, 1976.
48. G. Huet and B. Lang. Proving and applying program transformations expressed with second order patterns. *Acta Informatica*, 11:31–55, 1978.

49. Gérard Huet. *Constrained resolution: A complete method for higher order logic.* PhD thesis, Case Western University, Aug. 1972.
50. D.C. Jensen and T. Pietrzykowski. Mechanizing ω-order type theory through unification. *Theoretical Computer Science*, 3:123–171, 1976.
51. Frank Pfenning, Joëlle Despeyroux, and Carsten Schürmann. Primitive recursion for higher-order syntax. In R. Hindley, editor, *Proceedings of the 3rd Int. Conference on Typed Lambda Calculus and Applications (TLCA'97)*, volume 1210 of *Lecture Notes in Computer Science*, pages 147–163, Nancy, France, 1997. Springer-Verlag.
52. Patricia Johann and Michael Kohlhase. Unification in an extensional lambda calculus with ordered function sorts and constant overloading. In Alan Bundy, editor, *Conference on Automated Deduction*, volume 814 in Lecture Notes in Artificial Intelligence, pages 620–634. Springer-Verlag, 1994.
53. C. Kirchner and Ch. Ringeissen. Higher-order equational unification via explicit substitutions. In M. Hanus, J. Heering, and K. Meinke, editors, *Algebraic and Logic Programming, International Joint Conference ALP'97-HOA'97*, volume 1298 of *Lecture Notes in Computer Science*, pages 61–75. Springer-Verlag, 1997.
54. K. Kwon and G. Nadathur. An instruction set for higher-order hereditary harrop formulas (extended abstract). In *Proceedings of the Workshop on the lambda Prolog Programming Language, UPenn Technical Report MS-CIS-92-86*, 1992.
55. Peter Lee, Frank Pfenning, John Reynolds, Gene Rollins, and Dana Scott. Research on semantically based program-design environments: The ergo project in 1988. Technical Report CMU-CS-88-118, Computer Science Department, Carnegie Mellon University, 1988.
56. J. Levy. Linear second order unification. In H. Ganzinger, editor, *Rewriting Techniques and Applications*, volume 1103 of *Lecture Notes in Computer Science*, pages 332–346. Springer-Verlag, 1996.
57. Jordi Levy and Margus Veanes. On unification problems in restricted second-order languages. In *Annual Conf. of the European Ass. of Computer Science Logic (CSL98)*, Brno, Czech Republic, 1998.
58. Jordi Levy and Mateu Villaret. Linear second-order unification and context unification with tree-regular constraints. In *Proceedings of the 11th Int. Conf. on Rewriting Techniques and Applications (RTA 2000)*, volume 1833 of *Lecture Notes in Computer Science*, pages 156–171, Norwich, UK, 2000. Springer-Verlag.
59. R. Loader. The undecidability of λ-definability. To appear in Church memorial volume, 1994.
60. R. Loader. Higher order β matching is undecidable. Private communication, 2001.
61. C.L. Lucchesi. The undecidability of the unification problem for third order languages. Technical Report CSRR 2060, Department of applied analysis and computer science, University of Waterloo, 1972.
62. Stuart M. Shieber Mary Dalrymple and Fernando C.N. Pereira. Ellipsis and higher-order unification. *Linguistic and Philosophy*, 14:399–452, 1991.
63. R. Mayr and T. Nipkow. Higher-order rewrite systems and their confluence. *Theoretical Computer Science*, 192:3–29, 1998.
64. Raymond McDowell and Dale Miller. Reasoning with higher-order abstract syntax in a logical framework. *ACM Transactions on Computational Logic (TOCL)*, 3:80–136, 2002.
65. Dale Miller. Abstractions in logic programming. In P. Odifreddi, editor, *Logic and computer science*, pages 329–359. Academic Press, 1991.

66. Dale Miller. Unification under a mixed prefix. *Journal of Symbolic Computation*, 14:321–358, 1992.
67. Dale Miller and Gopalan Nadathur. A logic programming approach to manipulating formulas and programs. In *IEEE Symposium on Logic Programming*, pages 247–255, San Francisco, California, 1986.
68. Dale Miller and Gopalan Nadathur. Some uses of higher-order logic in computational linguistics. In *24th Annual Meeting of the Association for Computational Linguistics*, pages 247–255, 1986.
69. O. Müller and F. Weber. Theory and practice of minimal modular higher-order E-unification. In A. Bundy, editor, *Conference on Automated Deduction*, volume 814 in Lecture Notes in Artificial Intelligence, pages 650–664. Springer-Verlag, 1994.
70. Gopalan Nadathur. An explicit substitution notation in a λprolog implementation. In *First International Workshop on Explicit Substitutions*, 1998.
71. Gopalan Nadathur. *A Higher-Order Logic as the Basis for Logic Programming*. PhD thesis, University of Pennsylvania, Dec. 1986.
72. Gopalan Nadathur and Dale Miller. Higher-order horn clauses. *J. of the Association for Computing Machinery*, 37:777–814, 1990.
73. Gopalan Nadathur and Dale Miller. Higher-order logic programming. In D. M. Gabbay, C. J. Hogger, and J. A. Robinson, editors, *Handbook of logic in artificial intelligence and logic programming*, volume 5, pages 499–590. Clarendon Press, 1998.
74. Gopalan Nadathur and D.J. Mitchell. System description: A compiler and abstract machine based implementation of λprolog. In *Conference on Automated Deduction*, 1999.
75. T. Nipkow. Higher-order critical pairs. In *Logic in Computer Science*, pages 342–349, 1991.
76. T. Nipkow. Higher-order rewrite systems. In *Proceedings of the 6th Int. Conf. on Rewriting Techniques and Applications (RTA-95)*, volume 914 of *Lecture Notes in Computer Science*, pages 256–256, Keiserlautern, Germany, 1995. Springer-Verlag.
77. T. Nipkow and Ch. Prehofer. Higher-order rewriting and equational reasoning. In W. Bibel and P.H. Schmitt, editors, *Automated Deduction - A Basis for Applications*, volume 1, pages 399–430. Kluwer, 1998.
78. Tobias Nipkow and Zhenyu Qian. Modular higher-order e-unification. In *Proceedings of the 4th Int. Conf. on Rewriting Techniques and Applications (RTA)*, volume 488 of *Lecture Notes in Computer Science*, pages 200–214. Springer-Verlag, 1991.
79. Tobias Nipkow and Zhenyu Qian. Reduction and unification in lambda calculi with a general notion of subtype. *Journal of Automated Reasoning*, 12:389–406, 1994.
80. V. Padovani. Fourth-order matching is decidable. Manuscript, 1994.
81. V. Padovani. Decidability of all minimal models. In S. Berardi and M. Coppo, editors, *Types for Proof and Programs 1995*, volume 1158 in Lecture Notes in Computer Science, pages 201–215. Springer-Verlag, 1996.
82. V. Padovani. Filtrage d'ordre supérieur. Thèse de Doctorat, Université de Paris VII, 1996.
83. Larry C. Paulson. Isabelle: The next 700 theorem provers. In P. Odifreddi, editor, *Logic and computer science*, pages 361–385. Academic Press, 1991.
84. F. Pfenning. Partial polymorphic type inference and higher-order unification. In *Conference on Lisp and Functional Programming*, pages 153–163, 1988.

85. F. Pfenning. Logic programming in the LF logical framework. In G. Huet and G. Plotkin, editors, *Logical frameworks*, pages 149–181. Cambridge University Press, 1991.
86. F. Pfenning. Unification and anti-unification in the calculus of constructions. In *Logic in Computer Science*, pages 74–85, 1991.
87. F. Pfenning and I. Cervesato. Linear higher-order pre-unification. In *Logic in Computer Science*, 1997.
88. Frank Pfenning and Carsten Schürmann. System description: Twelf — A meta-logical framework for deductive systems. In *Proceedings of the 16th Int. Conference on Automated Deduction (CADE-16)*, volume 1632 of *Lecture Notes in Artificial Intelligence*. Springer-Verlag, 1999.
89. T. Pietrzykowski. A complete mecanization of second-order type theory. *J. of the Association for Computing Machinery*, 20:333–364, 1973.
90. G. Plotkin. Building-in equational theories. *Machine Intelligence*, 7:73–90, 1972.
91. C. Prehofer. Decidable higher-order unification problems. In A. Bundy, editor, *Conference on Automated Deduction*, volume 814 of *Lecture Notes in Artificial Intelligence*, pages 635–649. Springer-Verlag, 1994.
92. C. Prehofer. Higher-order narrowing. In *Logic in Computer Science (LICS'94*, pages 507–516, 1994.
93. C. Prehofer. Solving higher-order equations: From logic to programming. Doctoral thesis, Technische Universität München. Technical Report TUM-19508, Institut für Informatik, TUM, München, 1995.
94. C. Prehofer. *Solving Higher-Order Equations: From Logic to Programming.* Progress in theoretical coputer science. Birkhäuser, 1998.
95. S. Pulman. Higher-order unification and the interpretation of focus. *Linguistics and Philosophy*, 20:73–115, 1997.
96. D. Pym. Proof, search, and computation in general logic. Doctoral thesis, University of Edinburgh, 1990.
97. Z. Qian. Higher-order equational logic programming. In *Principle of Programming Languages*, pages 254–267, 1994.
98. Z. Qian and K. Wang. Higher-order equational E-unification for arbitrary theories. In K. Apt, editor, *Joint International Conference and Symposium on Logic Programming*, 1992.
99. Z. Qian and K. Wang. Modular AC unification of higher-order patterns. In J.-P. Jouannaud, editor, *International Conference on Constaints in Computational Logic*, volume 845 of *Lecture Notes in Computer Science*, pages 105–120. Springer-Verlag, 1994.
100. J.A. Robinson. New directions in mechanical theorem proving. In A.J.H. Morrell, editor, *International Federation for Information Processing Congress, 1968*, pages 63–67. North Holland, 1969.
101. J.A. Robinson. A note on mechanizing higher order logic. *Machine Intelligence*, 5:123–133, 1970.
102. H. Saïdi. Résolution d'équations dans le système T de Gödel. Mémoire de DEA, Université de Paris VII, 1994.
103. M. Schmidt-Schauß. Unification of stratified second-order terms. Technical Report 12, J.W.Goethe-Universität, Frankfurt, 1994.
104. M. Schmidt-Schauß. Decidability of bounded second order unification. Technical Report 11, J.W.Goethe-Universität, Frankfurt, 1999.
105. M. Schmidt-Schauß and K.U. Schulz. Solvability of context equations with two context variables is decidable. In H. Ganzinger, editor, *Conference on Automated*

Deduction, volume 1632 in Lecture Notes in Artificial Intelligence, pages 67–81, 1999.

106. A. Schubert. Linear interpolation for the higher order matching problem. In M. Bidoit and M. Dauchet, editors, *Theory and Practice of Software Development*, volume 1214 of *Lecture Notes in Computer science*, pages 441–452. Springer-Verlag, 1997.
107. A. Schubert. Second-order unification and type inference for Church-style polymorphism. In *Principle of Programming Languages*, pages 279–288, 1998.
108. W. Snyder. Higher-order E-unification. In M. E. Stickel, editor, *Conference on Automated Deduction*, volume 449 of *Lecture Notes in Artificial Intelligence*, pages 573–587. Springer-Verlag, 1990.
109. W. Snyder and J. Gallier. Higher order unification revisited: Complete sets of tranformations. *Journal of Symbolic Computation*, 8(1 & 2):101–140, 1989. Special issue on unification. Part two.
110. W. Snyder and J. Gallier. Higher-order unification revisited: Complete sets of transformations. *Journal of Symbolic Computation*, 8:101–140, 1989.
111. J. Springintveld. Algorithms for type theory. Doctoral thesis, Utrecht University, 1995.
112. J. Springintveld. Third-order matching in presence of type constructors. In M. Dezani-Ciancagliani and G. Plotkin, editors, *Typed Lambda Calculi and Applications*, volume 902 of *Lecture Notes in Computer Science*, pages 428–442. Springer-Verlag, 1995.
113. J. Springintveld. Third-order matching in the polymorphic lambda calculus. In G. Dowek, J. Heering, K. Meinke, and B. Möller, editors, *Higher-order Algebra, Logic and Term Rewriting*, volume 1074 of *Lecture Notes in Computer Science*, pages 221–237. Springer-Verlag, 1995.
114. Fernando C.N. Pereira Stuart, M. Shieber, and Mary Dalrymple. Interactions of scope and ellipsis. *Linguistics and Philosophy*, 19:527–552, 1996.
115. D.A. Wolfram. The clausal theory of types. Doctoral thesis, University of Cambridge, 1989.
116. M. Zaionc. The regular expression description of unifier set in the typed λ-calculus. *Fundementa Informaticae*, X:309–322, 1987.
117. M. Zaionc. Mechanical procedure for proof construction via closed terms in typed λ-calculus. *Journal of Automated Reasoning*, 4:173–190, 1988.

Combining Higher Order Abstract Syntax with Tactical Theorem Proving and (Co)Induction*

Simon J. Ambler, Roy L. Crole, and Alberto Momigliano

Department of Mathematics and Computer Science
University of Leicester, Leicester, LE1 7RH, U.K.
{S.Ambler,R.Crole,A.Momigliano}@mcs.le.ac.uk

Abstract. Combining Higher Order Abstract Syntax (HOAS) and induction is well known to be problematic. We have implemented a tool called Hybrid, within Isabelle HOL, which does allow object logics to be represented using HOAS, and reasoned about using tactical theorem proving in general and principles of (co)induction in particular. In this paper we describe Hybrid, and illustrate its use with case studies. We also provide some theoretical adequacy results which underpin our practical work.

1 Introduction

Many people are concerned with the development of computing systems which can be used to reason about and prove properties of programming languages. However, developing such systems is not easy. Difficulties abound in both practical implementation and underpinning theory. Our paper makes both a theoretical and practical contribution to this research area. More precisely, this paper concerns *how to reason about* object level logics with syntax involving variable binding—note that a programming language can be presented as an example of such an object logic. Our contribution is the provision of a mechanized tool, Hybrid, which has been coded within Isabelle HOL, and

- provides a form of *logical framework* within which the syntax of an object level logic can be adequately represented by *higher order abstract syntax* (HOAS);
- is consistent with *tactical theorem proving* in general, and principles of *induction and coinduction* in particular; and
- is *definitional* which guarantees consistency *within a classical type theory*.

We proceed as follows. In the introduction we review the idea of simple logical frameworks and HOAS, and the problems in combining HOAS and induction. In Section 2 we introduce our tool Hybrid. In Section 3 we provide some key technical definitions in Hybrid. In Section 4 we explain how Hybrid is used to represent and reason about object logics, by giving some case studies. In Section 5

* This work was supported by EPSRC grant number GR/M98555.

V.A. Carreño, C. Muñoz, S. Tahar (Eds.): TPHOLs 2002, LNCS 2410, pp. 13–30, 2002.

we give a mathematical description of Hybrid together with some underpinning theory. In Section 6 we review the articles related to our work. In Section 7 we comment on future plans.

Hybrid provides a form of logical framework. **Here we briefly recall some fundamental technical details of a basic logical framework, by using the vehicle of quantified propositional logic (*QPL*) as an object level logic—the notation will be used in Sections 4 and 5.** While this is a very small logic, it comes with a single binding construct which exemplifies the problems we are tackling. We let $V_1, V_2, \ldots$ be a countable set of (object level) variables. The *QPL* **formulae** are given by $Q ::= V_i \mid Q \supset Q \mid \forall V_i.\, Q$. Recall how to represent *QPL* in a simple logical framework—the framework here provides a form of HOAS, following [25]. A framework theory is specified by a signature of ground types which generate function types, and constants. The objects of the theory are given by $e ::= name \ \mid v_i \mid e\ e \mid \lambda\, v_i.\, e$ where *name* ranges over constants, and i ranges over $\mathbb{N}$ giving a countable set of (meta) variables v_i. From this, a standard type assignment system can be formulated, leading to a notion of canonical form. We refer the reader to [25] for the full definition. To represent *QPL* we take a ground type oo of *formulae*. We also give the constants of the theory, each of which corresponds to a *QPL* constructor. For *QPL* we specify $\mathsf{Imp} :: oo \Rightarrow oo \Rightarrow oo$ and $\mathsf{All} :: (oo \Rightarrow oo) \Rightarrow oo$. One can define a translation function $\ulcorner - \urcorner$ by the clauses

$$\ulcorner V_i \urcorner \stackrel{\text{def}}{=} v_i \qquad \ulcorner Q_1 \supset Q_2 \urcorner \stackrel{\text{def}}{=} \mathsf{Imp}\ \ulcorner Q_1 \urcorner \ulcorner Q_2 \urcorner \qquad \ulcorner \forall V_i.\, Q \urcorner \stackrel{\text{def}}{=} \mathsf{All}\ \ (\lambda\, v_i.\, \ulcorner Q \urcorner)$$

One can then show that this translation gives a "sensible" representation of *QPL* in the framework, meaning that the function $\ulcorner - \urcorner$ provides a bijection between *QPL* and canonical objects, and further $\ulcorner - \urcorner$ is compositional on substitution.

Although there are well known benefits in working with such higher order abstract syntax, there are also difficulties. In general, it is not immediately clear how to obtain a principle of induction over expressions or how to define functions on them by primitive recursion, although such principles do exist [27]. Worse still, apparently promising approaches can lead to inconsistencies [16]. The axiom of choice leads to loss of consistency, and exotic terms may lead to loss of adequacy. In our example, one would like to view the constants above as the constructors of a datatype $oo ::= var \mid \mathsf{Imp}\ (oo * oo) \mid \mathsf{All}\ (oo \Rightarrow oo)$ so that an induction principle is "immediate". Such datatype declarations would be perfectly legal in functional programming languages. In a theorem prover such as Isabelle HOL the constructors of a datatype are required to be injective [13]. However, the function $\mathsf{All} :: (oo \Rightarrow oo) \Rightarrow oo$ cannot be injective for cardinality reasons (Cantor's proof can be formalized within HOL) as Isabelle HOL provides a classical set theory. Moreover, the function space $oo \Rightarrow oo$ yields Isabelle HOL definable functions which do not "correspond" to terms of *QPL*. Such functions give rise to terms in oo which are unwanted, so-called *exotic* terms such as

$$\mathsf{All}\ (\lambda\, x.\, \mathsf{if}\ x = u\ \mathsf{then}\ u\ \mathsf{else}\ \mathsf{All}\ (\lambda\, z.\, z))$$

We show that it *is possible* to define a logical framework in which All is *injective on a subset* of $oo \Rightarrow oo$. The subset is sufficiently large to give an adequate representation of syntax—see Section 4.1.

2 Introducing Hybrid

Within Isabelle HOL, our goal is to **define a datatype for λ-calculus with constants over which we can deploy (co)induction principles, while representing variable binding through Isabelle's HOAS. We do this in a tool called Hybrid, which we introduce in this section.** Our starting point is the work [1] of Andrew Gordon, which we briefly review. It is well known (though rarely proved—see Section 5) that λ-calculus expressions are in bijection with (a subset of) de Bruijn expressions. Gordon defines a de Bruijn notation in which expressions have *named free variables* given by *strings*. He can write $T = \mathsf{dLAMBDA}\,\mathtt{v}\,t$ (where v is a string) which corresponds to an abstraction in which v is bound in t. The function dLAMBDA has a *definition* which converts T to the corresponding de Bruijn term which has an outer abstraction, and a subterm which is t in de Bruijn form, in which (free) occurrences of v are converted to bound de Bruijn indices. For example,

$$\mathsf{dLAMBDA}\,\mathtt{v}\,(\mathsf{dAPP}\,(\mathsf{dVAR}\,\mathtt{v})\,(\mathsf{dVAR}\,\mathtt{u})) = \mathsf{dABS}\,(\mathsf{dAPP}\,(\mathsf{dBND}\,0)\,(\mathsf{dVAR}\,\mathtt{u}))$$

Gordon demonstrates the utility of this approach. It provides a good mechanism through which one may work with named bound variables, but it does not exploit the built in HOAS which Isabelle HOL itself uses to represent syntax. The novelty of our approach is that we *do* exploit the HOAS at the meta (machine) level.

We introduce Hybrid by example. First, some basics. Of central importance is a Isabelle HOL datatype of de Bruijn expressions, where *bnd* and *var* are the natural numbers, and *con* provides names for constants

$$\boxed{expr ::= \mathsf{CON}\ con \mid \mathsf{VAR}\ var \mid \mathsf{BND}\ bnd \mid expr\ \$\$\ expr \mid \mathsf{ABS}\ expr}$$

Let $T_O = \Lambda V_1.\,\Lambda V_2.\,V_1\;V_3$ be a genuine, honest to goodness (object level) syntax[1] tree. Gordon would represent this by

$$T_G = \mathsf{dLAMBDA}\,\mathtt{v1}\,(\mathsf{dLAMBDA}\,\mathtt{v2}\,(\mathsf{dAPP}\,(\mathsf{dVAR}\,\mathtt{v1})\,(\mathsf{dVAR}\,\mathtt{v3})))$$

which equals

$$\mathsf{dABS}\,(\mathsf{dABS}\,(\mathsf{dAPP}\,(\mathsf{dBND}\,1)\,(\mathsf{dVAR}\,\mathtt{v3})))$$

Hybrid provides a binding mechanism with similarities to dLAMBDA. Gordon's T would be written as $\mathsf{LAM}\,v.\,t$ in Hybrid. This is simply a *definition for* a de Bruijn term. A *crucial difference* in our approach is that *bound variables in the object logic* are *bound variables in Isabelle HOL*. Thus the v in $\mathsf{LAM}\,v.\,t$

[1] We use a capital Λ and capital V to avoid confusion with meta variables v and meta abstraction λ.

is a metavariable (and not a string as in Gordon's approach). In Hybrid we also choose to denote object level free variables by terms of the form $\mathsf{VAR}\ i$; however, this has essentially no impact on the technical details—the important thing is the countability of free variables. In Hybrid the T_O above is rendered as $T_H = \mathsf{LAM}\, v_1.\,(\mathsf{LAM}\, v_2.\,(v_1\ \$\$\ \mathsf{VAR}\ 3))$. The LAM is an Isabelle HOL binder, and this expression is by *definition*

$$\mathsf{lambda}\ (\lambda\, v_1.\,(\mathsf{lambda}\ (\lambda\, v_2.\,(v_1\ \$\$\ \mathsf{VAR}\ 3))))$$

where λv_i is meta abstraction and one can see that the object level term is rendered in the usual HOAS format, where $\mathsf{lambda} :: (expr \Rightarrow expr) \Rightarrow expr$ is a defined function. Then Hybrid will reduce T_H to the de Bruijn term

$$\mathsf{ABS}\ (\mathsf{ABS}\ (\mathsf{BND}\ 1\ \$\$\ \mathsf{VAR}\ 3))$$

as in Gordon's approach. The key to this is, of course, the definition of lambda, which relies crucially on *higher order pattern matching.* We return to its definition in Section 3. In summary, Hybrid provides a form of HOAS where object level

- free variables correspond to Hybrid expressions of the form $\mathsf{VAR}\ i$;
- bound variables correspond to (bound) meta variables;
- abstractions $\Lambda V.\, E$ correspond to expressions $\mathsf{LAM}\, v.\, e = \mathsf{lambda}\ (\lambda\, v.\, e)$;
- applications $E_1\ E_2$ correspond to expressions $e_1\ \$\$\ e_2$.

3 Definition of Hybrid in Isabelle HOL

Hybrid consists of a small number of Isabelle HOL theories. One of these provides the dataype of de Bruijn expressions given in Section 2. The theories also contain definitions of various key functions and inductive sets. **In this section we outline the definitions, and give some examples, where e ranges over Isabelle HOL expressions. We show how Hybrid provides a form of HOAS, and how this can be married with induction.**

We are going to use a pretty-printed version of Isabelle HOL concrete syntax; a rule

$$\frac{H_1 \ldots H_n}{C}$$

will be represented as $[\![\ H_1; \ldots; H_n\ ]\!] \Longrightarrow C$. An Isabelle HOL type declaration has the form $s\ :: [\,t_1, \ldots t_n\,] \Rightarrow t$. The Isabelle metauniversal quantifier is $\bigwedge$, while Isabelle HOL connectives are represented via the usual logical notation. Free variables are implicitly universally quantified. Datatypes will be introduced using BNF grammars. The sign $\equiv$ (Isabelle metaequality) will be used for *equality by definition.*

Note that all of the infrastructure of Hybrid, which we now give, is specified *definitionally.* We do *not* postulate axioms, as in some approaches reviewed in Section 6, which require validation.

$\boxed{\mathsf{level} :: [\,bnd, expr\,] \Rightarrow bool}$ Recall that BND i corresponds to a bound variable in the λ-calculus, and VAR i to a free variable; we refer to **bound** and **free indices** respectively. We call a bound index i **dangling** if i or less Abs labels occur between the index i and the root of the expression tree. e is said to be at **level** $l \geq 1$, if enclosing e inside l Abs nodes ensures that the resulting expression has no dangling indices. These ideas are standard, as is the implementation of level.

$\boxed{\mathsf{proper} :: expr \Rightarrow bool}$ One has proper e == level 0 e. A proper expression is one that has no dangling indices and corresponds to a λ-calculus expression.

$\boxed{\mathsf{insts} :: bnd \Rightarrow (expr)list \Rightarrow expr \Rightarrow expr}$ We explain this function by example. Suppose that j is a bound index occurring in e, and $v_0, \ldots, v_m$ a list of metavariables of type $expr$. Let BND j be enclosed by a ABS nodes. If $a \leq j$ (so that j dangles) then insts replaces BND j by v_{j-a}. If j does not dangle, then insts leaves BND j alone. For example, noting $5 - 2 = 3$,

$$\mathsf{insts}\ 0\ v_0, \ldots, v_m\ \mathsf{ABS}\ (\mathsf{ABS}\ (\mathsf{BND}\ 0\ \$\$\ \mathsf{BND}\ 5)) = \mathsf{ABS}\ (\mathsf{ABS}\ (\mathsf{BND}\ 0\ \$\$\ v_3))$$

$\boxed{\mathsf{abst} :: [\,bnd, expr \Rightarrow expr\,] \Rightarrow bool}$ This predicate is defined by induction as a subset of $bnd * (expr \Rightarrow expr)$. The inductive definition is

$$\begin{aligned}
&\Longrightarrow \mathsf{abst}\ i\ (\lambda\, v.\, v)\\
&\Longrightarrow \mathsf{abst}\ i\ (\lambda\, v.\, \mathsf{VAR}\ n)\\
j < i &\Longrightarrow \mathsf{abst}\ i\ (\lambda\, v.\, \mathsf{BND}\ j)\\
[\![\, \mathsf{abst}\ i\ f; \mathsf{abst}\ i\ g\,]\!] &\Longrightarrow \mathsf{abst}\ i\ (\lambda\, v.\, f\, v\ \$\$\ g\, v)\\
\mathsf{abst}\ (\mathsf{Suc}\ i)\ f &\Longrightarrow \mathsf{abst}\ i\ (\lambda\, v.\, \mathsf{ABS}\ (f\, v))
\end{aligned}$$

This definition is best explained in terms of the next function.

$\boxed{\mathsf{abstr} :: [\,expr \Rightarrow expr\,] \Rightarrow bool}$ We set abstr e == abst 0 e. This function determines when an expression e of type $expr \Rightarrow expr$ is an **abstraction**. This is a key idea, and the notion of an abstraction is central to the formulation of induction principles. We illustrate the notion by example. Suppose that ABS e is proper; for example let e = ABS (BND 0 $$ BND 1). Then e is of level 1, and in particular there may be some bound indices which now dangle; for example BND 1 in ABS (BND 0 $$ BND 1). An abstraction is produced by replacing each occurrence of a dangling index with a metavariable (which can be automated with insts) and then abstracting the meta variable. Our example yields the abstraction $\lambda\, v.$ ABS (BND 0 $$ v).

$\boxed{\mathsf{lbnd} :: [\,bnd, expr \Rightarrow expr, expr\,] \Rightarrow bool}$ This predicate is defined as an inductive subset of $S \stackrel{\text{def}}{=} bnd * (expr \Rightarrow expr) * expr$. The inductive definition is

$$\begin{aligned}
&\Longrightarrow \mathsf{lbnd}\ i\ (\lambda\, v.\, v)\ (\mathsf{BND}\ i)\\
&\Longrightarrow \mathsf{lbnd}\ i\ (\lambda\, v.\, \mathsf{VAR}\ n)\ (\mathsf{VAR}\ n)\\
&\Longrightarrow \mathsf{lbnd}\ i\ (\lambda\, v.\, \mathsf{BND}\ j)\ (\mathsf{BND}\ j)\\
[\![\, \mathsf{lbnd}\ i\ f\ s; \mathsf{lbnd}\ i\ g\ t\,]\!] &\Longrightarrow \mathsf{lbnd}\ i\ (\lambda\, v.\, f\, v\ \$\$\ g\, v)\ (s\ \$\$\ t)\\
\mathsf{lbnd}\ (\mathsf{Suc}\ i)\ f\ s &\Longrightarrow \mathsf{lbnd}\ i\ (\lambda\, v.\, \mathsf{ABS}\ (f\, v))\ (\mathsf{ABS}\ s)
\end{aligned}$$

There is a default case (omitted above) which is called when the second argument does not match any of the given patterns. It is a theorem that this defines a function. The proof is by induction on the *rank* of the function where $\mathsf{rank}\, f = = \mathsf{size}\,(f\ (\mathsf{VAR}\ 0))$.

$\boxed{\mathsf{lbind} :: [\,bnd, expr \Rightarrow expr\,] \Rightarrow expr}$ Set $\mathsf{lbind}\ i\ e == \epsilon\, s.\, \mathsf{lbnd}\ i\ e\ s$ where ϵ is the description operator. Consider the abstraction $\lambda\, v.\, \mathsf{ABS}\ (\mathsf{BND}\ 0\ \$\$\ v)$. The arguments to lbind consist of a bound index, and an abstraction. The intuitive action of this function is that it replaces each bound occurrence of a binding variable in the body of an abstraction, with a bound index, so that a level 1 expression results. This is the reverse of the procedure defined in the paragraph concerning abstractions, where dangling indices were instantiated to metavariables using insts. In practice, lbind will be called on 0 at the top level. Thus one has

$$\mathsf{lbind}\ 0\ (\lambda\, v.\, \mathsf{ABS}\ (\mathsf{BND}\ 0\ \$\$\ v)) = \ldots = \mathsf{ABS}\ (\mathsf{BND}\ 0\ \$\$\ \mathsf{BND}\ 1)$$

$\boxed{\mathsf{lambda} :: (expr \Rightarrow expr) \Rightarrow expr}$ Set $\mathsf{lambda}\ e == \mathsf{ABS}\ (\mathsf{lbind}\ 0\ e)$. Its purpose is to transform an abstraction into the "corresponding" proper de Bruijn expression. Our running example yields

$$\mathsf{lambda}\ (\lambda\, v.\, \mathsf{ABS}\ (\mathsf{BND}\ 0\ \$\$\ v)) = \mathsf{ABS}\ (\mathsf{ABS}\ (\mathsf{BND}\ 0\ \$\$\ \mathsf{BND}\ 1))$$

It is easy to perform induction over a datatype of de Bruijn terms. However, we wish to be able to perform induction over the Hybrid expressions which we have just given. In order to do this, we want to view the functions CON, VAR, $$, and lambda as datatype constructors, that is, they should be injective, with disjoint images. In fact, we identify subsets of *expr* and *expr* ⇒ *expr* for which these properties hold. The subset of *expr* consists of those expressions which are *proper*. The subset of *expr* ⇒ *expr* consists of all those e for which $\mathsf{LAM}\, v.\, e\, v$ is proper. In fact, this means e is an abstraction, which is intuitive but requires proof—it is a Hybrid theorem. We can then prove that

$$[\![\ \mathsf{abstr}\ e;\ \mathsf{abstr}\ f\ ]\!] \Longrightarrow (\mathsf{Lam}\, x.\ e\ x = \mathsf{Lam}\, y.\ f\ y) = (e = f) \qquad \mathit{INJ}$$

which says that lambda is injective on the set of abstractions. This is crucial for the proof of an induction principle for Hybrid, which is omitted for reasons of space, but will appear in a journal version of this paper.

4 Hybrid as a Logical Framework

Recall that in Section 2 we showed that Hybrid supports HOAS. **In this section we show how Hybrid can be used as a logical framework to represent object logics, and further how we can perform tactical theorem proving.**

The system provides:

- A number of automatic tactics: for example proper_tac (resp. abstr_tac) will recognize whether a given term is indeed proper (resp. an abstraction).

– A suite of theorems: for example, the injectivity and distinctness properties of Hybrid constants, and induction principles over *expr* and $expr \Rightarrow expr$, as discussed in Section 3.

Note that the adequacy of our representations will be proved in a forthcoming paper.

4.1 Quantified Propositional Logic

We begin with an encoding of the quantified propositional logic introduced in Section 1. While the fragment presented there is functionally complete, we choose to work with the following syntax

$$Q ::= V_i \mid \neg Q \mid Q \wedge Q' \mid Q \vee Q' \mid Q \supset Q' \mid \forall V_i.\, Q \mid \exists V_i.\, Q$$

This will allow us to demonstrate the representation of object level syntax in detail, and show some properties of an algorithm to produce negation normal forms.

So far we have written *expr* for the type of Hybrid expressions. This was to simplify the exposition, and does not correspond directly to our code. There one sees that Hybrid actually provides a type *con expr* of de Bruijn expressions, where *con* is a type of names of constants. Typically, such names are for object logic constructors. Thus we can define a different type *con* for each object logic we are dealing with. In the case of *QPL*, we declare

$$con ::= cNOT \mid cIMP \mid cAND \mid cOR \mid cALL \mid cEX$$

followed by a Isabelle HOL type whose elements represent the object level formulae, namely $oo == con\ expr$. Readers should pause to recall the logical framework (and HOAS) of Section 1.

Now we show how to represent the object level formulae, as Hybrid expressions of type *oo*. The table below shows analogous constructs in $\mathcal{LF}$ and *Hybrid*.

	Constants	**Application**	**Abstraction**
$\mathcal{LF}$	name	$e_1\ e_2$	$\lambda\, v_i.\, e$
Hybrid	CON $cNAME$	e_1 \$\$ e_2	LAM $v_i.\, e$

In the next table, we show the representation of the object level formulae $Q \supset Q'$ and $\forall V_i.\, Q$ in $\mathcal{LF}$ and Hybrid, where $\ulcorner - \urcorner$ is a translation function

$\mathcal{LF}$	Imp $\ulcorner Q \urcorner \ulcorner Q' \urcorner$	All $(\lambda\, v.\, \ulcorner Q \urcorner)$
Hybrid	CON $cIMP$ \$\$ $\ulcorner Q \urcorner$ \$\$ $\ulcorner Q' \urcorner$	CON $cALL$ \$\$ LAM $v.\, \ulcorner Q \urcorner$
Abbrevs	$\ulcorner Q \urcorner$ Imp $\ulcorner Q' \urcorner$	All $v.\, \ulcorner Q \urcorner$

The bottom row introduces some Isabelle HOL binders as abbreviations for the middle row, where the Isabelle HOL definitions are

$$q \ \mathsf{Imp}\ q' == \mathsf{CON}\ cIMP\ \$\$\ q\ \$\$\ q' \text{ where } \mathsf{Imp} :: [\,oo, oo\,] \Rightarrow oo$$

and

$$\mathsf{All}\, v.\, q\, v == \mathsf{CON}\ cALL\ \$\$\ \mathsf{LAM}\, v.\, q\, v \text{ where } \mathsf{All} :: (oo \Rightarrow oo) \Rightarrow oo$$

The code for the representation of the remainder of QPL is similar:

$$\begin{array}{rclrcl}
\mathsf{And} & :: & [\, oo, oo\,] \Rightarrow oo & \mathsf{Or} & :: & [\, oo, oo\,] \Rightarrow oo \\
q\ \mathsf{And}\ q' & == & \mathsf{CON}\ cAND\ \$\$\ q\ \$\$\ q' & q\ \mathsf{Or}\ q' & == & \mathsf{CON}\ cOR\ \$\$\ q\ \$\$\ q' \\
\mathsf{Not} & :: & oo \Rightarrow oo & \mathsf{Ex} & :: & (oo \Rightarrow oo) \Rightarrow oo \\
\mathsf{Not}\ q & == & \mathsf{CON}\ cNOT\ \$\$\ q & \mathsf{Ex}\, v.\, q\ v & == & \mathsf{CON}\ cEX\ \$\$\ \mathsf{LAM}\, v.\, q\ v
\end{array}$$

The QPL formula $\forall V_1.\, \forall V_2.\, V_1 \supset V_2$ is represented by $\mathsf{All}\, v_1.\, \mathsf{All}\, v_2.\ v_1\ \mathsf{Imp}\ v_2$, although the "real" underlying form is

$$\mathsf{CON}\ cALL\ \$\$\ (\mathsf{LAM}\, v_1.\, \mathsf{CON}\ cALL\ \$\$\ \mathsf{LAM}\, v_2.\, (\mathsf{CON}\ cIMP\ \$\$\ v_1\ \$\$\ v_2)$$

These declarations almost induce a data-type, in the sense the above defined constants enjoy certain freeness properties, much as they would if they were datatype constructors. We can prove that they define *distinct* values; for example $\mathsf{All}\, v.\, q\ v \neq \mathsf{Ex}\, v.\, q\ v$. This is achieved by straightforward simplification of their definitions to the underlying representation. Injectivity of higher-order constructors (recall the end of Section 1) holds conditionally on their bodies being abstractions. In particular, recall from Section 3 the result INJ that the LAM binder is injective on the set of abstractions. Simplification will yield $[\![\ \mathsf{abstr}\ e; \mathsf{abstr}\ f; \mathsf{All}\, v.\, e\ \ v\ =\ \mathsf{All}\, v.\, f\ \ v\ ]\!] \Longrightarrow e = f$. Since the type oo of legal formulae is merely an abbreviation for Hybrid expressions, we need to introduce a "well-formedness" predicate, such that $\mathsf{isForm}\ \ulcorner Q \urcorner$ holds iff Q is a legal object level formula.[2] The inductive definition of isForm in Isabelle HOL is immediate as far as the propositional part of QPL is concerned, for example $[\![\ \mathsf{isForm}\ p; \mathsf{isForm}\ q\ ]\!] \Longrightarrow \mathsf{isForm}\ (\ p\ \mathsf{Imp}\ q)$. For the quantified part, we first remark that in a framework such as $\mathcal{LF}$ one would write

$$[\![\ \forall y.\, \mathsf{isForm}\ y \to \mathsf{isForm}\ (p\ y)\ ]\!] \Longrightarrow \mathsf{isForm}\ (\mathsf{All}\, v.\, p\ v)$$

This is not possible in Isabelle HOL, since the above clause, if taken as primitive, would induce a (set-theoretic) non-monotone operator, and cannot be part of an introduction rule in an inductive definition. Therefore, we instead descend into the scope of the quantifier replacing it with a fresh free variable and add the corresponding base case:

$$\begin{array}{r}
\mathsf{isForm}\ (\mathsf{VAR}\ i) \\
[\![\ \mathsf{abstr}\ p; \forall i.\, \mathsf{isForm}\ (p\ (\mathsf{VAR}\ i))\ ]\!] \Longrightarrow \mathsf{isForm}\ (\mathsf{All}\, v.\, p\ v)
\end{array}$$

We can now proceed to an encoding of an algorithm for negation normal form as an inductive relation, where we skip some of the propositional clauses,

[2] Please see remarks in Section 7 concerning internalizing such predicates as types.

and Φ abbreviates $\mathsf{abstr}\ p; \mathsf{abstr}\ q$;

$$
\begin{array}{rl}
& \mathsf{nnf}\ (\mathsf{VAR}\ i)\ (\mathsf{VAR}\ i) \\
& \mathsf{nnf}\ (\mathsf{Not}\ (\mathsf{VAR}\ i))\ (\mathsf{Not}\ (\mathsf{VAR}\ i)) \\
\mathsf{nnf}\ b\ d \Longrightarrow & \mathsf{nnf}\ (\mathsf{Not}\ (\mathsf{Not}\ b))\ d \\
[\![\ \mathsf{nnf}\ (\mathsf{Not}\ p)\ d; \mathsf{nnf}\ (\mathsf{Not}\ q) e\ ]\!] \Longrightarrow & \mathsf{nnf}\ (\mathsf{Not}\ (\ p\ \mathsf{And}\ q))\ (\ d\ \mathsf{Or}\ e) \\
& \ldots \\
[\![\ \Phi;\ \forall i.\, \mathsf{nnf}\ (\mathsf{Not}\ (p\ (\mathsf{VAR}\ i)))(q\ (\mathsf{VAR}\ i))\ ]\!] \Longrightarrow & \mathsf{nnf}\ (\mathsf{Not}\ (\mathsf{Ex}\, v.\, p\ v))\ (\mathsf{All}\, v.\, q\ v) \\
[\![\ \Phi;\ \forall i.\, \mathsf{nnf}\ (\mathsf{Not}\ (p\ (\mathsf{VAR}\ i)))(q\ (\mathsf{VAR}\ i))\ ]\!] \Longrightarrow & \mathsf{nnf}\ (\mathsf{Not}\ (\mathsf{All}\, v.\, p\ v))\ (\mathsf{Ex}\, v.\, q\ v) \\
[\![\ \Phi;\ \forall i.\, \mathsf{nnf}\ (\mathsf{Not}\ (p\ (\mathsf{VAR}\ i)))(q\ (\mathsf{VAR}\ i))\ ]\!] \Longrightarrow & \mathsf{nnf}\ (\mathsf{All}\, v.\, p\ v)\ (\mathsf{All}\, v.\, q\ v) \\
[\![\ \Phi;\ \forall i.\, \mathsf{nnf}\ (\mathsf{Not}\ (p\ (\mathsf{VAR}\ i)))(q\ (\mathsf{VAR}\ i))\ ]\!] \Longrightarrow & \mathsf{nnf}\ (\mathsf{Ex}\, v.\, p\ v)\ (\mathsf{Ex}\, v.\, q\ v)
\end{array}
$$

Note how, in the binding cases, we explicitly state in Φ which second-order terms are abstractions; this allows us to exploit the injectivity of abstractions to derive the appropriate elimination rules.

It is possible to show in a fully automatic way that the algorithm yields negation normal forms (whose definition is omitted), that is $\mathsf{nnf}\ q\ q' \Longrightarrow \mathsf{isNnf}\ q'$. Moreover it is a functional relation: $\mathsf{nnf}\ q\ q_1 \Longrightarrow \forall q_2.\, \mathsf{nnf}\ q\ q_2 \rightarrow q_1 = q_2$. The latter proof exploits a theorem regarding extensionality of abstractions, namely:

$$[\![\ \mathsf{abstr}\ e; \mathsf{abstr}\ f; \forall i.\, e\ (\mathsf{VAR}\ i) = f\ (\mathsf{VAR}\ i)\ ]\!] \Longrightarrow e = f$$

Note that the above is taken as an *axiom* in the *Theory of Contexts* [17].

4.2 Operational Semantics in the Lazy Lambda Calculus

The object logic in this section is yet another λ-calculus, Abramsky's lazy one [2]. We describe some properties of its operational semantics. In particular, we give HOAS encodings of some notions such as divergence and simulation which are naturally rendered *coinductively*—this can only be approximated in other approaches, as we discuss in Section 6.

To represent the lazy λ-calculus, the type *con* will contain the names *cAPP* and *cABS*, used to represent object level application and abstraction. We then define the constants below from these names, where $lexp == con\ expr$.

$$
\begin{array}{rlrl}
@ & :: [\, lexp, lexp\,] \Rightarrow lexp & \mathsf{Fun}\,.\ & :: (lexp \Rightarrow lexp) \Rightarrow lexp \\
p\ @\ q & == \mathsf{CON}\ cAPP\ \$\$\ p\ \$\$\ q & \mathsf{Fun}\, x.\, f\ x & == \mathsf{CON}\ cABS\ \$\$\ \mathsf{LAM}\, x.\, f\ x
\end{array}
$$

The definition of the well-formedness predicate isExp is analogous to the one in Section 4 and is omitted.

The benefits of obtaining object-level substitution via metalevel β-conversion are exemplified in the encoding of call-by-name evaluation (on closed terms) via the inductive definition of $\gg\ :: [\, lexp, lexp\,] \Rightarrow bool$.

$$
\begin{array}{rl}
[\![\ \mathsf{abstr}\ e;\ \forall i.\, \mathsf{isExp}\ (e\ (\mathsf{VAR}\ i))\ ]\!] \Longrightarrow & \mathsf{Fun}\, x.\, e\ x \gg \mathsf{Fun}\, x.\, e\ x \\
[\![\ e1 \gg \mathsf{Fun}\, x.\, e\ x;\ \mathsf{abstr}\ e;\ \mathsf{isExp}\ e2;\ \ (e\ e2) \gg v\ ]\!] \Longrightarrow & (e1\ @\ e2) \gg v
\end{array}
$$

Standard properties such as uniqueness of evaluation and value soundness have direct proofs based only on structural induction and the introduction and elimination rules.

Divergence can be defined co-inductively as the predicate $\mathsf{divrg} :: lexp \Rightarrow bool$:

$$[\![\ \mathsf{isExp}\ e1;\ \mathsf{isExp}\ e2;\ \mathsf{divrg}\ e1\ ]\!] \Longrightarrow \mathsf{divrg}\ (e1\ @\ e2)$$
$$[\![\ e1 \gg \mathsf{Fun}\,x.\,e\ x;\ \mathsf{abstr}\ e;\ \mathsf{isExp}\ e2;\ \mathsf{divrg}\ (e\ e2)\ ]\!] \Longrightarrow \mathsf{divrg}\ (e1\ @\ e2)$$

We can give a fully automated co-inductive proof of the divergence of combinators such as $\Omega \stackrel{\text{def}}{=} (\mathsf{Fun}\,x.\,x\ @\ x)\ @\ (\mathsf{Fun}\,x.\,x\ @\ x)$, once we have added the $\mathsf{abstr\_tac}$ tactic to the built-in simplifier. Moreover, there is a direct proof that convergence and divergence are exclusive and exhaustive.

Applicative (bi)simulation $\leq\ ::\ [\,lexp, lexp\,] \Rightarrow bool$ is another interesting (co-inductive) predicate with the single introduction rule

$$[\![\ \forall t.\ r \gg \mathsf{Fun}\,x.\,t\ x \wedge \mathsf{abstr}\ t \rightarrow (\exists u.\ s \gg \mathsf{Fun}\,x.\,u\ x \wedge \mathsf{abstr}\ u\ \wedge$$
$$(\forall p. \mathsf{isExp}\ p \rightarrow\ (t\ p) \leq\ (u\ p)))\ ]\!] \Longrightarrow r \leq s$$

The HOAS style here greatly simplifies the presentation and correspondingly the metatheory. Indeed, with the appropriate instantiation of the coinductive relation, the proofs that simulation is a pre-order and bisimulation an equivalence relation are immediate. Other verified properties include Kleene equivalence is a simulation, and divergent terms are the least element in this order.

4.3 The Higher-Order π-Calculus

We present an encoding of the monadic higher-order π-calculus [24], with structural congruence and reaction rules; see [11] for a review of other styles of encodings for the first-order case. The syntax is given by the following, where $a, \overline{a}$ ranges over names, X over agent variables:

$$\alpha ::= \tau \mid a(X) \mid \overline{a}\langle P\rangle$$
$$P ::= X \mid \Sigma_{i\in I}\alpha_i.\,P \mid (P_1 \mid P_2) \mid (\nu a)P$$

The encoding introduces appropriate type abbreviations (namely pi and $name$) and constants; we concentrate on restriction, input and output:

$$\mathsf{New}\ ::\ (name \Rightarrow pi) \Rightarrow pi$$
$$(\mathsf{New}\ a)p\ a == \mathsf{CON}\ cNU\ \$\$\ \mathsf{LAM}\,a.\,p\ a$$
$$\mathsf{In}\ ::\ [\,name, (pi \Rightarrow pi)\,] \Rightarrow pi$$
$$\mathsf{In}\ a\ (p) == \mathsf{CON}\ cIN\ \$\$\ \mathsf{lambda}\ p$$
$$\mathsf{Out}\ ::\ [\,name, pi, pi\,] \Rightarrow pi$$
$$\mathsf{Out}\ \overline{a}\langle p\rangle q == \mathsf{CON}\ cOUT\ \$\$\ a\ \$\$\ p\ \$\$\ q$$

Replication is defined from these constants and need not be primitive:

$$!\,p == (\mathsf{New}\ a)(D|\mathsf{Out}\ \overline{a}\langle p|D\rangle)\ \text{ where } D == \mathsf{In}\ a\ (\lambda\,x.\,(x|\mathsf{Out}\ \overline{a}\langle x\rangle))$$

We then inductively define well-formed processes $\mathsf{isProc}\ p$, whose introduction rules are, omitting non-binding cases,

$$\mathsf{isProc}\ (\mathsf{VAR}\ i)$$
$$[\![\ \mathsf{abstr}\ p;\ \forall a.\ \mathsf{isName}\ a \rightarrow \mathsf{isProc}\ (p\ a)\ ]\!] \Longrightarrow \mathsf{isProc}\ (\mathsf{New}\ a)p\ a$$
$$[\![\ \mathsf{abstr}\ p;\ \mathsf{isName}\ a;\ \forall i.\ \mathsf{isProc}\ (p\ (\mathsf{VAR}\ i))\ ]\!] \Longrightarrow \mathsf{isProc}\ \mathsf{In}\ a\ (p)$$

Restriction, being represented as a function of type $name \Rightarrow pi$ is well-formed if its body $(p\ a)$ is, under the assumption $\mathsf{isName}\ a$; this differs from input, which contains a function of type $pi \Rightarrow pi$.

Process abstraction is defined by $\mathsf{procAbstr}\ p \;==\; \mathsf{abstr}\ p \wedge (\forall q.\, \mathsf{isProc}\ q \rightarrow \mathsf{isProc}\ (p\ q))$. Next, structural congruence is introduced and we proceed to encode reaction rules as the inductive definition $\longmapsto\ ::\ [\,pi \Rightarrow pi\,] \Rightarrow bool$. Note that the presence of the structural rule make this unsuitable for search. Here is a sample:

$$[\![\ \mathsf{procAbstr}\ p; \mathsf{isProc}\ q; \mathsf{isProc}\ p'; \mathsf{isName}\ a\ ]\!] \Longrightarrow \mathsf{In\ a\ (p)|Out\ \bar{a}\langle q\rangle p'} \longmapsto \mathsf{(p\ q)|p'}$$
$$[\![\ \mathsf{abstr}\ p; \mathsf{abstr}\ p'; (\forall a.\, \mathsf{isName}\ a \rightarrow \mathsf{(p\ a)} \longmapsto \mathsf{(p'\ a)})\ ]\!] \Longrightarrow \mathsf{(New\ a)p\ a} \longmapsto \mathsf{(New\ a)p'\ a}$$

A formalization of late operational semantics following [18] is possible and will be treated within a separate paper.

5 A Theory of Hybrid

The goal of this section is to describe a mathematical model of Hybrid and then show that the model provides an adequate representation of the λ-calculus. The work here serves to illustrate and under-pin *Hybrid*. We proceed as follows. In Section 5.1 we set up an *explicit bijection* between the set of alpha equivalence classes of lambda expressions, and the set of proper de Bruijn terms. This section also allows us to introduce notation. In Section 5.2 we prove adequacy for an object level λ-calculus by proving an equivalent result for (proper) object level de Bruijn expressions, and appealing to the bijection.

First, some notation *for the object level.* λ-calculus expressions are inductively defined by $E ::= V_i \mid E\ E \mid \Lambda V_i.E$. We write $E[E'/V_i]$ for capture avoiding substitution, $E \sim_\alpha E'$ for α-equivalence, and $[E]_\alpha$ for alpha equivalence classes. We write $\mathcal{LE}$ for the set of expressions, and $\mathcal{LE}/{\sim_\alpha}$ for the set of all alpha equivalence classes. The set of de Bruijn expressions is denoted by $\mathcal{DB}$, and generated by $D ::= \bar{\imath} \mid i \mid D\ \$\ D \mid A(D)$ where $\bar{\imath}$ is a free index. We write $\mathcal{DB}(l)$ for the set of expressions at level l, and $\mathcal{PDB}$ for the set of proper expressions. Note that $\mathcal{PDB} = \mathcal{DB}(0) \subset \mathcal{DB}(1) \ldots \subset \mathcal{DB}(l) \ldots \subset \mathcal{DB}$ and $\mathcal{DB} = \bigcup_{l<\omega} \mathcal{DB}(l)$.

5.1 A Bijection between λ-Calculus and Proper de Bruijn

Theorem 1. *There is a bijection* $\theta\colon \mathcal{LE}/{\sim_\alpha} \rightleftarrows \mathcal{PDB}\colon \phi$ *between the set* $\mathcal{LE}/{\sim_\alpha}$ *of alpha equivalence classes of λ-calculus expressions, and the set* $\mathcal{PDB}$ *of proper de Bruijn expressions,*

In order to prove the theorem, we first establish in Lemma 1 and Lemma 2 the existence of a certain family of pairs of functions $[\![-]\!]_L\colon \mathcal{LE} \rightleftarrows \mathcal{DB}(|L|)\colon (\!|-|\!)_L$. Here, **list** L of length $|L|$ is one whose elements are object level variables V_i. We say that L is **ordered** if it has an order reflected by the indices i, and no repeated elements. The first element has the largest index. We write ϵ for the empty list.

Lemma 1. *For any* L*, there exists a function* $[\![-]\!]_L\colon \mathcal{LE} \rightarrow \mathcal{DB}(|L|)$ *given recursively by*

- $[\![V_i]\!]_L \stackrel{\text{def}}{=}$ *if* $V_i \notin L$ *then* $\bar{i}$ *else posn* $V_i\,L$ *where posn* $V_i\,L$ *is the position of* V_i *in* L*, counting from the head, which has position* 0.
- $[\![E_1\ E_2]\!]_L \stackrel{\text{def}}{=} [\![E_1]\!]_L \ \$\ [\![E_2]\!]_L$
- $[\![\Lambda V_i.\, E]\!]_L \stackrel{\text{def}}{=} A([\![E]\!]_{V_i,L})$

Lemma 2. *For any ordered* L*, there exists a function* $(\!|-|\!)_L\colon \mathcal{DB}(|L|) \to \mathcal{LE}$ *given recursively by*

- $(\!|\bar{i}|\!)_L \stackrel{\text{def}}{=} V_i$
- $(\!|j|\!)_L \stackrel{\text{def}}{=} elem\ j\ L$ *the* j*th element of* L
- $(\!|D_1 \ \$\ D_2|\!)_L \stackrel{\text{def}}{=} (\!|D_1|\!)_L\ (\!|D_2|\!)_L$
- $(\!|A(D)|\!)_L \stackrel{\text{def}}{=} \Lambda V_{M+1}.\, (\!|D|\!)_{V_{M+1},L}$ *where* $M \stackrel{\text{def}}{=} Max(D;L)$

$Max(D;L)$ *denotes the maximum of the free indices which occur in* D *and the index* j*, where* V_j *is the head of* L.

The remainder of the proof involves establishing facts about these functions. It takes great care to ensure all details are correct, and many sub-lemmas are required! We prove that each pair of functions "almost" gives rise to an isomorphism, namely that $[\![(\!|D|\!)_L]\!]_L = D$ and that $(\!|[\![E]\!]_L|\!)_{L'} \sim_\alpha E[L'/L]$. Let $q\colon \mathcal{LE} \to \mathcal{LE}/\!\sim_\alpha$ be the quotient function. We define $\theta([E]_\alpha) \stackrel{\text{def}}{=} [\![E]\!]_\epsilon$ and $\phi \stackrel{\text{def}}{=} q \circ (\!|-|\!)_\epsilon$, and the theorem follows by simple calculation. Although our proofs take great care to distinguish α-equivalence classes from expressions, we now write simply E instead of $[E]_\alpha$, which will allow us to write $[\![-]\!]_L\colon \mathcal{LE}/\!\sim_\alpha \leftrightarrows \mathcal{DB}(|L|)\colon (\!|-|\!)_L$

5.2 Adequacy of Hybrid for the λ-Calculus

It would not be possible to provide a "complete" proof of the adequacy of the Hybrid machine tool for λ-calculus. A compromise would be to work with a mathematical description of Hybrid, but even that would lead to extremely long and complex proofs. Here we take a more narrow approach—we work with what amounts to a description of a fragment of Hybrid as a simply typed lambda calculus—presented as a theory in a logical framework as described in Section 1. Before we can state the adequacy theorem, we need a theory in the logical framework which we regard as a "model" of Hybrid. The theory has ground types *expr*, *var* and *bnd*. The types are generated by $\sigma ::= expr \mid var \mid bnd \mid \sigma \Rightarrow \sigma$. We declare constants

$$
\begin{array}{rlrl}
i :: & var & \mathsf{BND} :: & bnd \Rightarrow expr \\
i :: & bnd & \$\$ :: & expr \Rightarrow expr \Rightarrow expr \\
\mathsf{VAR} :: & var \Rightarrow expr & \mathsf{ABS} :: & expr \Rightarrow expr
\end{array}
$$

where i ranges over the natural numbers. The objects are generated as in Section 1. Shortage of space means the definitions are omitted—here we aim to give a flavour of our results. Our aim is to prove

Theorem 2 (Representational Adequacy).

- *There is an injective function* $|| - ||_\epsilon : \mathcal{LE}/\!\sim_\alpha \to \mathcal{CLF}_{expr}(\emptyset)$ *which is*
- *compositional on substitution, that is*

$$||E[E'/V_i]||_\epsilon = \mathsf{subst}\ ||E'||_\epsilon\ i\ ||E||_\epsilon$$

where $\mathsf{subst}\ e'\ i\ e$ *is the function on canonical expressions in which* e' *replaces occurrences of* $\mathsf{VAR}\ i$ *in* e*. The definition of meta substitution, and the set* $\mathcal{CLF}_{expr}(\Gamma)$ *of canonical objects in typing environment* Γ *is omitted.*

To show adequacy we establish the existence of the following functions

$$\mathcal{LE}/\!\sim_\alpha \xrightarrow{||-||_L} \mathcal{CLF}_{expr}(\beta(L)) \xleftarrow{\Sigma_L} \mathcal{DB}(|L|)$$

The function β is a bijection between object level variables V_i and meta variables v_i. Thus $\beta(V_i) = v_i$. The notation $\beta(L)$ means "map" β along L. We write $\beta(L)$ for a typing environment in the framework with all variables of type expression. From now on, all lists L are ordered. The function Σ_L is defined by the clauses

- $\Sigma_L\, \bar{i} \stackrel{\text{def}}{=} \mathsf{VAR}\ i$
- $\Sigma_L\, i \stackrel{\text{def}}{=} elem\, i\, \beta(L)$
- $\Sigma_L\, (D_1 \ \$\ D_2) \stackrel{\text{def}}{=} \Sigma_L\, D_1 \ \$\$\ \Sigma_L\, D_2$
- $\Sigma_L\, A(D) \stackrel{\text{def}}{=} \mathsf{lambda}\ (\lambda\, v_{M+1}.\, \Sigma_{V_{M+1},L}\, D) = \mathsf{LAM}\, v_{M+1}.\, \Sigma_{V_{M+1},L}\, D$ where $M \stackrel{\text{def}}{=} Max(D; L)$ (see Lemma 2).

The definition of $||-||_L$ is $\Sigma_L \circ [\![-]\!]_L$. Note that this is a well defined function on an α-equivalence class because $[\![-]\!]_L$ is. In view of Theorem 1, and the definition of $|| - ||_\epsilon$, it will be enough to show the analogous result for the function Σ_ϵ. The proof strategy is as follows. We show that there is a function insts which maps de Bruijn expressions to canonical objects, possesses a left inverse, and is compositional with respect to substitution. Then we prove that the action of Σ_L is simply that of the insts function. Proofs of results will appear in a journal version of this paper.

6 Related Work

Our work could not have come about without the contributions of others. Here, following [28] we classify some of these contributions according to the mathematical construct chosen to model abstraction. The choice has dramatic consequences on the associated notions of recursion and proof by induction.

- De Bruijn syntax [4]: we do not review this further.
- Name-carrying syntax: here abstractions are pairs "(name, expression)" and the mechanization works directly on parse trees, which are quotiented by α-conversion [22,29,9]. While recursion/induction is well-supported, the detail

that needs to be taken care of on a case-by-case basis tends to be overwhelming. To partially alleviate this [1] *defines* name-carrying syntax in terms of an underlying type of De Bruijn λ-expressions, which is then used as a metalogic where equality is α-convertibility. Very recently, in [10,28] Gabbay and Pitts have introduced a novel approach, based on the remarkable observation that a first-order theory of α-conversion and binding is better founded on the notion of name *swapping* rather than renaming. The theory can either be presented axiomatically as formalizing a primitive notion of swapping and *freshness* of names from which binding can be derived, or as a non-classical set-theory with an internal notion of *permutation* of atoms. Such a set-theory yields a natural notion of structural induction and recursion over α-equivalence classes of expressions, but it is incompatible with the axiom of choice. The aim of Pitts and Gabbay's approach is to give a satisfying foundation of the informal practice of reasoning modulo renaming, more than formulate an alternative logical framework for metareasoning (although this too is possible). An ML-like programming language, *FreshML*, is under construction geared towards metaprogramming applications.

- Abstractions as functions from *names* to expressions: mentioned in [1] (and developed in [12]) it was first proposed in [5], as a way to have binders as functions on inductive data-types, while coping with the issue of *exotic* expressions stemming from an inductive characterization of the set of names. The most mature development is Honsell et al.'s framework [17], which explicitly embraces an *axiomatic* approach to metareasoning with HOAS. It consists of a higher-order logic inconsistent with unique choice, but extended with a set of axioms, called the *Theory of Contexts,* parametric to a HOAS signature. Those axioms include the reification of key properties of names akin to *freshness.* More crucially, higher-order induction and recursion schemata on expressions are also assumed. The consistency of such axioms with respect to functor categories is left to a forthcoming paper. The application of this approach to object logics such as the π-calculus [18] succeeds not only because the possibility to "reflect" on names is crucial for the metatheory of operations such as mismatch, but also because here hypothetical judgments, which are only partially supported in such a style [5] are typically not needed. Moreover β-conversion can implement object-level substitution, which is in this case simply "name" for bound variable in a process. The latter may not be possible in other applications such as the ambient calculus. This is also the case for another case-study [23], where these axioms seem less successful. In particular, co-induction is available, but the need to code substitution explicitly makes some of the encoding fairly awkward.
- Abstractions as functions from *expressions* to expressions [26,15]. We can distinguish here two main themes to the integration of HOAS and induction: one where they coexist in the same language and the other where inductive reasoning is conducted at an additional metalevel. In the first one, the emphasis is on trying to allow (primitive) recursive definitions on functions of higher type while preserving adequacy of representations; this has

been realized for the simply-typed case in [7] and more recently for the dependently-typed case in [6]. The idea is to separate at the type-theoretic level, via an S4 modal operator, the *primitive* recursive space (which encompasses functions defined via case analysis and iteration) from the *parametric* function space (whose members are those convertible to expressions built only via the constructors).

On the other side, the *Twelf* project [27] is built on the idea of devising an explicit (meta)metalogic for reasoning (inductively) about logical frameworks, in a fully automated way. $\mathcal{M}_2$ is a constructive first-order logic, whose quantifiers range over possibly open LF object over a signature. In the metalogic it is possible to express and inductively prove metalogical properties of an object logic. By the adequacy of the encoding, the proof of the existence of the appropriate LF object(s) guarantees the proof of the corresponding object-level property. It must be remarked that *Twelf* usage model is explicitly non-interactive (i.e. not programmable by tactics). Moreover, co-inductive definitions have to be rewritten in an inductive fashion, exploiting the co-continuity of the said notion, when possible.

Miller and McDowell [20] introduce a metameta logic, $FO\lambda^{\Delta I\!N}$, that is based on intuitionistic logic augmented with definitional reflection [14] and induction on natural numbers. Other inductive principles are *derived* via the use of appropriate measures. At the metameta level, they reason about object-level judgments formulated in second-order logic. They prove the consistency of the method by showing that $FO\lambda^{\Delta I\!N}$ enjoys cut-elimination [19]. $FO\lambda^{\Delta I\!N}$ approach [20] is interactive; a tactic-based proof editor is under development, but the above remark on co-induction applies.

7 Conclusions and Future Work

The induction principles of Hybrid involve universal quantifications over free variables when instantiating abstractions. It remains future work to determine the real utility of our principles, and how they compare to more standard treatments. Informal practice utilises the some/any quantifier that has emerged formally in [10]. McKinna and Pollack discuss some of these issues in [21].

Several improvements are possible:

We will internalize the well-formedness predicates as abstract types in Isabelle HOL, significantly simplifying judgments over object logics. For example, the subset of lazy λ-calculus expressions identified by predicate isExp will become a type, say $tExp$, so that evaluation will be typed as $\gg\; :: [\,tExp, tExp\,] \Rightarrow bool$. We will specialize the abstr predicate to the defined logic so that it will have type $(tExp \Rightarrow tExp) \Rightarrow bool$.

We envisage eventually having a system, similar in spirit to Isabelle HOL's datatype package, where the user is only required to enter a binding signature for a given object logic; the system will provide an abstract type characterizing the logic, plus a series of theorems expressing freeness of the constructors of such a type and an induction principle on the shape of expressions analogous to the one mentioned in Section 3.

We are in the process of further validating our approach by applying our methods to the compiler optimization transformations for Benton & Kennedy's MIL-lite language [3].

We intend to develop further the theory of Hybrid. Part of this concerns presenting the full details of the material summarized here. There are also additional results, which serve to show how Hybrid relates to λ-calculus. For example, we can prove that if $\mathsf{abstr}\ e$, then there exists $[\Lambda\, V_i.\, E]_\alpha \in \mathcal{LE}/{\sim_\alpha}$ such that $||\Lambda\, V_i.\, E||_\epsilon = \mathsf{LAM}\, v_i.\, e\, v_i$. On a deeper level, we are looking at obtaining categorical characterisations of some of the notions described in this paper, based on the work of Fiore, Plotkin, and Turi in [8].

A full journal version of this paper is currently in preparation, which in particular will contain a greatly expanded section on the theory of Hybrid, and provide full details of the case studies.

Acknowledgments

We would like to thank Andy Gordon for useful discussions and having provided the HOL script from [1]. We thank Simon Gay for discussions and ideas concerning the π-calculus. Finally, we are very grateful for the financial support of the UK EPSRC.

References

1. A. Gordon. A mechanisation of name-carrying syntax up to alpha-conversion. In J.J. Joyce and C.-J.H. Seger, editors, *International Workshop on Higher Order Logic Theorem Proving and its Applications*, volume 780 of *Lecture Notes in Computer Science*, pages 414–427, Vancouver, Canada, Aug. 1993. University of British Columbia, Springer-Verlag, published 1994.
2. S. Abramsky. The lazy lambda calculus. In D. Turner, editor, *Research Topics in Functional Programming*, pages 65–116. Addison-Wesley, 1990.
3. N. Benton and A. Kennedy. Monads, effects and transformations. In *Proceedings of the 3rd International Workshop in Higher Order Operational Techniques in Semantics*, volume 26 of *Electronic Notes in Theoretical Computer Science*. Elsevier, 1998.
4. N. de Bruijn. Lambda-calculus notation with nameless dummies: A tool for automatic formula manipulation with application to the Church-Rosser theorem. *Indag. Math.*, 34(5):381–392, 1972.
5. J. Despeyroux, A. Felty, and A. Hirschowitz. Higher-order abstract syntax in Coq. In M. Dezani-Ciancaglini and G. Plotkin, editors, *Proceedings of the International Conference on Typed Lambda Calculi and Applications*, pages 124–138, Edinburgh, Scotland, Apr. 1995. Springer-Verlag LNCS 902.
6. J. Despeyroux and P. Leleu. Metatheoretic results for a modal λ-calculus. *Journal of Functional and Logic Programming*, 2000(1), 2000.
7. J. Despeyroux, F. Pfenning, and C. Schürmann. Primitive recursion for higher-order abstract syntax. In R. Hindley, editor, *Proceedings of the Third International Conference on Typed Lambda Calculus and Applications (TLCA'97)*, pages 147–163, Nancy, France, Apr. 1997. Springer-Verlag LNCS.

8. M. Fiore, G.D. Plotkin, and D. Turi. Abstract Syntax and Variable Binding. In G. Longo, editor, *Proceedings of the 14th Annual Symposium on Logic in Computer Science (LICS'99)*, pages 193–202, Trento, Italy, 1999. IEEE Computer Society Press.
9. J. Ford and I.A. Mason. Operational Techniques in PVS – A Preliminary Evaluation. In *Proceedings of the Australasian Theory Symposium, CATS '01*, 2001.
10. M. Gabbay and A. Pitts. A new approach to abstract syntax involving binders. In G. Longo, editor, *Proceedings of the 14th Annual Symposium on Logic in Computer Science (LICS'99)*, pages 214–224, Trento, Italy, 1999. IEEE Computer Society Press.
11. S. Gay. A framework for the formalisation of pi-calculus type systems in Isabelle/HOL. In *Proceedings of the 14th International Conference on Theorem Proving in Higher Order Logics (TPHOLs 2001*, LNCS. Springer-Verlag, 2001.
12. A.D. Gordon and T. Melham. Five axioms of alpha-conversion. In J. von Wright, J. Grundy, and J. Harrison, editors, *Proceedings of the 9th International Conference on Theorem Proving in Higher Order Logics (TPHOLs'96)*, volume 1125 of *Lecture Notes in Computer Science*, pages 173–190, Turku, Finland, August 1996. Springer-Verlag.
13. E.L. Gunter. Why we can't have SML style `datatype` declarations in HOL. In L.J.M. Claese and M.J.C. Gordon, editors, *Higher Order Logic Theorem Proving and Its Applications*, volume A–20 of *IFIP Transactions*, pages 561–568. North-Holland Press, Sept. 1992.
14. L. Hallnas. Partial inductive definitions. *Theoretical Computer Science*, 87(1):115–147, July 1991.
15. R. Harper, F. Honsell, and G. Plotkin. A framework for defining logics. *Journal of the Association for Computing Machinery*, 40(1):143–184, Jan. 1993.
16. M. Hofmann. Semantical analysis for higher-order abstract syntax. In G. Longo, editor, *Proceedings of the 14th Annual Symposium on Logic in Computer Science (LICS'99)*, pages 204–213, Trento, Italy, July 1999. IEEE Computer Society Press.
17. F. Honsell, M. Miculan, and I. Scagnetto. An axiomatic approach to metareasoning on systems in higher-order abstract syntax. In *Proc. ICALP 2001*, volume 2076 in LNCS, pages 963–978. Springer-Verlag, 2001.
18. F. Honsell, M. Miculan, and I. Scagnetto. π-calculus in (co)inductive type theories. *Theoretical Computer Science*, 2(253):239–285, 2001.
19. R. McDowell. *Reasoning in a Logic with Definitions and Induction.* PhD thesis, University of Pennsylvania, 1997.
20. R. McDowell and D. Miller. Reasoning with higher-order abstract syntax in a logical framework. *ACM Transaction in Computational Logic*, 2001. To appear.
21. J. McKinna and R. Pollack. Some Type Theory and Lambda Calculus Formalised. To appear in Journal of Automated Reasoning, Special Issue on Formalised Mathematical Theories (F. Pfenning, Ed.),
22. T.F. Melham. A mechanized theory of the π-calculus in HOL. *Nordic Journal of Computing*, 1(1):50–76, Spring 1994.
23. M. Miculan. Developing (meta)theory of lambda-calculus in the theory of contexts. In S. Ambler, R. Crole, and A. Momigliano, editors, *MERLIN 2001: Proceedings of the Workshop on MEchanized Reasoning about Languages with variable bINding*, volume 58 of *Electronic Notes in Theoretical Computer Scienc*, pages 1–22, November 2001.
24. J. Parrow. An introduction to the pi-calculus. In J. Bergstra, A. Ponse, and S. Smolka, editors, *Handbook of Process Algebra*, pages 479–543. Elsevier Science, 2001.

25. F. Pfenning. Computation and deduction. Lecture notes, 277 pp. Revised 1994, 1996, to be published by Cambridge University Press, 1992.
26. F. Pfenning and C. Elliott. Higher-order abstract syntax. In *Proceedings of the ACM SIGPLAN'88 Symposium on Language Design and Implementation*, pages 199–208, Atlanta, Georgia, June 1988.
27. F. Pfenning and C. Schürmann. System description: Twelf — A metalogical framework for deductive systems. In H. Ganzinger, editor, *Proceedings of the 16th International Conference on Automated Deduction (CADE-16)*, pages 202–206, Trento, Italy, July 1999. Springer-Verlag LNAI 1632.
28. A. M. Pitts. Nominal logic: A first order theory of names and binding. In N. Kobayashi and B. C. Pierce, editors, *Theoretical Aspects of Computer Software, 4th International Symposium, TACS 2001, Sendai, Japan, October 29-31, 2001, Proceedings*, volume 2215 of *Lecture Notes in Computer Science*, pages 219–242. Springer-Verlag, Berlin, 2001.
29. R. Vestergaard and J. Brotherson. A formalized first-order conflence proof for the λ-calculus using one sorted variable names. In A. Middelrop, editor, *Proceedings of RTA 2001*, volume 2051 of *LNCS*, pages 306–321. Springer-Verlag, 2001.

Efficient Reasoning about Executable Specifications in Coq

Gilles Barthe and Pierre Courtieu

INRIA Sophia-Antipolis, France
{Gilles.Barthe,Pierre.Courtieu}@sophia.inria.fr

Abstract. We describe a package to reason efficiently about executable specifications in Coq. The package provides a command for synthesizing a customized induction principle for a recursively defined function, and a tactic that combines the application of the customized induction principle with automatic rewriting. We further illustrate how the package leads to a drastic reduction (by a factor of 10 approximately) of the size of the proofs in a large-scale case study on reasoning about JavaCard.

1 Introduction

Proof assistants based on type theory, such as Coq [9] and Lego [15], combine an expressive specification language (featuring inductive and record types) and a higher-order predicate logic (through the Curry-Howard Isomorphism) to reason about specifications. Over the last few years, these systems have been used extensively, in particular for the formalization of programming languages. Two styles of formalizations are to be distinguished:

- *the functional style*, in which specifications are written in a functional programming style, using pattern-matching and recursion;
- *the relational style*, in which specifications are written in a logic programming style, using inductively defined relations;

In our opinion, the functional style has some distinctive advantages over its relational counterpart, especially for formalizing complex programming languages. In particular, the functional style offers support for testing the specification and comparing it with a reference implementation, and the possibility to generate programs traces upon which to reason using e.g. temporal logic. Yet, it is striking to observe that many machine-checked accounts of programming language semantics use inductive relations. In the case of proof assistants based on type theory, two factors contribute to this situation:

- firstly, type theory requires functions to be total and terminating (in fact, they should even be provably so, using a criterion that essentially captures functions defined by structural recursion), therefore specifications written in a functional style may be more cumbersome than their relational counterpart;

V.A. Carreño, C. Muñoz, S. Tahar (Eds.): TPHOLs 2002, LNCS 2410, pp. 31–46, 2002.

- secondly, a proof assistant like Coq offers, through a set of inversion [10,11] and elimination tactics, effective support to reason about relational specifications. In contrast, there is no similar level of support for reasoning about executable functions.

Here we address this second issue by giving a package that provides effective support for reasoning about complex recursive definitions, see Section 2. Further, we illustrate the benefits of the package in reasoning about executable specifications of the JavaCard platform [3,4], and show how its use yields compact proofs scripts, up to 10 times shorter than proofs constructed "by hand", see Section 3. Related work is discussed in Section 3. We conclude in Section 5.

2 Elimination Principles for Functions

2.1 Elimination on Inductive Types and Properties

One of the powerful tools provided by proof assistants like Coq for reasoning on inductive types is the *elimination principle*, which is usually generated automatically from the definition of the type. We see here the definition in Coq of the type nat and the type of its associated principle:

```
Inductive nat : Set := O : nat | S : nat→nat.
nat_ind:
(P: nat→Prop) (P O) →((n:nat)(P n)→(P (S n))) →(n:nat)(P n).
```

The logical meaning of an elimination principle attached to an inductive type T is that closed normal terms of type T have a limited set of possible forms, determined by the constructors of T. It also captures the recursive structure of the type. It is important to notice that elimination principles are nothing else than recursive functions, for example the definition of nat_ind is the following:

```
[P: nat→Prop; f:(P O); f0:((n:nat)(P n)→(P (S n)))]
 Fixpoint F [n:nat] : (P n) :=
<[x:nat](P x)> Cases n of O ⇒f
                        | (S n0) ⇒(f0 n0 (F n0))
  end}
```

The expression <[x:nat](P x)> is a specific type scheme for dependently typed case expressions (dependent types are necessary to represent predicates like P). It is used to build case branches having different types, provided that each one corresponds to the scheme applied to the corresponding pattern. For example the first branch is correct because ([x:nat](P x) O)=(P O). Similarly the second case is accepted because (f0 n0 (F n0)) has type (P (S n0)), which is equal to ([x:nat](P x) (S n0)).

The type of the whole case expression is obtained by applying the scheme to the term on which the case analysis is done, here n, i.e. (P n).

Reasoning by elimination on an inductive *property* P can be understood intuitively as reasoning by induction on the proof of P. We see here the principle associated to the relation le:

```
Inductive le [n : nat] : nat→Prop :=
     le_n : (le n n)
   | le_S : (m:nat)(le n m)→(le n (S m)).

le_ind: (n:nat; P:(nat→Prop))
          (P n) → ((m:nat)(le n m)→(P m)→(P (S m)))
        → (m:nat)(le n m)→(P m).
```

Suppose we are trying to prove `(P m)` for all m such that `H:(le n m)`, we can apply the theorem `le_ind`, which leads to two subgoals corresponding to the possible constructors of `le`. This way of reasoning, which has proved to be very useful in practice, makes relational specifications a popular solution.

2.2 Elimination Following the Shape of a Function

Basic Idea. When we choose a functional style of specification, recursive functions are used instead of inductive relations and elimination principles are usually not automatically available. However, once a function is defined (i.e. proved terminating which is automatic in Coq for structural recursion) it is relatively natural to build an induction principle, which follows its shape. It allows *reasoning by induction on the possible branches* of the definition of the function. Since the new term follows the same recursive calls as the function, we know that this new term defines a correct induction scheme and that it will be accepted by Coq. For example the following function:

```
Fixpoint isfourtime [n:nat] : bool :=
 Cases n of | O ⇒ true
            | (S S S S m) ⇒ (isfourtime m)
            | _ ⇒ false
 end.
```

yields an induction principle with a nested case analysis of depth 5. Redoing all the corresponding steps each time we want to consider the different paths of the function would be long and uninteresting (see Section 3 for more examples).

We have designed a Coq package allowing automation of the inductive/case analysis process. According to the Coq paradigm, our package builds *proof terms* in the Curry-Howard philosophy. Surprisingly the presence of proof terms is helpful: since functions and proofs are all represented by λ-terms, we can build proof terms by transforming the term of the function we consider. Since moreover we focus on functions defined by structural recursion (see Section 4.2 for extension to more general recursion schemes), termination of the function (and consequently the correctness of the elimination principles we generate) is ensured by the type checking of Coq.

The proof of a property $\forall x.P$ using `isfourtime` as a model will be a term Q obtained by transforming `isfourtime` (notice how implicit cases have been expanded):

```
Fixpoint Q [n:nat]: P :=
<[x:nat](P x)> Cases n of
```

```
 | O ⇒(?: (P O))
 | (S O) ⇒(?: (P (S O)))
 | (S S O) ⇒(?: (P (S S O)))
 | (S S S O) ⇒(?: (P (S S S O)))
 | (S S S S m) ⇒((?:(x:nat)(P x)→(P (S S S S x))) m (Q m))
end.
```

where the five (?:H) stand for properties H yet to be proved (subgoals). In Subsection 2.5 we concentrate on the structure of this incomplete term and its automatic generation from the definition of the function.

Once built, this incomplete term can be applied to a particular goal with the **Refine** tactic, or completed into a general reusable induction principle. To achieve the latter, we replace the ?'s by abstracted variables (Hypx in the example below). We also abstract the predicate (P) and finally obtain a general principle. On our example this leads to:

```
[P:(nat→Prop); Hyp0:(P O); Hyp1:(P (S O));
 Hyp2:(P (S (S O))); Hyp3:(P (S (S (S O))));
 Hyp4:(x:nat) (P x) →(P (S (S (S (S x)))))]
Fixpoint Q [n:nat]: P :=
<[x:nat](P x)> Cases n of
  | O ⇒Hyp0
  | (S O) ⇒Hyp1
  | (S S O) ⇒Hyp2
  | (S S S O) ⇒Hyp3
  | (S S S S m) ⇒Hyp4 m (Q m)
  end
    : (P:(nat→Prop)) (P O) →(P (S O)) →(P (S (S O)))
    →(P (S (S (S O)))) →((x:nat) (P x) →(P (S (S (S (S x))))))
    →(n:nat)(P n)
```

This term is well-typed and thus defines a correct induction principle.

Capturing the Environment of Each Branch. In the examples above subgoals contain nothing more than induction hypotheses, but generally this will not be enough. Indeed, we need to capture for each subgoal the *environment* induced by the branches of **Cases** expressions which have been chosen. We illustrate this situation with the following function (which computes for every n the largest $m \leq n$ such that $m = 4k$ for some k):

```
Fixpoint findfourtime [n:nat]: nat :=
 Cases n of
 | O ⇒O
 | (S m) ⇒
  Cases (isfourtime n) of
   | true ⇒n
   | false ⇒(findfourtime m)
  end
 end.
```

In this case it is necessary to remember that in the first branch `n=O`, in the second branch `n=O` and `(isfourtime n)=true` and in the third branch `(isfourtime n)=false`. Otherwise it will be impossible to prove for example the following property:

$$P \equiv \forall \texttt{n:nat.(isfourtime (findfourtime n)) = true.}$$

Finally the proof of the elimination principle generated from `findfourtime` has the form:

```
Fixpoint Q2 [n:nat]: (P n) :=
(<[x:bool](isfourtime n)=x →(P n)>
Cases n of
 | O ⇒(?:(P O))
 | (S m) ⇒
  <[b:bool](isfourtime (S m))=b→(P (S m))>
  Cases (isfourtime (S m)) of
   | true ⇒(?:(isfourtime (S m))=true→(P (S m)))
   | false ⇒(?:(P m)→(isfourtime (S m))=false→(P (S m))) (Q2 m)
  end (refl_equal bool (isfourtime (S m))))
end
```

where `(reflequal bool (isfourtime (S m)))` is the proof of the different equalities induced by the case analysis. Recall that `reflequal` is the constructor of the inductive type `eq` representing the equality, so `{(reflequal A x)}` is of type `(eq A x x)`.

These proofs are gathered outside the case expression so as to obtain a well-typed term. This point is subtle: equality proofs have the form `(reflequal bool (isfourtime (S m)))`, which is of type `(isfourtime (S m))=(isfourtime (S m))`. We see therefore that to be accepted by the case typing rule explained above, we must apply this term to a term of type `(isfourtime (S m))=(isfourtime (S m)) → (P (S m))`. This is the type of the whole case expression, but not of each branch taken separately. So moving the `reflequal` expression inside the `Cases` would result in an ill-typed term.

2.3 Contents of the Package

The main components of the package are:

- A Coq *command* **`Functional Scheme`** which builds a general induction principle from the definition of a function f. It is a theorem of the form:

  $$(P : (\forall \boldsymbol{x_i} : \boldsymbol{T_i}.Prop))(H_1 : PO_1)...(H_n : PO_n) \rightarrow \forall \boldsymbol{x_i} : \boldsymbol{T_i}.(P\ \boldsymbol{x_i})$$

  where the PO_i's are the proof obligations generated by the algorithm described above, and $\boldsymbol{x_i}$'s correspond to the arguments of f. To make an elimination, the user just applies the theorem. The advantage of this method is that the structure of the function (and its type checking) is not duplicated each time we make an elimination;
- A *tactic* **`Analyze`** which applies the above algorithm to a particular goal. This tactic allows for more automation, see for that section 2.6.

2.4 Using the Package on an Example

We now prove the following property:

$$P \equiv \forall \texttt{n:nat.(findfourtime n)} \leq \texttt{n}$$

Using the Tactic **Analyze**. In order to benefit from the rewriting steps of the tactic, we first unfold the definition of the function, and then apply the tactic:

```
Lemma P:(n:nat)(le (findfourtime n) n).
Intro n. Unfold findfourtime.
Analyze findfourtime params n.
```

At this point we have the following three subgoals corresponding to the three branches of the definition of `findfourtime`, notice the induction hypothesis on the last one:

```
1: (le O O)
2: [(isfourtime (S m))=true] ⊢ (le (S m) (S m))
3: [(le (findfourtime m) m); (isfourtime (S m))=false]
      ⊢ (le (findfourtime m) (S m))
```

Each of these subgoals is then proved by the **Auto** tactic.

Using the General Principle. We set:

```
Functional Scheme findfourtime_ind:= Induction for findfourtime.
```

Then the proof follows the same pattern as above:

```
Lemma Q':(n:nat)(le (findfourtime n) n).
Intro n. Unfold findfourtime.
Elim n using findfourtime_ind.
```

Again we have three subgoals:

```
1: (le O O)
2: [eq:(isfourtime (S m))=true]
     ⊢ (le (if (isfourtime (S m))
                then (S m) else (findfourtime m)) (S m))
3:[(le (findfourtime m) m); eq:(isfourtime (S m))=false]
     ⊢ (le (if (isfourtime (S m))
                then (S m) else (findfourtime m)) (S m))
```

We see that some rewriting steps must be done by hand to obtain the same result than with the tactic. Finally the whole script is the following:

```
Lemma Q':(n:nat)(le (findfourtime n) n).
Intro n. Unfold findfourtime.
Elim n using findfourtime_ind;Intros;Auto;Rewrite eq;Auto.
Save.
```

In more complex cases, like in section 3.2, the rewriting operations done by the **Analyze** tactic make the script significantly smaller than when using the general principle.

2.5 Inference System

In this subsection, we give an inference system to build an elimination principle for a particular property G from a function t. The main goal of this system is to reproduce the structure of the term t, including **Case**, **Fix**, **let** and λ-abstractions, until it meets an application, a constant or a variable. The remaining terms are replaced by proof obligations whose type is determined by the **Case**, **Fix**, **let** and λ-abstractions from above. The result of this algorithm is an incomplete term, like `Q2` above, from which we can either apply the **Refine** tactic or build a general principle.

Grammar. Coq's internal representation of the expressions of the Calculus of Inductive Constructions (CIC) is given in Figure 1. Although our package deals with mutually recursive functions, we omit them from the presentation for clarity reasons. The construction $Ind(x:T)\{\boldsymbol{T}\}$, where $\boldsymbol{T}$ stands for a vector of terms (types actually), is an inductive type whose n-th constructor is denoted by $C_{n,\boldsymbol{T}}$ (in the sequel we often omit $\boldsymbol{T}$). In the **Fix** expression X is bound in f and corresponds to the function being defined, t is its type and f is the body of the function. In the **Case** expression t_i's are functions taking as many arguments as the corresponding constructors take, see [16,17].

$$
\begin{aligned}
\text{Variables}: V &::= x, y \ldots \\
\text{Sorts}: S &::= Set \mid Prop \mid Type \\
\text{Terms}: T &::= V \mid S \mid \lambda V:T.T \mid \forall V:T.T \mid (T\ T) \\
&\mid \mathbf{Fix}(V,T,T) \mid \mathbf{Case}\ T\ \mathbf{of}\ \boldsymbol{T}\ \mathbf{end} \mid \mathbf{let}\ T = T\ \mathbf{in}\ T \\
&\mid Ind(x:T)\{\boldsymbol{T}\} \mid C_{n,\boldsymbol{T}}
\end{aligned}
$$

Fig. 1. Syntax of terms of CCI

Judgment. Judgments are of the form: $t, X, G, \Gamma_1, \Gamma_2 \vdash P$ where t is the function to use as a model, X is the variable corresponding to the recursive function (bound by **Fix** in t, used to find recursive calls), G is the property to be proved (the goal), Γ_1 is the list of bound variables (initially empty), Γ_2 is the list of equalities corresponding to the case analysis, and P is the proof term of the principle. When proving a property G, we build a term P containing proof obligations represented by $(?:T)$ where T is the awaited property (i.e. type) that must be proved.

Rules are given in Figure 2. In the last rule, $(\boldsymbol{X\ t_i})$ represents all recursive calls found in t. In the present implementation, no nested recursion is allowed. In the rule (Case), we note $(C_i\ \boldsymbol{x_i})$ the fully applied constructor C_i.

$$\frac{t, X,\ G,\ x : T \cup \Gamma_1,\ \Gamma_2 \vdash P}{\lambda x : T.t,\ X,\ G,\ \Gamma_1,\ \Gamma_2 \vdash \lambda x : T.P}(\text{Lambda})$$

$$\frac{\forall i.\{t_i, X,\ G,\ \boldsymbol{x_i} : \boldsymbol{T_i} \cup \Gamma_1,\ E = (C_i\ \boldsymbol{x_i}) \cup \Gamma_2 \vdash P_i\}}{\begin{array}{l}\texttt{Case}\ E\ \textbf{of}\ \boldsymbol{t_i}\ \textbf{end},\ X,\ G,\ \Gamma_1,\ \Gamma_2 \\ \qquad\qquad \vdash (\texttt{Case}\ E\ \textbf{of}\ \boldsymbol{P_i}\ \textbf{end}\ (\texttt{reflequal}\ T_E\ E))\end{array}}(\text{Case})$$

$$\frac{t, X,\ G,\ \boldsymbol{x_i} : \boldsymbol{T_i} \in \boldsymbol{u} \cup \Gamma_1,\ u = v \cup \Gamma_2 \vdash P}{\texttt{let}\ u = v\ \texttt{in}\ t,\ X,\ G,\ \Gamma_1,\ \Gamma_2 \vdash \texttt{let}\ u = v\ \texttt{in}\ P}(\text{Let})$$

$$\frac{f, X, G, \Gamma_1, \Gamma_2 \vdash P}{\texttt{Fix}(X, T, f),\ \_,\ G,\ \Gamma_1,\ \Gamma_2 \vdash \texttt{Fix}(X, G, P)}(\text{Fix})$$

$$\frac{(\boldsymbol{X}\ \boldsymbol{t_i}) \in t}{t, X, G, \Gamma_1, \Gamma_2 \vdash ((? : \forall \Gamma_1.\boldsymbol{X}\ \boldsymbol{t_i} \to \Gamma_2 \to G)\ \Gamma_1)}(\text{Rec})$$
$$\text{if } t \neq \texttt{Fix},\ \texttt{Case}, \texttt{let} \text{ or } \lambda x : T.t.$$

Fig. 2. Elimination algorithm

2.6 More Automation

We can use the equalities generated by the tactic to perform rewriting steps automatically, thereby reducing the user interactions to achieve the proof.

- First, we can rewrite in the generated hypothesis of each branch. We give a modified version of the (Case) rule in figure 3, where $\Gamma_2[E \leftarrow C_i(\boldsymbol{x_i})]$ is the set of equalities Γ_2 where E has been replaced by $C_i(\boldsymbol{x_i})$ in the right members.

$$\frac{\forall i.\{t_i[E \leftarrow C_i(\boldsymbol{x_i})], X,\ G,\ \boldsymbol{x_i} : \boldsymbol{T_i} \cup \Gamma_1,\ E = (C_i\ \boldsymbol{x_i}) \cup \Gamma_2[E \leftarrow C_i(\boldsymbol{x_i})] \vdash P_i\}}{\begin{array}{l}\texttt{Case}\ E\ \textbf{of}\ \boldsymbol{t_i}\ \textbf{end},\ X,\ G,\ \Gamma_1,\ \Gamma_2 \\ \qquad\qquad \vdash (\texttt{Case}\ E\ \textbf{of}\ \boldsymbol{P_i}\ \textbf{end}\ (\texttt{reflequal}\ T_E\ E))\end{array}}(\text{Case})$$

Fig. 3. (Case) rule with rewriting

- Second, we can propagate these rewriting steps in the goal itself. This is possible in the tactic **`Analyze`**, where we call the **`Rewrite`** tactic of Coq, which performs substitutions in the goal, with the set of equalities Γ_2.
  As Coq provides a mechanism for folding/unfolding constants, it is possible that rewriting steps become possible later during the proofs of the generated subgoals. This is specially true when dealing with complex specifications

where unfolding all definitions is not comfortable. Therefore we also provide a rewriting database that contains all the equalities of Γ_2 for each branch. The database can be used later with the **AutoRewrite** tactic.

- Third, we can optimize the **Analyze** tactic in the particular case of a non recursive function applied to constructor terms, i.e. terms of which head symbol is a constructor of an inductive type. For example, suppose we have a goal of the form:

  ```
  (n:nat)(P (f O (S (S (S n)))))
  ```

  Where `f` is a non recursive function. It is clear that we do not want to consider all possible constructors for the arguments of `f`. For example the case `(f (S ...) ...)` is not interesting for us. The idea is to focus on the case branch corresponding to the constructors. We use a simple trick to achieve this: we apply the function to its arguments and reduce. The resulting term is a new function containing only the relevant case analysis. We apply the tactic to this term instead of the initial function.
  Note that this optimization is hardly useful for recursive functions because some recursive hypothesis can be lost when focusing on a particular branch.

3 Applications to JavaCard

In this section, we illustrate the benefits of our package in establishing correctness results for the JavaCard platform. Note that for the clarity of presentation, the Coq code presented below is a simplified account of [3,4].

3.1 Background

JavaCard. is a dialect of Java tailored towards programming multi-application smartcards. Once compiled, JavaCard applets, typically electronic purses and loyalty applets, are verified by a bytecode verifier (BCV) and loaded on the card, where they can be executed by a JavaCard Virtual Machine (JCVM).

Correctness of the JavaCard Platform. The JCVM comes in two flavours: a defensive JCVM, which manipulates typed values and performs type-checking at run-time, and an offensive JCVM, which manipulates untyped values and relies on successful bytecode verification to eliminate type verification at run-time. Following a strategy streamlined in [14], we want to prove that the offensive and defensive VMs coincide on those programs that pass bytecode verification. This involves:

- formalizing both VMs, and show that both machines coincide on those programs whose execution on the defensive VM does not raise a type error;
- formalizing a BCV as a dataflow analysis of an abstract VM that only manipulates types and show that the defensive VM does not raise a type error for those programs that pass bytecode verification.

In both cases, we need to establish a correspondence between two VMs. More precisely, we need to show that the offensive and abstract VMs are sound (non-standard) abstract interpretations of the defensive VMs, in the sense that under suitable constraints "abstraction commutes with execution".

CertiCartes [3,4] is an in-depth feasibility study in the formal verification of the JavaCard platform. CertiCartes contains formal executable specifications of the three VMs (defensive, abstract and offensive) and of the BCV, and a proof that the defensive and offensive VM coincide on those programs that pass bytecode verification. The bulk of the proof effort is concerned with the soundness of the offensive and abstract VMs w.r.t. the defensive VM. In our initial work, such proofs were performed by successive unfoldings of the case analyses arising in the definition of the semantics of each bytecode, leading to cumbersome proofs which were hard to produce, understand and modify. In contrast, the **Analyze** tactic leads to (up to 10 times) smaller proofs that are easier to perform and understand, as illustrated in the next subsections.

3.2 Applications to Proving the Correspondence between Virtual Machines

In order to establish that the offensive VM is a sound abstraction of the defensive VM, one needs to prove that abstraction commutes with execution for every bytecode of the JavaCard instruction set. In the particular case of method invocation, one has to show:

```
nargs:nat
nm: class_method_idx
state: dstate
cap:jcprogram
===========================================
let res=(dinvokevirtual nargs nm state cap) in
let ostate=(alpha_off state) in
let ores=(oinvokevirtual nargs nm ostate cap)
in (INVOKEVIRTUAL_conds res) →(alpha_off res) = ores
```

where:

- the abstraction function `alpha_off` maps a defensive state into an offensive one;
- the predicate `INVOKEVIRTUAL_conds` ensures that the result state `res` is not a type error;
- the functions `dinvokevirtual` and `oinvokevirtual` respectively denote the defensive and offensive semantics of virtual method invokation.

The definition of `dinvokevirtual` is:

```
Definition dinvokevirtual :=
[nargs:nat][nm:class_method_idx][state:dstate][cap:jcprogram]
```

```
(* The initial state is decomposed *)
Cases state of
(sh, (hp, nil)) ⇒(AbortCode state_error state) |
(sh, (hp, (cons h lf))) ⇒

(* nargs must be greater than zero *)
 Cases nargs of
 O    ⇒(AbortCode args_error state) |
 (S _) ⇒

  (* Extraction of the object reference (the nargsth element) *)
  Cases (Nth_func (opstack h) nargs) of
  error ⇒(AbortCode opstack_error state) |
  (value x) ⇒

   (* Tests if this element is a reference *)
   Cases x of
   ((Prim _), vx) ⇒(AbortCode type_error state) |
   ((Ref _), vx) ⇒

   (* tests if the reference is null *)
   (if (test_NullPointer vx)
   then (ThrowException NullPointer state cap)
   else

   (* Extraction of the referenced object *)
   Cases (Nth_func hp (absolu vx)) of
   error ⇒(AbortMemory heap_error state) |
   (value nhp) ⇒

   (* Get the corresponding class *)
   Cases (Nth_elt (classes cap) (get_obj_class_idx nhp)) of
   (value c) ⇒

   (* Get the corresponding method *)
   Cases (get_method c nm) of
   (* Successful method call *)
   (value m) ⇒(new_frame_invokevirtual nargs m nhp state cap) |
   error ⇒(AbortCap methods_membership_error state)
   end |
   _ ⇒(AbortCap class_membership_error state)
   end end) end end end end.
```

In our initial work on CertiCartes, such statements were proved "by hand", by following the successive case analyses arising in the definition of the semantics of each bytecode. The proofs were hard to produce, understand and modify. In contrast, the **Analyze** tactic leads to smaller proofs that are easier to perform and understand. For example, the following script provides the first steps of the proof:

```
Simpl.
Unfold dinvokevirtual.
Analyze dinvokevirtual params nargs nm state cap.
```

At this point, nine different subgoals are generated, each of them corresponding to a full case analysis of the function. In the case of a successful method call, the corresponding subgoal is:

```
...
_eg_8 : (get_method c nm)=(value m)
_eg_7 : (Nth_elt (classes cap) (get_obj_class_idx nhp))
        =(value c)
_eg_7 : (Nth_func hp (absolu vx))=(value nhp)
_eg_6 : (test_NullPointer vx)=false
_eg_5 : t=(Ref t0)
_eg_4 : x=((Ref t0),vx)
_eg_3 : (Nth_func (opstack h) (S n))=(value ((Ref t0),vx))
_eg_2 : nargs=(S n)
_eg_1 : state=(sh,(hp,(cons h lf)))
...
H : (INVOKEVIRTUAL_conds
(new_frame_invokevirtual (S n) m nhp (sh,(hp,(cons h lf))) cap))
===========================================
(alpha_off
(new_frame_invokevirtual (S n) m nhp (sh,(hp,(cons h lf))) cap))
=(oinvokevirtual (S n) nm (alpha_off (sh,(hp,(cons h lf)))) cap)
```

This goal, as all other goals generated by the tactic, can in turn be discharged using rewriting and some basic assumptions on the representation of JavaCard programs.

3.3 Applications to Proving Memory Invariants

In order to establish that the abstract VM is a sound abstraction of the defensive VM, one needs to establish a number of invariants on the memory of the virtual machine, for example that executing the abstract virtual machine does not create illegal JavaCard types such as arrays of arrays. Proving such a property is tedious and involves several hundreds of case analyses, including a topmost case analysis on the instruction set. In the case of the `load` bytecode, one is left with the goal:

```
s : astate
t : type
l : locvars_idx
================================================
(legal_types s) →(legal_types (tload t l s))
```

Whereas a proof "by hand" leaves the user with 29 subgoals to discharge, an application of the **Analyze** tactic leaves the user with 3 subgoals to discharge, namely the three interesting cases corresponding to the successful execution of the `tload` bytecode. In such examples, the use of our tactic reduces the size of proof scripts by a factor of 10.

4 Extensions and Related Work

4.1 Related Work

Our work is most closely related to Slind's work on reasoning about terminating functional programs, see e.g. [20] where Slind presents a complete methodology to define a function from a list of equations and a *termination relation*. This relation is used to generate *termination conditions*, which need to be discharged to prove that the rewrite system defined by the equations terminates. From the proofs of the termination conditions, Slind automatically synthesizes an induction principle for the function. Slind's induction principle is closely related to ours but there are some differences:

- Slind's work focuses on proof-assistants based on higher-order logic, in particular on Isabelle [18] and HOL [13], that do not feature proof terms. In our framework, objects, properties and proofs are all terms of the Calculus of Inductive Constructions, and hence we need to provide a proof term for the customized principle attached to the function. As pointed out in Section 2, the construction of the proof term can be done directly by transforming the term corresponding to the function;
- Slind's work provides for each total and terminating function `f` a generic induction principle that can be used to prove properties about `f`. In contrast, the tactic **`Analyze`** starts from the property to be proven, say ϕ, and proceeds to build directly a proof of ϕ by following the shape of the definition of `f`. Eventually the tactic returns some proof obligations that are required to complete the proof of ϕ. This allows our package to perform rewriting and other operations during the building of the proof, leading to a drastic increase in automation. Since similar proof terms are constructed every time one applies the tactic on the function `f`, it could be argued that such a feature is costly in terms of memory. However, we have not experienced any efficiency problem although our JavaCard development is fairly large;
- Slind's work provides support for well-founded and nested recursion, which our package does not handle currently.

Our work is also closely related to the induction process defined by Boyer and Moore. In Nqthm [8], the directive `INDUCT (f v1...vn)` can be used during a proof to perform the same case analysis and recursion steps as in the definition of `f`. An interesting feature that we do not handle in this paper is the ability of *merging* different functions into one single induction principle.

These works, as well as ours, aims at providing efficient support for reasoning about executable specifications. An alternative strategy consists in developing tools for executing specifications that may be relational, at least in part. For example, Elf [19] combines type theory and logic programming and hence provides support for executing relational specifications. Further, there have been some efforts to develop tools for executing specifications written in Coq, Isabelle or related systems. In particular some authors have provided support for executing relational specifications: for example Berghofer and Nipkow [5] have developed a tool to execute Isabelle theories in a functional language, and used it in the context of their work on Java and JavaCard. Conversely, some authors have provided support to translate executable specifications into input for proof assistants: for example, Terrasse [21,22] has developed a tool to translate Typol specifications into Coq. However, we believe that executable specifications are better suited for automating proofs, especially because they lend themselves well to automated theorem-proving and rewriting techniques.

A third possible strategy would consist of relating relational and functional specifications. For example, one could envision a tool that would derive automatically from a functional specification (1) its corresponding relational specification (2) a formal proof that the two specifications coincide. Such a tool would allow the combination of the best of both worlds (one could be using elimination/inversion principles to reason about relational specifications, and then transfer the results to functional ones); yet we are not aware of any conclusive experience in this direction.

4.2 Extensions

As emphasized in the introduction, the main motivation behind the package is to develop effective tool support to reason about the JavaCard platform. The following two extensions are the obvious next steps towards this objective:

- *support for conditional rewriting rules.* We would like to introduce functions using the format of conditional rewriting rules of [2]. Indeed, this format, which can be automatically translated into Coq (provided the guard conditions are satisfied), lead to more readable specifications, especially for partial functions.
- *support for inversion principles.* We also would like to provide inversion principles like those that exist for inductive types [10]. Such principles do not use induction; instead they proceed by case analysis and use the fact that constructors of inductive types are injective. For example, inversion principles for functions should allow us to deduce:

  `x=0 ∨ x=(S S S S y)∧((isfourtime y)=true)`

  from `(isfourtime x)=true`.
- *support for merging.* Merging [8] allows to generate a single induction principle for several functions. This technique allows to define very accurate induction principles, that are very useful in fully automated tactics.

From a more general perspective, it would be interesting to provide support for nested and well-founded recursion. Furthermore, a higher automation of equational reasoning is required in the proofs of commuting diagrams. One possibility is to rely on an external tool to generate rewriting traces that are then translated in Coq. There are several candidates for such tools, including Elan [6] and Spike [7], and ongoing work to interface them with proof assistants such as Coq, see e.g. [1].

5 Conclusion

We have described a toolset for reasoning about complex recursive functions in Coq. A fuller account of the package, including a tutorial, can be found in [12].

In addition, we have illustrated the usefulness of our package in reasoning about the JavaCard platform. This positive experience supports the claim laid in [2] that proofs of correctness of the JavaCard platform can be automated to a large extent and indicates that proof assistants could provide a viable approach to validate novel type systems for future versions of Javacard.

Thanks

We would like to thank the anonymous referees for there constructive remarks, Yves Bertot for its precious advice, ideas and examples, Guillaume Dufay for the feedback on using the tactic intensively, and Konrad Slind for the precisions on TFL/Isabelle.

References

1. C. Alvarado and Q.-H. Nguyen. ELAN for equational reasoning in COQ. In J. Despeyroux, editor, *Proceedings of LFM'00*, 2000. Rapport Technique INRIA.
2. G. Barthe, G. Dufay, M. Huisman, and S. Melo de Sousa. Jakarta: a toolset to reason about the JavaCard platform. In I. Attali and T. Jensen, editors, *Proceedings of e-SMART 2001*, volume 2140 of *Lecture Notes in Computer Science*, pages 2–18. Springer-Verlag, 2001.
3. G. Barthe, G. Dufay, L. Jakubiec, and S. Melo de Sousa. A formal correspondence between offensive and defensive JavaCard virtual machines. In A. Cortesi, editor, *Proceedings of VMCAI 2002*, volume 2294 of *Lecture Notes in Computer Science*, pages 32–45. Springer-Verlag, 2002.
4. G. Barthe, G. Dufay, L. Jakubiec, B. Serpette, and S. Melo de Sousa. A Formal Executable Semantics of the JavaCard Platform. In D. Sands, editor, *Proceedings of ESOP 2001*, volume 2028 of *Lecture Notes in Computer Science*, pages 302–319. Springer-Verlag, 2001.
5. S. Berghofer and T. Nipkow. Executing higher order logic. In P. Callaghan, Z. Luo, J. McKinna, and R. Pollack, editors, *Proceedings of TYPES 2000*, volume LNCS 2277 of *Lecture Notes in Computer Science*. Springer-Verlag, 2002.
6. P. Borovanský, H. Cirstea, H. Dubois, C. Kirchner, H. Kirchner, P.-E. Moreau, C. Ringeissen, and M. Vittek. *The Elan V3.4. Manual*, 2000.
7. A. Bouhoula. Automated theorem proving by test set induction. *Journal of Symbolic Computation*, 23:47–77, 1997.
8. R.S. Boyer and J.S. Moore. *A Computational Logic Handbook*. Academic Press, 1988.
9. Coq Development Team. *The Coq Proof Assistant User's Guide. Version 7.2*, January 2002.
10. C. Cornes. *Conception d'un langage de haut niveau de representation de preuves: Récurrence par filtrage de motifs; Unification en présence de types inductifs primitifs; Synthèse de lemmes d'inversion.* PhD thesis, Université de Paris 7, 1997.

11. C. Cornes and D. Terrasse. Automating inversion and inductive predicates in Coq. In S. Berardi and M. Coppo, editors, *Proceedings of Types'95*, volume 1158 of *Lecture Notes in Computer Science*, pages 85–104. Springer-Verlag, 1995.
12. P. Courtieu. Function Schemes in Coq: Documentation and tutorial. See `http://www-sop.inria.fr/lemme/Pierre.Courtieu/funscheme.html`.
13. M.J.C. Gordon and T.F. Melham, editors. *Introduction to HOL: A theorem proving environment for higher-order logic.* Cambridge University Press, 1993.
14. J.-L. Lanet and A. Requet. Formal Proof of Smart Card Applets Correctness. In J.-J. Quisquater and B. Schneier, editors, *Proceedings of CARDIS'98*, volume 1820 of *Lecture Notes in Computer Science*, pages 85–97. Springer-Verlag, 1998.
15. Z. Luo and R. Pollack. LEGO proof development system: User's manual. Technical Report ECS-LFCS-92-211, LFCS, University of Edinburgh, May 1992.
16. C. Paulin-Mohring. Inductive definitions in the system Coq. Rules and properties. In M. Bezem and J.F. Groote, editors, *Proceedings of TLCA'93*, volume 664 of *Lecture Notes in Computer Science*, pages 328–345. Springer-Verlag, 1993.
17. C. Paulin-Mohring. *Définitions Inductives en Theorie des Types d'Ordre Superieur.* Habilitation à diriger les recherches, Université Claude Bernard Lyon I, 1996.
18. L. Paulson. *Isabelle: A generic theorem prover*, volume 828 of *Lecture Notes in Computer Science.* Springer-Verlag, 1994.
19. F. Pfenning. Elf: a meta-language for deductive systems. In A. Bundy, editor, *Proceedings of CADE-12*, volume 814 of *Lecture Notes in Artificial Intelligence*, pages 811–815. Springer-Verlag, 1994.
20. K. Slind. *Reasoning about Terminating Functional Programs.* PhD thesis, TU Münich, 1999.
21. D. Terrasse. Encoding natural semantics in Coq. In V. S. Alagar, editor, *Proceedings of AMAST'95*, volume 936 of *Lecture Notes in Computer Science*, pages 230–244. Springer-Verlag, 1995.
22. D. Terrasse. *Vers un environnement d'aide au développement de preuves en Sémantique Naturelle.* PhD thesis, Ecole Nationale des Ponts et Chaussées, 1995.

Verified Bytecode Model Checkers

David Basin, Stefan Friedrich, and Marek Gawkowski

Albert-Ludwigs-Universität Freiburg, Germany
{basin,friedric,gawkowsk}@informatik.uni-freiburg.de

Abstract. We have used Isabelle/HOL to formalize and prove correct an approach to bytecode verification based on model checking that we have developed for the Java Virtual Machine. Our work builds on, and extends, the formalization of the Java Virtual Machine and data flow analysis framework of Pusch and Nipkow. By building on their framework, we can reuse their results that relate the run-time behavior of programs with the existence of well-typings for the programs. Our primary extensions are to handle polyvariant data flow analysis and its realization as temporal logic model checking. Aside from establishing the correctness of our model-checking approach, our work contributes to understanding the interrelationships between classical data flow analysis and program analysis based on model checking.

1 Introduction

The security of Java and the Java Virtual Machine (JVM), and in particular of its bytecode verifier, has been the topic of considerable study. Research originally focused on abstract models of the JVM [4,19] and more recently on machine checked proofs of the correctness of bytecode verifier models [2,12,13,15] and type inference algorithms based on data flow analysis [11]. There are two research directions that are the starting point for our work. First, the work of Pusch [15] and Nipkow [11] on models and proofs in Isabelle/HOL for verifying bytecode verifiers based on data flow algorithms. Second, the work of Posegga and Vogt [14], later extended in [1], on reducing bytecode verification to model checking. These two directions are related: both are based on abstract interpretation and solving fixpoint equations. However, whereas data flow analysis computes a type for a method and checks that the method's instructions are correctly applied to data of that type, the model-checking approach is more declarative; here one formalizes the instruction applicability conditions as formulae in a temporal logic (e.g. LTL) and uses a model checker to verify that an abstraction of the method (corresponding to the abstract interpreter of the data flow approach) satisfies these applicability conditions.

In this paper we explore the interrelationships between these two research directions. We present the first machine-checked proof of the correctness of the model-checking approach to bytecode verification, and in doing so we build upon the Isabelle/HOL formalizations of the JVM [15] and the abstract verification framework that Nipkow developed for verifying data flow algorithms in [11].

V.A. Carreño, C. Muñoz, S. Tahar (Eds.): TPHOLs 2002, LNCS 2410, pp. 47–66, 2002.

This framework formalizes the notion of a well-typing for bytecode programs and proves that a bytecode verifier is correct (i.e., accepts only programs free of runtime type errors) when it accepts only programs possessing a well-typing. We will show that every bytecode program whose abstraction globally satisfies the instruction applicability conditions (which can be established by model checking) in fact possesses such a well-typing, i.e. our goal is to validate the model-checking approach by proving a theorem of the form

$$(\mathit{abstraction}(\mathit{Method}) \models_{LTL} \Box \mathit{app\_conditions}(\mathit{Method})) \\ \Longrightarrow (\exists\, \phi.\ \mathit{welltyping}(\phi, \mathit{Method}))\,.$$

We achieve this by modifying and extending the framework of Pusch and Nipkow to support polyvariant data flow analysis and model checking.

Our development can be subdivided into six areas. Based on (1) *preliminary definitions* and (2) *semilattice-theoretic foundations* we define (3) the *JVM model*, which includes the JVM type system, the JVM abstract semantics, and the definition of the JVM state type. In (4), the *model-checking framework*, we define Kripke structures and traces, which we later use to formalize model checking. Afterwards, we define the translation of bytecode programs into (5) *finite transition systems*. We use the abstract semantics of JVM programs to define the transition relations and the JVM type system to build LTL formulae for model checking. Finally, we state and prove in (6) our *main theorem*.

Overall, most of the theories we use are adopted from the work of Nipkow [11], and hence our work can, to a large part, be seen as an instance of Nipkow's abstract correctness framework. The table below provides an overview of how our formalization differs from Nipkow's.

Theories	Status
JBasis (1), Type (1), Decl (1), TypeRel (1), State (1), WellType (1), Conform (1), Value(1), Semilat (2), Err (2), Opt (2), Product(2), JType (3), BVSpec (3)	unchanged
Listn (2)	the stack model is changed
JVMInstructions (1), JVMExecInstr (1), JVMExec (1) JVMType (3), Step (3),	modified due to the changes in the stack model
Semilat2 (2), Kripke (4), LTL (4), ModelChecker (5), JVM_MC (6)	new

Our original motivation for undertaking this development was a pragmatic one: we had developed a model-checking based tool [1] for verifying bytecode and we wanted to establish the correctness of the approach taken. We see our contributions, however, as more general. Our development provides insight into the relationship between monovariant and polyvariant data flow analysis and model checking, in particular, what differences are required in their formalization. Monovariant analysis associates one program state type, which contains information about the stack type and register type, to each control point (as in Sun's verifier or more generally Kildall algorithm, which was analyzed by Nipkow), or one such type per subroutine to each control point. In contrast, polyvariant analysis allows multiple program state types per control point, depending on the number of control-flow paths that lead to this control point [9]. In the formalization of Pusch and Nipkow [15,11], monovariant data flow analysis is used, which is adequate since they do not consider subroutines and interfaces. In our approach, we use model checking, which performs a polyvariant data flow analysis. The result is not only that we can base our bytecode verification tool on standard model checkers, but also that our bytecode verifier accepts more (type-correct) programs than bytecode verifiers performing monovariant data flow analysis. For instance, our tool can successfully type check programs where, for a given program point and for two different execution paths, the operand stack has two different sizes or unused stack elements have different (and incompatible) types.

Despite switching to a polyvariant, model-checking approach, we were able to reuse a surprising amount of Nipkow's formal development, as the above table indicates. Our main change was to allow stack elements of incompatible types, which we achieved by generalizing the notion of the JVM state type. In doing so, we enlarged the set of programs that fulfill the well-typing definition for bytecode programs as formalized by Pusch [15]. An additional change was required from the model-checking side: to allow the computation of the supremum of two stacks at each program point, we had to specialize our notion of type correct program by imposing additional constraints concerning the stack size on programs. Overall, the changes we made appear generally useful. Polyvariant data flow analysis constitutes a generalization of monovariant data flow analysis and it should be possible (although we have not formally checked this) to formalize the monovariant data flow analysis using our JVM model with an accordingly modified notion of well-typing.

Finally, note that we formalize model checking not algorithmically, as done in [16], but declaratively. Consequently our formalization of model checking is independent of the implemented model-checking algorithm. In our work with Posegga [1], for instance, we implemented the backends both for symbolic and for explicit state model checkers. As our correctness theorem here is a statement about the correctness of the model-checking approach to bytecode verification it is valid for both these backends. Hence the plural in our title "verified bytecode model checkers."

2 Background

We present background concepts necessary for our formalization. The sections 2.1, 2.2, and 2.4–2.6 describe (unmodified) parts of Nipkow's formalization and are summarized here for the sake of completeness.

2.1 Basic Types

We employ basic types and definitions of Isabelle/HOL. Types include *bool*, *nat*, *int*, and the polymorphic types $\alpha\,set$ and $\alpha\,list$. We employ a number of standard functions on (cons) lists including a conversion function set from lists to sets, infix operators # and @ to build and concatenate lists, and a function size to denote the length of a list. $xs!i$ denotes the i-th element of a list xs and $xs[i := x]$ overwrites i-th element of xs with x. Finally, we use Isabelle/HOL records to build tuples and functional images over sets: $(\!|\ \mathsf{a} :: \alpha,\ \mathsf{b} :: \beta\ |\!)$ denotes the record type containing the components a and b of types α and β respectively and $\mathsf{f}\,`\,A$ denotes the image of the function f over the set A.

2.2 Partial Orders and Semilattices

A partial order is a binary predicate of type $\alpha\ ord = \alpha \rightarrow \alpha \rightarrow bool$. We write $x \leq_r y$ for $r\,x\,y$ and $x <_r y$ for $x \leq_r y \wedge x \neq y$. $r :: \alpha\ ord$ is a **partial order** iff r is reflexive, antisymmetric und transitive. We formalize this using the predicate $\mathsf{order} :: \alpha\, ord \rightarrow bool$. $\top :: \alpha$ is the **top** element with respect to the partial order r iff the predicate $\mathsf{top} :: \alpha\ ord \rightarrow \alpha \rightarrow bool$, defined by $\mathsf{top}\, r\, T \equiv \forall x.\, x \leq_r T$ holds. Given the types $\alpha\ binop = \alpha \rightarrow \alpha \rightarrow \alpha$ and $\alpha\ sl = \alpha\ set \times \alpha\ ord \times \alpha\ binop$ and the supremum notation $x +_f y = f\,x\,y$, we say that $(A,\ r,\ f) :: \alpha\ sl$ is a **(supremum) semilattice** iff the predicate $\mathsf{semilat} :: \alpha\ sl \rightarrow bool$ holds

$$
\begin{aligned}
\mathsf{semilat}(A,\ r,\ f) \equiv\ & \mathsf{order}\, r\ \wedge\ \mathsf{closed}\, A\, f\ \wedge \\
& (\forall x\, y \in A.\ x \leq_r x +_f y)\ \wedge\ (\forall x\, y \in A.\ y \leq_r x +_f y)\ \wedge \\
& (\forall x\, y\, z \in A.\ x \leq_r z\ \wedge\ y \leq_r z \longrightarrow x +_f y \leq_r z),
\end{aligned}
$$

where $\mathsf{closed}\, A\, f\ \equiv\ \forall x\, y \in A.\ x +_f y \in A$.

2.3 Least Upper Bounds of Sets

The above definitions are for reasoning about the supremum of binary operators $f :: \alpha\ binop$. We define a new theory, Semilat2 to reason about the supremum of sets.

To build the suprema over sets, we define the function $\mathsf{lift\_sup} :: \alpha\ sl \rightarrow \alpha \rightarrow \alpha \rightarrow \alpha \rightarrow \alpha$, written $\mathsf{lift\_sup}\,(A,\ r,\ f)\, T\, x\, y = x \uplus_{(A,\, r,\, f),\, T}\ y$, that lifts the binary operator $f :: \alpha\ binop$ over the type α:

$$
\begin{aligned}
x \uplus_{(A,\, r,\, f),\, T}\ y \equiv\ & \mathtt{if}\ (\mathsf{semilat}\,(A,\ r,\ f)\ \wedge\ \mathsf{top}\, r\, T)\ \mathtt{then} \\
& \qquad \mathtt{if}\ (x \in A\ \wedge y \in A)\mathtt{then}\ (x +_f\ y)\ \mathtt{else}\ T \\
& \mathtt{else}\ \mathsf{arbitrary}
\end{aligned}
$$

To support reasoning about the least upper bounds of sets, we introduce the **bottom** element $B :: \alpha$, which is defined by $\mathsf{bottom}\ r\ B\ \equiv\ \forall x.\, B \leq_r x$. We use the Isabelle/HOL function $\mathsf{fold} :: (\beta \rightarrow \alpha \rightarrow \alpha) \rightarrow \alpha \rightarrow \beta\ set \rightarrow \alpha$ to build the least upper bounds of sets and in the following we write $\bigsqcup_{(A,\, r,\, f),\, T,\, B} A'$ for $\mathsf{fold}\,(\lambda\, x\, y.\, x \uplus_{(A,\, r,\, f),\, T}\ y)\, B\, A'$. We have shown that $\bigsqcup_{(A,\, r,\, f),\, T,\, B} A'$ is a least upper bound over A'.

Lemma 1. *Given* $\mathsf{semilat}\,(A,\ r,\ f)$, $\mathsf{finite}\, A$, $T \in A$, $\mathsf{top}\, r\, T$, $B \in A$, $\mathsf{bottom}\, r\, B$ *and given an arbitrary function* g*, where* $g\ B = B$, $\forall x \in A.\ g\ x \in A$*, and* $\forall x\, y \in A.\, g\,(x +_f\ y) \leq_r (g\, x +_f\ g\, y)$*, then*

$$
(g\,`\, A'' \subseteq A') \longrightarrow g\,(\bigsqcup_{(A,\, r,\, f),\, T,\, B} A'') \leq_r (\bigsqcup_{(A,\, r,\, f),\, T,\, B} A')
$$

2.4 The Error Type and *err*-Semilattices

The theory Err defines an error element to model the situation where the supremum of two elements does not exist. We introduce both a datatype and a corresponding construction on sets:

$$\textbf{datatype}\ \alpha\ \mathit{err} \equiv \mathsf{Err} \mid \mathsf{OK}\ \alpha \qquad \mathsf{err}\ A \equiv \{\mathsf{Err}\} \cup \{\mathsf{OK}\ a \mid a \in A\}$$

Orderings r on α can be lifted to $\alpha\ \mathit{err}$ by making Err the top element:

$$\mathsf{le}\ r\ (\mathsf{OK}\ x)\ (\mathsf{OK}\ y) = x\ \leq_r\ y$$

We now employ the following lifting function

$$\begin{array}{l} \mathsf{lift2} :: (\alpha \to \beta \to \gamma\ \mathit{err}) \to \alpha\ \mathit{err} \to \beta\ \mathit{err} \to \gamma\ \mathit{err} \\ \mathsf{lift2}\ f\ (\mathsf{OK}\ x)\ (\mathsf{OK}\ y) = f\ x\ y \\ \mathsf{lift2}\ f\ \_ \qquad\quad \_ \qquad\quad = \mathsf{Err} \end{array}$$

to define a new notion of an *err*-semilattice, which is a variation of a semilattice with a top element. It suffices to say how the ordering and the supremum are defined over non-top elements and hence we represent a semilattice with top element Err as a triple of type *esl*: $\alpha\ \mathit{esl}\ =\ \alpha\ \mathit{set}\ \times\ \alpha\ \mathit{ord}\ \times\ \alpha\ \mathit{ebinop}$, where $\alpha\ \mathit{ebinop}\ =\ \alpha\ \to\ \alpha\ \to\ \alpha\ \mathit{err}$. We introduce also conversion functions between the types *sl* and *esl*:

$$\begin{array}{ll} \mathsf{esl} :: \alpha\ \mathit{sl} \to \alpha\ \mathit{esl} & \mathsf{sl} :: \alpha\ \mathit{esl} \to \alpha\ \mathit{err}\ \mathit{sl} \\ \mathsf{esl}\ (A, r, f) = (A, r, \lambda x\, y.\ \mathsf{OK}\ (f\ x\ y)) & \mathsf{sl}\ (A, r, f) = (\mathsf{err}\ A, \mathsf{le}\ r, \mathsf{lift2}\ f) \end{array}$$

Finally we define $L :: \alpha\ \mathit{esl}$ to be an *err*-**semilattice** iff sl L is a semilattice. It follows that esl L is an *err*-semilattice if L is a semilattice.

2.5 The Option Type

Theory Opt introduces the type *option* and the set opt as duals to the type *err* and the set err:

$$\textbf{datatype}\ \alpha\ \mathit{option} \equiv \mathsf{None} \mid \mathsf{Some}\ \alpha \qquad \mathsf{opt}\ A \equiv \{\mathsf{None}\} \cup \{\mathsf{Some}\ a \mid a \in A\}$$

The theory also defines an ordering where None is the bottom element and a supremum operation:

$$\mathsf{le}\ r\ (\mathsf{Some}\ x)\ (\mathsf{Some}\ y) = x \leq_r y \qquad \mathsf{sup}\ f\ (\mathsf{Some}\ x)\ (\mathsf{Some}\ y) = \mathsf{Some}(f\ x\ y)$$

Note that the function $\mathsf{esl}\ (A, r, f) = (\mathsf{opt}\ A, \mathsf{le}\ r, \mathsf{sup}\ f)$ maps *err*-semilattices to *err*-semilattices.

2.6 Products

Theory Product provides what is known as the *coalesced* product, where the top elements of both components are identified.

$$\begin{aligned}
&\mathsf{esl} :: \alpha\ esl \rightarrow \beta\ esl \rightarrow (\alpha \times \beta)\ esl\\
&\mathsf{esl}\ (A, r_A, f_A)\ (B, r_B, f_B) = (A \times B, \mathsf{le}\ r_A\ r_B, \mathsf{sup}\ f_A\ f_B)\\
&\mathsf{sup} :: \alpha\ ebinop \rightarrow \beta\ ebinop \rightarrow (\alpha \times \beta)\ ebinop\\
&\mathsf{sup}\ f\ g = \lambda\,(a_1, b_1)\,(a_2, b_2).\ \mathsf{Err.sup}\ (\lambda\, x\, y.\ (x, y))\ (a_1 +_f a_2)\ (b_1 +_g b_2)
\end{aligned}$$

The ordering function $\mathsf{le} :: \alpha\ ord \rightarrow \beta\ ord \rightarrow (\alpha \times \beta)\ ord$ is defined as expected.

Note that $\times$ is used both on the type and set level. Nipkow has shown that if both L_1 and L_2 are *err*-semilattices, so is $\mathsf{esl}\ L_1\ L_2$.

2.7 Lists of Fixed Length

For our application we must model the JVM stack differently from Nipkow. Our formalization is modified to allow the polyvariant analysis and thus requires associating multiple stack-register types to each control point within the program.

To facilitate this modification, we define the theory Listn, of fixed length lists over a given set. In HOL, this is formalized as a set rather than a type:

$$\mathsf{list}\ n\ A \equiv \{xs \mid \mathsf{size}\ xs = n \wedge \mathsf{set}\ xs \subseteq A\}$$

This set can be turned into a semilattice in a componentwise manner, essentially viewing it as an n-fold Cartesian product:

$$\begin{aligned}
&\mathsf{sl} :: nat \rightarrow \alpha\ sl \rightarrow \alpha\ list\ sl &\qquad& \mathsf{le} :: \alpha\ ord \rightarrow \alpha\ list\ ord\\
&\mathsf{sl}\ n\ (A, r, f) = (\mathsf{list}\ n\ A, \mathsf{le}\ r, \mathsf{map2}\ f) && \mathsf{le}\ r = \mathsf{list_all2}\ (\lambda\, x\, y.\ x \leq_r y)
\end{aligned}$$

Here the auxiliary functions $\mathsf{map2} :: (\alpha \rightarrow \beta \rightarrow \gamma) \rightarrow \alpha\ list \rightarrow \beta\ list \rightarrow \gamma\ list$ and $\mathsf{list\_all2} :: (\alpha \rightarrow \beta \rightarrow bool) \rightarrow \alpha\ list \rightarrow \beta\ list \rightarrow bool$ are defined as expected. We write $xs \leq_{[r]} ys$ for $xs \leq_{(\mathsf{le}\ r)} ys$ and $xs +_{[f]} ys$ for $xs +_{(\mathsf{map2}\ f)} ys$. Observe that if L is a semilattice then so is $\mathsf{sl}\ n\ L$.

To combine lists of different lengths, we use the following function:

$$\begin{aligned}
&\mathsf{sup} :: (\alpha \rightarrow \beta \rightarrow \gamma\ err) \rightarrow \alpha\ list \rightarrow \beta\ list \rightarrow \gamma\ list\ err\\
&\mathsf{sup}\ f\ xs\ ys = \mathtt{if}\ (\mathsf{size}\ xs = \mathsf{size}\ ys)\ \mathtt{then}\ \mathsf{OK}\ (\mathsf{map2}\ f\ xs\ ys)\ \mathtt{else}\ \mathsf{Err}
\end{aligned}$$

Note that in our JVM formalization, the supremum of two lists xs and ys of equal length returns a result of the form $\mathsf{OK}\ zs$ with $zs!i = \mathsf{Err}$ in the case that the supremum of two corresponding elements $xs!i$ and $ys!i$ equals Err. This differs from sup in Nipkow's formalization, which returns Err in this case. Below we present the *err*-semilattice upto_esl of all lists up to a specified length:

$$\begin{aligned}
&\mathsf{upto_esl} :: nat \rightarrow \alpha\ sl \rightarrow \alpha\ list\ esl\\
&\mathsf{upto_esl}\ n = \lambda\,(A, r, f).\ (\cup_{i \leq n}\ \mathsf{list}\ i\ A, \mathsf{le}\ r, \mathsf{sup}\ f)
\end{aligned}$$

We have shown that if L is a semilattice then $\mathsf{upto\_esl}\ n\ L$ is an *err*-semilattice.

3 The JVM Model

In this section we show how the JVM can be formalized for the purpose of polyvariant data flow analysis. In Section 3.1, our formalization adopts, unchanged, Nipkow's formalization of the JVM type system. Using this formalization, we define our modified program state type and construct a semilattice whose carrier set consists of elements of this type. Based on the modified notion of the program state type, we redefine, in Section 3.2, the syntax and abstract semantics of JVM programs and consequently also redefine the definitions of a JVM abstract execution function and of well-typed methods in Section 3.3.

3.1 Type System and Well-Formedness

The theory Types defines the types of our JVM. Our machine supports the void type, integers, null references, and class types (based on the type *cname* of class names):

$$\textbf{datatype}\ ty\ \equiv\ \mathsf{Void} \mid \mathsf{Integer} \mid \mathsf{Boolean} \mid \mathsf{NullT} \mid \mathsf{Class}\ cname$$

The theory Decl defines class declarations and programs. Based on the type *mname* of method names and *vname* of variable names, we model a JVM program $\mathsf{P} :: (cname \times (cname \times (vname \times ty)\ list \times ((mname \times ty\ list) \times ty \times \gamma)\ list))\ list$ as a list of class files. Each class file records its class name, the name of its super class, a list of field declarations, and a list of method declarations. The type *cname* is assumed to have a distinguished element Object. Our program formalization gives rise to a subclass relation subcls1 and subclasses induce a subtype relation $\mathsf{subtype} :: \gamma\ prog \rightarrow ty \rightarrow ty \rightarrow bool$, where $\gamma\ prog$ is a type association defined as follows:

$$\begin{aligned}
\gamma\ mdecl &= (mname \times ty\ list) \times ty \times \gamma \\
\gamma\ class &= cname \times (vname \times ty)\ list \times \gamma\ mdecl\ list \\
\gamma\ cdecl &= cname \times \gamma\ class \\
\gamma\ prog &= \gamma\ cdecl\ list
\end{aligned}$$

Corresponding to the subtype P relation we have supremum on types $\mathsf{sup} :: ty \rightarrow ty \rightarrow ty\ err$. As abbreviations, we define $\mathsf{types}\ P = \{\tau \mid \mathsf{is\_type}\ P\ \tau\}$ and $\tau_1 \sqsubseteq_P \tau_2 = \tau_1 \leq_{\mathsf{subtype}\ P} \tau_2$. Below, we use the predicate $\mathsf{is\_class}\ P\ C$ to express that the class C is in the program P.

Well-formedness of JVM programs is defined by context conditions that can be checked statically prior to bytecode verification. We formalize this using a predicate $\mathsf{wf\_prog} :: \gamma\ wf\_mb \rightarrow \gamma\ prog \rightarrow bool$, where $\gamma\ wf\_mb = \gamma\ prog \rightarrow cname \rightarrow \gamma\ mdecl \rightarrow bool$. The predicate wf_mb expresses that a method $m :: \gamma\ mdecl$ in class C from program P is a well-typing. Informally, $\mathsf{wf\_prog}\ wf\_mb\ P$ means that: $\mathsf{subcls1}\ P$ is univalent (i.e. $\mathsf{subcls1}\ P$ represents a single inheritance hierarchy) and acyclic, and both $(\mathsf{subcls1}\ P)^{-1}$ and $(\mathsf{subtype}\ P)^{-1}$ are well-founded. The following lemma holds for all well-formed programs P:

Lemma 2. $\forall\, wf\_mb\ P.$
wf_prog $wf\_mb\ P \longrightarrow$ semilat (sl (types P, subtype P, sup P)) $\wedge$ finite (types P)

We will use the semilattice sl (types P, subtype P, sup P) to construct a semilattice with the carrier set of program states.

The JVM is a stack machine where each activation record consists of a stack for expression evaluation and a list of local variables (called **registers**). The abstract semantics, which operates with types as opposed to values, records the type of each stack element and each register. At different program points, a register may hold incompatible types, e.g. an integer or a reference, depending on the computation path that leads to that point. This facilitates the reuse of registers and is modeled by the HOL type *ty err*, where OK τ represents the type τ and Err represents the inconsistent type. In our JVM formalization, the elements of the stack can also be reused. Thus the configurations of our abstract JVM are pairs of lists, which model an expression stack and a list of registers:

$$state\_type \;=\; ty\ err\ list \;\times\; ty\ err\ list$$

Note that this type differs from the type *ty list* $\times$ *ty err list* presented in [11], where the stack only holds the values that can actually be used.

We now define the type of the program state as *state_type option err*, where OK None indicates a program point that is not reachable (dead code), OK(Some s) is a normal configuration s, and Err is an error. In the following, we use the type association *state* $=$ *state_type option err*. Turning *state* into a semilattice structure is easy because all of its constituent types are (*err*-)semilattices. The set of operand stacks forms the carrier set of an *err*-semilattice because the supremum of stacks of different size is Err; the set of lists of registers forms the carrier set of a semilattice because the number of registers is fixed:

$$\begin{array}{ll}
\mathsf{stk_esl} & ::\ \gamma\ prog \rightarrow nat \rightarrow ty\ err\ list\ esl \\
\mathsf{stk_esl}\ P\ maxs & \equiv \mathsf{upto_esl}\ maxs\ (\mathsf{sl}\ (\mathsf{types}\ P,\ \mathsf{subtype}\ P,\ \mathsf{sup}\ P)) \\
\mathsf{reg_sl} & ::\ \gamma\ prog \rightarrow nat \rightarrow ty\ err\ list\ sl \\
\mathsf{reg_sl}\ P\ maxr & \equiv \mathsf{Listn.sl}\ maxr\ (\mathsf{sl}\ (\mathsf{types}\ P,\ \mathsf{subtype}\ P,\ \mathsf{sup}\ P))
\end{array}$$

Since any error on the stack must be propagated, the stack and registers are combined in a coalesced product using Product.esl and then embedded into *option* and *err* to form the semilattice sl:

$$\begin{array}{ll}
\mathsf{sl} & ::\ \gamma\ prog \rightarrow nat \rightarrow nat \rightarrow state\ sl \\
\mathsf{sl}\ P\ maxs\ maxr & \equiv \mathsf{Err.sl}\ (\mathsf{Opt.esl}\ (\mathsf{Product.esl}\ (\mathsf{stk_esl}\ P\ maxs) \\
& \qquad\qquad\qquad\qquad (\mathsf{Err.esl}\ \mathsf{reg_sl}\ P\ maxr)))
\end{array}$$

In the following we abbreviate sl P *maxs maxr* as sl, states P *maxs maxr* as states, le P *maxs maxr* as le, and sup P *maxs maxr* as sup. Using the properties about (*err*-) semilattices, it is easy to prove the following lemma:

Lemma 3.
(1). $\forall\, P\ maxs\ maxr.$ sl $=$ (states, le, sup),
(2). $\forall\, wf\_mb\ P\ maxs\ maxr.$ (wf_prog $wf\_mb\ P$) $\longrightarrow$ semilat (states, le, sup),

(3). $\forall\, wf\_mb\; P\; maxs\; maxr.$ (wf_prog $wf\_mb\; P$) $\longrightarrow$ finite (states),
(4). $\forall\, P\; maxs\; maxr.$ Err $\in$ states,
(5). $\forall\, P\; maxs\; maxr.$ top le Err,
(6). $\forall\, P\; maxs\; maxr.$ (OK None) $\in$ states, *and*
(7). $\forall\, P\; maxs\; maxr.$ bottom le (OK None).

3.2 Program Syntax and Abstract Semantics

The theory JVMInstructions defines the JVM instruction set. In our JVM formalization, the polymorphic instructions Load, Store, and CmpEq are replaced with instructions that have their counterparts in the Sun JVM instruction set, i.e. one such instruction for each base type (see the explanation at the end of Section 3.3).

datatype *instr* =	iLoad *nat*	| bLoad *nat*	| aLoad *nat*
| Invoke *cname mname*	| iStore *nat*	| bStore *nat*	| aStore *nat*
| Getfield *vname cname*	| iIfcmpeq *nat*	| LitPush *val*	| Dup
| Putfield *vname cname*	| bIfcmpeq *nat*	| New *cname*	| Dup_x1
| Checkcast *cname*	| aIfcmpeq *nat*	| Return	| Dup_x2
| Pop	| Swap	| IAdd	| Goto *int*

We instantiate the polymorphic type of program $\gamma\, prog$ using the type $nat \times nat \times instr\ list$, which reflects that the bytecode methods contain information about the class file attributes `max_stack` and `max_locals`, and the type of the method body. We model the type-level execution of a single instruction with step′ :: $instr \times jvm\_prog \times state\_type \times state\_type$, where $jvm\_prog = (nat \times nat \times instr\, list)\, list$. Below we show the definition of step′ for selected instructions:

$$
\begin{aligned}
\text{step}'\ (\text{iLoad}\ n,\ P,\ (st, reg)) &= ((reg!n)\#st, reg)\\
\text{step}'\ (\text{iStore}\ n,\ P,\ (\tau_1\#st_1, reg)) &= (st_1, reg[n := \tau_1])\\
\text{step}'\ (\text{iAdd},\ P,\ ((\text{OK Integer})\#(\text{OK Integer})\#st_1, reg)) &= ((\text{OK Integer})\#st_1, reg)\\
\text{step}'\ (\text{Putfield}\ fn\ C,\ P,\ (\tau_v\#\tau_o\#st_1, reg)) &= (st_1, reg)\\
\text{step}'\ (\text{iIfcmpeq}\ b,\ P,\ (\tau_1\#\tau_2\#st_1, reg)) &= (st_1, reg)\\
\text{step}'\ (\text{Return},\ P,\ (st, reg)) &= (st, reg)
\end{aligned}
$$

Note that the execution of the Return instruction is modeled by a self-loop. This will be useful when we model the traces of the transition system as a set of infinite sequences of program states. Finite sequences can occur only in ill-formed programs, where an instruction has no successor. The main omissions in our model are the same in [11]: both build on the type safety proof of Pusch [15], which does not cover exceptions, object initialization, and `jsr`/`ret` instructions.

3.3 Bytecode Verifier Specification

Now we define a predicate app′ :: $instr \times jvm\_prog \times nat \times ty \times state\_type \rightarrow bool$, which expresses the applicability of the function step′ to a given configuration $(st,\ reg)$ :: $state\_type$. We use this predicate later to formalize bytecode verification as a model-checking problem.

$$
\begin{array}{lll}
\mathsf{app}'\,(i,\,P,\,maxs,\,rT,\,(st,\,reg)) \equiv \mathtt{case}\ i\ \mathtt{of} & & \\
\quad (\mathsf{iLoad}\ n) & \Rightarrow & (n < \mathsf{size}\ st) \wedge (reg!n\ =\ (\mathsf{OK\ Integer})) \wedge (\mathsf{size}\ st < maxs) \\
\mid (\mathsf{iStore}\ n) & \Rightarrow & \exists\,\tau\ st_1.\,(n < \mathsf{size}\ st) \wedge (st = \tau \# st_1) \wedge (\tau = (\mathsf{OK\ Integer})) \\
\mid \mathsf{iAdd} & \Rightarrow & \exists\, st_1.\, st = (\mathsf{OK\ Integer})\#(\mathsf{OK\ Integer})\#\mathsf{st}_1 \\
\mid (\mathsf{Putfield}\ fn\ C) & \Rightarrow & \exists\,\tau_v\ \tau_o\ \tau_f\ st_2.\, st = (\mathsf{OK}\ \tau_v)\#(\mathsf{OK}\ \tau_o)\#\mathsf{st}_2 \wedge (\mathsf{is\_class}\ P\ C) \wedge \\
 & & (\mathsf{field}\ (P, C)\ fn\ =\ \mathsf{Some}\,(C,\ \tau_f)) \wedge \tau_v \sqsubseteq \tau_f \wedge \tau_0 \sqsubseteq \mathsf{Class}\ C \\
\mid (\mathsf{ilfcmpeq}\ b) & \Rightarrow & \exists\, st_2.\, st = (\mathsf{OK\ Integer})\#(\mathsf{OK\ Integer})\#\mathsf{st}_2 \\
\mid \mathsf{Return} & \Rightarrow & \exists\,\tau\ \mathsf{st}_1.\, st = \tau\# st_1 \wedge \tau \sqsubseteq \mathsf{rT}
\end{array}
$$

Further, we introduce a successor function succs :: $instr \rightarrow nat \rightarrow nat\ list$, which computes the possible successor instructions of a program point with respect to a given instruction.

$$
\begin{array}{l}
\mathsf{succs}\ i\ p \equiv \mathtt{case}\ i\ \mathtt{of}\ \mathsf{Return} \Rightarrow [p] \mid \mathsf{Goto}\ b \Rightarrow [p+b] \mid \mathsf{ilfcmpeq}\ b \Rightarrow [p+1,\ p+b] \\
\mid \mathsf{blfcmpeq}\ b \Rightarrow [p+1,\ p+b] \mid \mathsf{ilfcmpeq}\ b \Rightarrow [p+1,\ p+b] \mid \_ \Rightarrow [p+1]
\end{array}
$$

We use succs to construct the transition function of a finite transition system. To reason about the boundedness of succs, we define the predicate bounded :: $(nat \rightarrow nat\ list) \rightarrow list \rightarrow bool$, where bounded $f\ n\ \equiv\ \forall p < n.\, \forall q \in \mathsf{set}\,(f\ p).\, q < n$.

We can now define a well-typedness condition for bytecode methods. The predicate wt_method formalizes that, for a given program P, class C, method parameter list pTs, method return type rT, maximal stack size $maxs$, and maximal register index $maxr$, a bytecode instruction sequence bs is well-typed with respect to a given method type ϕ.

$$method\_type\ =\ state\_type\ option\ list$$

$$
\begin{array}{l}
\mathsf{le\_state\_opt}\ ::\ jvm\_prog \rightarrow state\_type\ option \rightarrow state\_type\ option \rightarrow bool \\
\mathsf{le\_state\_opt}\ P\ \equiv\ \mathsf{Opt.le}\ (\lambda\,(st_1,\ reg_1)\,(st_2,\ reg_2).\ st_1 \\
\qquad\qquad \leq_{[\mathsf{Err.le}\,(\mathsf{subtype}\ P)]} st_2 \wedge reg_1 \leq_{[\mathsf{Err.le}\,(\mathsf{subtype}\ P)]} reg_2)
\end{array}
$$

$$
\begin{array}{l}
\mathsf{step}\ ::\ instr \rightarrow jvm\_prog \rightarrow state\_type\ option \rightarrow state\_type\ option \\
\mathsf{step}\ i\ P\ \equiv\ \lambda\, i\ P\ s.\ \mathtt{case}\ s\ \mathtt{of}\ \mathsf{None}\ \Rightarrow\ \mathsf{None} \mid \mathsf{Some}\ s'\ \Rightarrow\ \mathsf{step}'\,(i,\ P,\ s')
\end{array}
$$

$$
\begin{array}{l}
\mathsf{app}\ ::\ instr \rightarrow jvm\_prog \rightarrow nat \rightarrow ty \rightarrow state\_type\ option \rightarrow bool \\
\mathsf{app}\ i\ P\ maxs\ rT\ s\ \equiv\ \mathtt{case}\ s\ \mathtt{of}\ \mathsf{None}\ \Rightarrow\ \mathtt{True} \mid \mathsf{Some}\ s'\ \Rightarrow\ \mathsf{app}'\,(i,\ P,\ maxs,\ rT,\ s')
\end{array}
$$

$$
\begin{array}{l}
\mathsf{wt\_method}\ ::\ jvm\_prog \rightarrow cname \rightarrow ty\ list \rightarrow ty \rightarrow nat \rightarrow nat \rightarrow instr\ list \\
\qquad\qquad \rightarrow method\_type \rightarrow bool \\
\mathsf{wt\_method}\ P\ C\ pTs\ rT\ maxs\ maxl\ bs\ \phi\ \equiv \\
(0 < \mathsf{size}\ bs) \wedge \\
(\mathsf{Some}\,([],\ (\mathsf{OK}\,(\mathsf{Class}\ C))\#(\mathsf{map\ OK}\ pTs)@(\mathsf{replicate}\ maxl\ \mathsf{Err})))\ \leq_{\mathsf{le\_state\_opt}}\ \phi!0 \wedge \\
(\forall pc.\, pc < \mathsf{size}\ bs\ \longrightarrow \\
\quad (\mathsf{app}\ (bs!pc)\ P\ maxs\ rT\ (\phi!pc)) \wedge \\
\quad (\forall pc' \in \mathsf{set}\,(\mathsf{succs}\,(bs!pc)\ pc).\,(pc' < \mathsf{size}\ bs)\ \wedge\ (\mathsf{step}\ (bs!pc)\ P\ (\phi!pc) \\
\qquad\qquad \leq_{\mathsf{le\_state\_opt}}\ \phi!pc'))
\end{array}
$$

This definition of a well-typed method is in the style of that given by Pusch [15].

The last element of our JVM model is the definition of the abstract execution function exec:

$\mathsf{exec} :: \mathit{jvm\_prog} \to \mathit{nat} \to \mathit{ty} \to \mathit{instr\ list} \to \mathit{nat} \to \mathit{state} \to \mathit{state}$
$\mathsf{exec}\ P\ \mathit{maxs}\ rT\ \mathit{bs}\ \mathit{pc} \equiv \lambda s.\ \mathtt{case}\ s\ \mathtt{of}\ \mathsf{Err} \Rightarrow \mathsf{Err} \mid \mathsf{OK}\ s' \Rightarrow$
$\quad \mathtt{if}\ \mathsf{app}\ (\mathit{bs}!\mathit{pc})\ P\ \mathit{maxs}\ rT\ s'\ \mathtt{then}\ \mathsf{OK}\,(\mathsf{step}\ (\mathit{bs}!\mathit{pc})\ P\ s')\ \mathtt{else}\ \mathsf{Err}$

Abbreviating exec P *maxs* rT *bs* *pc* as exec, we now prove:

Lemma 4. $\forall\, \mathit{wf\_mb}\ P\ \mathit{maxs}\ \mathit{maxr}\ \mathit{bs}\ \mathit{pc}.\ \forall\, s_1\ s_2 \in \mathsf{states}.$
$(\mathsf{wf\_prog}\ \mathit{wf\_mb}\ P \wedge \mathsf{semilat\ sl}) \longrightarrow \mathsf{exec}\,(s_1 +_{\mathsf{sup}} s_2) \leq_{\mathsf{le}} (\mathsf{exec}\, s_1) +_{\mathsf{sup}} (\mathsf{exec}\, s_2)$

This lemma states a "semi-homomorphism" property of the exec function with respect to the le relation. To prove it we must show that it holds in our formalization of the JVM that if two arbitrary program states x, $y \in A$ are well-typed with respect to an instruction at a given program point, then so is $x +_f y$. This would have been impossible to prove using the Nipkow's formalization of the JVM with polymorphic instructions; hence we have replaced the polymorphic instructions in the JVM instruction set with collections of monomorphic ones.

4 Model Checking

Bytecode verification by model checking is based on the idea that the bytecode of a method represents a transition system which, when suitably abstracted (e.g. by replacing data with their types) is finite and that one can uniformly produce, from bytecode, a temporal logic formula that holds iff the bytecode is well-typed with respect to a given type system. In [1] we presented a system based on this idea. For a bytecode method M our system generates a transition system K (in the input language of either the SPIN or the SMV model checker) and a correctness property ψ and uses the model checker to prove that ψ is globally invariant (i.e. $\Box\,\psi$ or $AG\ \psi$ depending on the model checker).

Here we verify the correctness of this approach. Namely we first formalize the translation of a bytecode method M to the transition system K; second we formalize the way in which ψ is generated, and third we formalize what it means for ψ to globally hold for K, i.e. $K \models \Box\,\psi$. We show that when model checking succeeds then there exists a type ϕ that is a well-typing for the method M.

4.1 Kripke Structures and Method Abstraction

Kripke Structures. In the following we define Kripke structures, which we use to formalize the abstract transition system of a method. A Kripke structure K consists of a non-empty set of states, a set of initial states, which is a subset of the states of K, and a transition relation next over the states of K. A trace of K is an infinite sequence of states such that the first state is an initial state and successive states are in the transition relation of K. A state is reachable in K if it is contained in a trace of K. We also define a suffix function on traces, which is needed to define the semantics of LTL-formulas. These definitions are standard and their formalization in Isabelle/HOL is straightforward.

$$\alpha\ kripke = (\!|\ \mathsf{states} :: \alpha\ set, \mathsf{init} :: \alpha\ set, \mathsf{next} :: (\alpha \times \alpha)\ set\ |\!)$$

$$\begin{aligned}&\mathsf{is\_kripke} :: \alpha\ kripke \to bool\\ &\mathsf{is\_kripke}\ K \ \equiv\ \mathsf{states}\ K \neq \emptyset \wedge \mathsf{init}\ K \subseteq \mathsf{states}\ K \wedge \mathsf{next}\ K \subseteq \mathsf{states}\ K \times \mathsf{states}\ K\end{aligned}$$

$$\alpha\ trace = nat \to \alpha$$

$$\begin{aligned}&\mathsf{is\_trace} :: \alpha\ kripke \to \alpha\ trace \to bool\\ &\mathsf{is\_trace}\ K\ t \ \equiv\ t\ 0 \in \mathsf{init}\ K \wedge \forall i.\ (t\ i,\ t\ (\mathsf{Suc}\ i)) \in \mathsf{next}\ K\end{aligned}$$

$$\begin{aligned}&\mathsf{traces} :: \alpha\ kripke \to \alpha\ trace\ set\\ &\mathsf{traces}\ K \ \equiv\ \{t \mid \mathsf{is\_trace}\ K\ t\}\end{aligned}$$

$$\begin{aligned}&\mathsf{reachable} :: \alpha\ kripke \to \alpha \to bool\\ &\mathsf{reachable}\ K\ q \ \equiv\ \exists t\, i.\ \mathsf{is\_trace}\ K\ t \wedge q = t\ i\end{aligned}$$

$$\begin{aligned}&\mathsf{suffix} :: \alpha\ trace \to nat \to \alpha\ trace\\ &\mathsf{suffix}\ t\ i \ \equiv\ \lambda j.\ t\,(i+j)\end{aligned}$$

Method Abstraction. With the above definitions at hand, we can formalize the abstraction of bytecode methods as a finite Kripke system over the type $abs\_state = nat \times state\_type$. We generate the transition system using the function abs_method, which for a given program P, class name C, method parameter list pTs, return type rT, maximal register index $maxr$, and bytecode instruction sequence bs, yields a Kripke structure of type *abs_state kripke*.

$$\begin{aligned}&\mathsf{abs\_method} :: jvm\_prog \to cname \to ty \to list \to ty \to nat \to instr\ list\\ &\qquad\qquad\qquad\qquad\qquad\qquad\qquad \to abs\_state\ kripke\end{aligned}$$

$$\begin{aligned}&\mathsf{abs\_method}\ P\ C\ pTs\ rT\ maxr\ bs \ \equiv\\ &\qquad (\!|\ \mathsf{states}\ =\ \mathsf{UNIV},\\ &\qquad\ \ \mathsf{init}\ =\ \mathsf{abs\_init}\ C\ pTs\ maxr,\\ &\qquad\ \ \mathsf{next}\ =\ (\textstyle\bigcup_{pc \in \{p.\, p < (\mathsf{size}\ bs)\}} \mathsf{abs\_instr}\ (bs!pc)\ P\ pc)\ |\!)\end{aligned}$$

The set of states of the transition system is modeled by the set UNIV of all elements of type *abs_state*. The initial abstract state set abs_init contains one element, which models the method entry where the program counter is set to 0 and the stack is empty. At the method entry, the list of local variables contains the this reference OK (Class C), the methods parameters map OK pTs, and $maxr$ uninitialized registers replicate $maxr$ Err.

$$\begin{aligned}&\mathsf{abs\_init} :: cname \to ty\ list \to nat \to abs\_state\ set\\ &\mathsf{abs\_init}\ C\ pTs\ maxr \ \equiv\\ &\qquad\qquad \{\,(0,\ ([],\ (\mathsf{OK}\ (\mathsf{Class}\ C))\#((\mathsf{map}\ \mathsf{OK}\ pTs))@(\mathsf{replicate}\ maxr\ \mathsf{Err})))\,\}\end{aligned}$$

The relation next is generated by the function abs_instr, which uses step′ and succs, which together make up the transition relation of our transition system. As we have shown in a lemma, the next relation is finite because both the type system and the number of storage locations (stack and local variables) of the abstracted method are finite.

$$\begin{aligned}&\mathsf{abs\_instr} :: \mathit{instr} \to \mathit{jvm\_prog} \to \mathit{nat} \to (\mathit{abs\_state} * \mathit{abs\_state})\ \mathit{set}\\&\mathsf{abs\_instr}\ i\ P\ pc \equiv \{((pc', q), (pc'', q')) \mid pc'' \in \mathsf{set}\,(\mathsf{succs}\ i\ pc') \wedge\\&\qquad (q' = \mathsf{step}'\ (i, P, q)) \wedge pc = pc'\}\end{aligned}$$

4.2 Temporal Logic and Applicability Conditions

Temporal Logic. The syntax of LTL formulae is given by the datatype $\alpha\ ltl$.

$$\begin{aligned}\textbf{datatype}\ \alpha\ ltl \equiv\ & \mathsf{Tr} \mid \mathsf{Atom}\ (\alpha \to bool) \mid \mathsf{Neg}\ (\alpha\ ltl) \mid \mathsf{Conj}\ (\alpha\ ltl)\ (\alpha\ ltl)\\ & \mid \mathsf{Next}\ (\alpha\ ltl) \mid \mathsf{Until}\ (\alpha\ ltl)\ (\alpha\ ltl)\end{aligned}$$

As is standard, the modalities eventually, $\Diamond :: \alpha\ ltl \to \alpha\ ltl$, and globally, $\Box :: \alpha\ ltl \to \alpha\ ltl$, can be defined as syntactic sugar.

$$\Diamond f \equiv \mathsf{Until}\ \mathsf{Tr}\ f \qquad \Box f \equiv \mathsf{Neg}\ (\Diamond\ (\mathsf{Neg}\ f))$$

We give the semantics of LTL formulae using a satisfiability predicate

$$\_ \models \_ :: \alpha\ trace \to \alpha\ kripke \to \alpha\ ltl \to bool$$

where

$$\begin{aligned}t &\models \mathsf{Tr} &&= \mathsf{True}\\ t &\models \mathsf{Atom}\ p &&= p\,(t\,0)\\ t &\models \mathsf{Neg}\ f &&= \neg(t \models f)\\ t &\models \mathsf{Conj}\ f\ g &&= (t \models f) \wedge (t \models g)\\ t &\models \mathsf{Next}\ f &&= \mathsf{suffix}\ t\ 1 \models f\\ t &\models \mathsf{Until}\ f\ g &&= \exists j.\ (\mathsf{suffix}\ t\ j \models g) \wedge\\ &&&\qquad (\forall i.\ i < j \longrightarrow \mathsf{suffix}\ t\ i \models f)\end{aligned}$$

Furthermore, we say that property f is globally satisfied in the transition system K if it is satisfied for all traces of K.

$$\begin{aligned}\_ &\models \_ :: \alpha\ kripke \to \alpha\ ltl \to bool\\ K &\models f \equiv \forall t \in \mathsf{traces}\ K.\ t \models f\end{aligned}$$

Applicability Conditions. The applicability conditions are expressed by an LTL formula of the form $\Box\ \mathsf{Atom}\ f$. We extract the formula f from the bytecode, using the functions spc_instr and spc_method.

$$\begin{aligned}&\mathsf{spc\_instr} :: instr \to jvm\_prog \to nat \to nat \to nat \to ty \to abs\_state \to bool\\&\mathsf{spc\_instr}\ i\ P\ pc'\ h\ maxs\ rT \equiv\\&\quad \lambda(pc, q).\, pc = pc' \longrightarrow ((\mathsf{size}\ (fst\ q)) = h) \wedge (\mathsf{app}'\ (i, P, maxs, rT, q))\\&\mathsf{spc\_method} :: jvm\_prog \to nat\ list \to ty \to nat \to instr\ list \to abs\_state \to bool\\&\mathsf{spc\_method}\ P\ hs\ rT\ maxs\ bs\ q \equiv\\&\quad \forall pc < (\mathsf{size}\ bs).\ (\mathsf{spc\_instr}\ (bs!pc)\ P\ pc\ (hs!pc)\ maxs\ rT\ q)\end{aligned}$$

The function spc_instr expresses the instruction applicability condition for a program state at a program point pc'. Note that besides the conditions formalized in app$'$, we also require that, in all states associated with this program point, the stack has a fixed predefined size h. We employ this to fit the polyvariant model-checking approach into the monovariant framework. In order to prove that there is a state type, we must show that the supremum of all possible stack types associated with the program point is different from the error element Err, and hence we require this restriction.

The function spc_method builds f as the conjunction of the applicability conditions of all program points in the bytecode bs (the list hs contains, for each program point in bs, a number that specifies the stack size at that program point). The resulting formula expresses the well-typedness condition for a given program state and hence, $K \models \Box\,\mathsf{Atom}\ f$ formalizes, for a Kripke structure K resulting from a method abstraction, the global type safety of the method. This formula, together with the additional constraints explained below, makes up the definition of a bytecode-modelchecked method, which is formalized using the predicate bcm_method.

$$\begin{aligned}
&\mathsf{bcm_method} :: \mathit{jvm_prog} \rightarrow \mathit{cname} \rightarrow \mathit{ty\ list} \rightarrow \mathit{ty} \rightarrow \mathit{nat} \rightarrow \\
&\quad \mathit{nat} \rightarrow \mathit{instr\ list} \rightarrow \mathit{bool}
\end{aligned}$$

$$\begin{aligned}
&\mathsf{bcm_method}\ P\ C\ pTs\ rT\ maxr\ maxs\ bs \ \equiv\ \exists\, hs. \\
&\quad (\mathsf{abs_method}\ P\ C\ pTs\ rT\ maxr\ bs) \models \Box(\mathsf{Atom}\ (\mathsf{spc_method}\ P\ hs\ rT\ maxs\ bs)) \\
&\quad \wedge \mathsf{size}\ hs\ =\ \mathsf{size}\ bs\ \wedge \\
&\quad (\mathsf{traces}\ (\mathsf{abs_method}\ P\ C\ pTs\ rT\ maxr\ bs))\ \neq\ \{\}\ \wedge \\
&\quad (\forall\, (x,\ y)\ \in\ (\mathsf{next}\ (\mathsf{abs_method}\ P\ C\ pTs\ rT\ maxr\ bs)). \\
&\qquad \exists\, t\, i.\ (\mathsf{is_trace}\ (\mathsf{abs_method}\ G\ C\ pTs\ rT\ maxr\ bs)\ t)\ \wedge\ (t\ i,\ t\ (\mathsf{Suc}\ i)) \\
&\qquad =\ (x,\ y))\ \wedge 0\ <\ (\mathsf{size}\ bs)\ \wedge \\
&\quad \mathsf{bounded}\ (\lambda\, n.\ \mathsf{succs}\ (bs!n)\ n)\ (\mathsf{size}\ bs)
\end{aligned}$$

The first conjunct expresses the global applicability conditions for the given bytecode bs. The second conjunct requires that a stack size is specified for every program point. The third conjunct states that the transition system is nonempty, i.e. that there exists at least one non-trivial model for a given LTL formula. The fourth conjunct characterizes a liveness property of the abstract interpreter. We use this assumption to show that a given program point pc, which is reachable in finitely many steps, is also contained in an (infinite) trace $t \in \mathit{traces}\, K$. The last two conjuncts express well-formedness properties of the bytecode, namely, that the list of instructions is not empty and that all instructions have a successor.

4.3 Main Theorem and Proof

Our main theorem, stating the correctness of bytecode verification by model checking can now be given.

Theorem 1.

$$\begin{aligned}
&(\mathsf{wf_prog}\ \mathit{wf_mb}\ P)\ \wedge\ (\mathsf{is_class}\ P\ C)\ \wedge\ (\forall\, x \in \mathsf{set}\ pTs.\ \mathsf{is_type}\ P\ x)\ \longrightarrow \\
&(\mathsf{bcm_method}\ P\ C\ pTs\ rT\ maxr\ maxs\ bs)\ \longrightarrow \\
&\exists\, \phi.\ \mathsf{wt_method}\ P\ C\ pTs\ rT\ maxs\ maxr\ bs\ \phi
\end{aligned}$$

This theorem states the correctness only for the bytecode bs for a single method, of a single class C, of a program P; however, it can easily be extended to a more general statement for the entire program P. We have actually proved in Isabelle the more general statement but, to avoid unnecessary complexity, we restrict ourselves here to this simpler statement.

This theorem has four assumptions: (A1) the program P is well-formed; (A2) the class C belongs to the program P; (A3) the method signature is well-formed (i.e., all method parameters have declared types); and (A4) that the bytecode bs is successfully model-checked. Under these assumptions, the conclusion states that bs has a well-typing given by the method type ϕ. The impact of this theorem is that the existence of such a well-typing implies that bs is free of runtime type errors.

In the following, to increase readability, we abbreviate abs_method P C pTs rT $maxr$ bs as abs_method and bcm_method P C pTs rT $maxr$ $maxs$ bs as bcm_method.

Proof Intuition. Our proof of the existence of a method type ϕ is constructive and builds ϕ from the transition system K pointwise for each program point pc as follows. First, we define a function _ at _ :: $(\alpha \times \beta)$ *kripke* $\rightarrow \alpha \rightarrow \beta$ *set*, where K at $pc \equiv \{ a \mid \text{reachable } K\ (pc, a) \}$ denotes the program configurations reachable on any trace that belong to the program point pc. Second, we build the supremum sup_{pc} of the set K at pc. The third step uses the fact (which must be proved) that each supremum sup_{pc} is of the form OK x_{pc} (i.e., it is not Err) and hence x_{pc} is the method type at the program point pc. In essence, we perform a data flow analysis here in our proof to build a method type.

The bulk of our proof is to show that this method type is actually a well-typing for the model checked bytecode. The main challenge here is to show that each method instruction is well-typed with respect to the instruction's state type (applicability) and that the successors of an instruction are well-typed with respect to the state type resulting from the execution of the instruction (stability). We establish this by induction on the set K at pc; this corresponds to an induction over all traces that contribute to the state set associated with the pcth program point. Since model checking enforces applicability and thus stability for every trace, and since exec is semi-continuous in the sense that

$$\text{exec}\,(\bigsqcup_{\text{sl}} (\text{OK}\,\text{`}\,\text{Some}\,\text{`}\,\text{abs\_method at}\ pc)) \leq_{\text{le}} \bigsqcup_{\text{sl}} (\text{exec}\,\text{`}\,(\text{OK}\ \text{`Some'}(\text{abs\_method at}\ pc))),$$

then the pointwise supremum of the state types gives rise to a method type that is a well-typing.

Some Proof Details. We now describe the construction of the method type more formally and afterwards the proof that it is a well-typing.

The first two steps of the construction are formalized using the function $\mathsf{ok\_method\_type} :: \mathit{jvm\_prog} \to \mathit{cname} \to \mathit{ty\ list} \to \mathit{ty} \to \mathit{nat} \to \mathit{nat} \to \mathit{instrlist} \to \mathit{nat} \to \mathit{state}$, where we abbreviate $\bigsqcup_{\mathsf{sl,Err,OK(None)}}$ as $\bigsqcup_{\mathsf{sl}}$.

$$\begin{array}{l}\mathsf{ok\_method\_type}\ P\ C\ pTs\ rT\ maxr\ maxs\ bs\ pc\ \equiv\\ \quad \mathtt{let}\ sl = \mathsf{sl}\ P\ maxs\ (1 + (\mathsf{size}\ pTs) + maxr);\\ \qquad K = \mathsf{abs\_method}\ P\ C\ pTs\ rT\ maxr\ bs;\\ \quad \mathtt{in}\ \bigsqcup_{sl}\ \mathsf{OK}\text{`}\mathsf{Some}\text{`}\ (K\ \mathsf{at}\ pc)\end{array}$$

Here, the supremum function over sets, $\bigsqcup_{sl}$, builds from K at pc the supremum sup_{pc} in the semilattice sl. To increase readability, we will use ok_method_type as an abbreviation for ok_method_type $P\ C\ pTs\ rT\ maxr\ maxs\ bs$.

For the third step, we extract the method type using the function $\mathsf{ok\_val} :: \alpha\ err \to \alpha$, where $\mathsf{ok\_val}\ e = \mathtt{case}\ e\ \mathtt{of}\ (\mathsf{OK}\ x) \Rightarrow x \mid \_ \Rightarrow \mathsf{arbitrary}$. Thus the method type at program point pc can be computed by the function $\mathsf{method\_type} \equiv \mathsf{ok\_val} \circ \mathsf{ok\_method\_type}$ and we chose as the overall method type

$$\phi = \mathsf{map}\ (\mathsf{method\_type}\ P\ C\ pTs\ rT\ maxr\ maxs\ bs)\ [0, \ldots, \mathsf{size}\ bs - 1]\ .$$

Before explaining why ϕ is a well-typing, we state two lemmata central to the proof. As mentioned above, the correctness of our construction requires that for each program point pc, there is a state type x such that $\mathsf{ok\_method\_type}\ pc = \mathsf{OK}\,x$. The following lemma states this, and moreover that this type is applicable at program point pc.

Lemma 5. ***Well-Typed Supremum***

$$\begin{array}{l}0 < \mathsf{size}\ bs\ \wedge\ pc < \mathsf{size}\ bs\ \wedge\ \mathsf{wf\_prog}\ wf\_mb\ P\ \wedge\ \mathsf{is\_class}\ P\ C\\ \wedge\ (\forall x \in \mathsf{set}\ pTs.\ \mathsf{is\_type}\ P\ x)\ \wedge\\ \mathsf{size}\ hs\ =\ \mathsf{size}\ bs\ \wedge\ \mathsf{bounded}\ (\lambda n.\ \mathsf{succs}\ (bs!n)\ n)\ (\mathsf{size}\ bs)\ \wedge\\ (\mathsf{abs\_method} \models \Box\ \mathsf{Atom\ spc\_method})\\ \longrightarrow \exists x.\ (\mathsf{ok\_method\_type}\ pc = \mathsf{OK}\ x)\ \wedge\ (x = \mathsf{None} \vee (\exists\, st\ reg.\ x = \mathsf{Some}\,(st, reg)\ \wedge\\ \qquad \mathsf{app}'\ (bs!pc,\ P,\ maxs,\ rT,\ (st, reg))\ \wedge\ (\mathsf{size}\ st\ =\ hs!pc)))\end{array}$$

We prove this lemma by induction on the set of program states that belong to the program point pc. This proof uses the fact that if a property holds globally in a Kripke structure then it holds for every reachable state of the structure. The step case requires the following lemma, which states that when two states (st_1, reg_1) and (st_2, reg_2) of the type *state_type* satisfy the predicate app′ for the instruction $bs!pc$ and the sizes of their stacks are equal, then the supremum of these states also satisfies the applicability condition for this instruction and the size of supremum's stack is the size of the stack of the two states. Here we abbreviate lift JType.sup P as sup′.

Lemma 6. ***Well-Typed Induction Step***

$$\begin{array}{l}pc < (\mathsf{size}\ bs)\ \wedge\ \mathsf{semilat\ sl}\ \wedge\ \mathsf{size}\ hs = \mathsf{size}\ bs \longrightarrow\\ \mathsf{size}\ st_1 = hs!pc\ \wedge\ \mathsf{app}'\ (bs!pc,\ P,\ maxs,\ rT,\ (st_1, reg_1))\ \wedge\\ \mathsf{size}\ st_2 = hs!pc\ \wedge\ \mathsf{app}'\ (bs!pc,\ P,\ maxs,\ rT,\ (st_2, reg_2)) \longrightarrow\\ \mathsf{size}\ (st_1 +_{[\mathsf{sup}']} st_2) = hs!pc \wedge \mathsf{app}'\ (bs!pc,\ P,\ maxs,\ rT,\ (st_1 +_{[\mathsf{sup}']} st_2,\ reg_1 +_{[\mathsf{sup}']} reg_2))\end{array}$$

We now sketch the proof that ϕ is a well-typing. To simplify notation, we abbreviate the initial state type $([], (\mathsf{OK}\ (\mathsf{Class}\ C))\#\ (\mathsf{map}\ \mathsf{OK}\ pTs)@$ $(\mathsf{replicate}\ maxl\ \mathsf{Err}))$ as (st_0, reg_0).

Following the definition of wt_method, we must show, that under the four assumptions (A1)–(A4), three properties hold: First, the method body must contain at least one instruction, i.e. $0 < \mathsf{size}\ bs$. This follows directly from the definition of bcm_method.

Second, the start of the method must be well-typed, that is $\mathsf{Some}\,(st_0,\ reg_0)$ $\leq_{\mathsf{le\_state\_opt}}\ \phi!0$. Since the set of traces is not empty, the initial state type $\mathsf{Some}\,(st_0,\ reg_0)$ is contained in the set $\mathsf{Some}\,`\,(\mathsf{abs\_method}\ \mathsf{at}\ 0)$ and hence it follows that

$$\mathsf{OK}\,(\mathsf{Some}\,(st_0, reg_0)) \leq_{\mathsf{le}} \bigsqcup_{\mathsf{sl}}\,(\mathsf{OK}\,`\,\mathsf{Some}\,`\,(\mathsf{abs\_method}\ \mathsf{at}\ 0))\,,$$

which is (by the definition of ok_method_type) the same as $\mathsf{OK}\,(\mathsf{Some}\,(st_0, reg_0))$ $\leq_{\mathsf{le}}\ \mathsf{ok\_method\_type}\ 0$. By lemma 5 we know that the right-hand side of the inequality is an OK value and thus we can strip off OK yielding $\mathsf{Some}\,(st_0,\ reg_0)$ $\leq_{\mathsf{le\_state\_opt}}\ \mathsf{ok\_val}(\mathsf{ok\_method\_type}\ 0)$, which is (by the choice of ϕ and the definition of ok_method_type) the desired result.

Finally, we must show that all instructions of the method bs are well-typed, i.e.,

$$\begin{array}{lr} \forall (pc < \mathsf{size}\ bs).\ \forall\, pc' \in \mathsf{set}\,(\mathsf{succs}\,(bs!pc)\ pc). & \\ \quad pc' < \mathsf{size}\ bs\ \wedge & \text{(boundedness)} \\ \quad \mathsf{app}\,(bs!pc)\ P\ maxs\ rT\ (\phi!pc)\ \wedge & \text{(applicability)} \\ \quad \mathsf{step}\,(bs!pc)\ P\ (\phi!pc)\ \leq_{\mathsf{le\_state\_opt}} (\phi!pc') & \text{(stability)} \end{array}$$

This splits into three subgoals (boundedness, applicability, and stability). For all three we fix a pc, where $pc < \mathsf{size}\ bs$ and a successor pc' of pc. Boundedness holds trivially, since from bcm_method it follows that succs is bounded.

To show the applicability of the instruction at the program point pc, by the definition of method_type, we must show that

$$\mathsf{app}\,(bs!pc)\ P\ maxs\ rT\ (\mathsf{ok\_val}\,(\mathsf{ok\_method\_type}\ pc))\,,$$

which we prove by case analysis. The first case is when ok_method_type pc is OK None, which in our formalization means that pc is not reachable (i.e. there is dead code). Then applicability holds by the definition of app. The second case is when the transition system abs_method generates program configurations that belong to the program point pc. We must then prove (according to the definition of app)

$$\mathsf{app}'\,(bs!pc,\ P,\ maxs,\ rT,\ (st,\ reg))\,,$$

which follows from Lemma 5. This lemma also guarantees that the third case, where ok_method_type pc is Err, does not occur.

The majority of the work is in showing stability. Due to the choice of ϕ and the definition of method_type we must prove

$$\mathsf{step}\ (bs!pc)\ P\ (\mathsf{ok\_val}\ (\mathsf{ok\_method\_type}\ pc)) \\ \le_{\mathsf{le\_state\_opt}} (\mathsf{ok\_val}\ (\mathsf{ok\_method\_type}\ pc')) .$$

By the definition of $\le_{\mathsf{le}}$ it suffices to show the inequality

$$\mathsf{OK}(\mathsf{step}\ (bs!pc)\ P\ (\mathsf{ok\_val}\ (\mathsf{ok\_method\_type}\ pc))) \le_{\mathsf{le}} \mathsf{OK}(\mathsf{ok\_val}\ (\mathsf{ok\_method\_type}\ pc'))$$

Lemma 5 states the applicability of ok_val (ok_method_type pc) to the instruction at program point pc on the left-hand side of the inequality and hence, by the definition of exec, we can reduce our problem to

$$\mathsf{exec}(\mathsf{OK}(\mathsf{ok\_val}\ (\mathsf{ok\_method\_type}\ pc))) \le_{\mathsf{le}} \mathsf{OK}(\mathsf{ok\_val}\ (\mathsf{ok\_method\_type}\ pc')) .$$

Moreover, this lemma allows us to conclude that ok_method_type delivers OK values for pc and pc' and thus the argument of exec is equal to ok_method_type pc and the right-hand side of the inequality is equal to ok_method_type pc'. By unfolding the definition of ok_method_type, the inequality simplifies to

$$\mathsf{exec}\ (\bigsqcup_{\mathsf{sl}}(\mathsf{OK}\ \text{`}\ \mathsf{Some}\ \text{`}\ (\mathsf{abs\_method\,at}\ pc))) \le_{\mathsf{le}} \bigsqcup_{\mathsf{sl}}(\mathsf{OK}\ \text{`}\ \mathsf{Some}\ \text{`}\ (\mathsf{abs\_method\,at}\ pc')) . \qquad (*)$$

In Lemma 1 we proved that if a function g is a semi-homomorphism and $g\ \text{`}\ A'' \subseteq A'$, then $g\,(\bigsqcup_{\mathsf{sl}} A'') \le_{\mathsf{le}} \bigsqcup_{\mathsf{sl}} A'$. Inequation (*) is an instance of the conclusion of this lemma. We can prove the (corresponding instance of the) first premise of this lemma, showing that exec is a semi-homomorphism, using Lemma 4. Thus, it suffices to prove the (corresponding instance of the) second premise, i.e, to prove

$$(\mathsf{exec}\ \text{`}\ (\mathsf{OK}\ \text{`}\ \mathsf{Some}\ \text{`}\ (\mathsf{abs\_method\ at}\ pc))) \subseteq (\mathsf{OK}\ \text{`}\ \mathsf{Some}\ \text{`}\ (\mathsf{abs\_method\ at}\ pc')) .$$

We assume for an arbitrary state type $(st,\ reg)$ that $(st,\ reg) \in \mathsf{abs\_method\ at}\ pc$ and prove that

$$\exists (st',\ reg') \in \mathsf{abs\_methodat} pc'.\ \mathsf{exec}(\mathsf{OK}(\mathsf{Some}(st,\ reg))) = \mathsf{OK}(\mathsf{Some}\ (st',\ reg')) .$$

From this assumption it follows that $(pc, (st,\ reg))$ is a reachable state of abs_method, which together with (A4) entails that the applicability conditions hold for $(st,\ reg)$. Hence, by the definition of exec, we can reduce our goal to

$$\exists (st',\ reg') \in \mathsf{abs\_method\ at}\ pc'.\ \mathsf{OK}(\mathsf{step}\ (bs!pc)\ P\ \mathsf{Some}(st,\ reg))) = \mathsf{OK}(\mathsf{Some}\ (st',\ reg'))$$

and further, by the definition of step and by stripping off OK, to

$$\exists (st',\ reg') \in \mathsf{abs\_method\ at}\ pc'.\ \mathsf{step}'\ (bs!pc, P, (st, reg)) = (st', reg') .$$

However, this is equivalent to

$$\exists (st',\ reg') \in \mathsf{abs\_method\ at}\ pc'.\ ((st, reg), (st', reg')) \in \mathsf{next}(\mathsf{abs\_method}) ,$$

which follows directly from the liveness property that is part of (A4).

5 Conclusion

We have formalized and proved in Isabelle/HOL the correctness of our approach to model checking based bytecode verification. We were fortunate in that we could build on the framework of Pusch and Nipkow. As such, our work also constitutes a fairly large scale example of theory reuse, and the generality of their formalisms, in particular Nipkow's verification framework, played a major role in this regard. As mentioned in the introduction, the changes we made to the verification framework appear generally useful. There are some loose ends remaining though as future work. First, we would like to formalize the monovariant approach in our framework. Second, our approach to polyvariant analysis required slight changes in the stack component of the state type and this requires a new proof that programs possessing a well-typing are type sound. This should be straightforward but also remains as future work. Finally, our approach supports polyvariant analysis in that it admits bytecode with incompatible types at different program points. However, for each program point, our formalization requires the stack size to be a constant. As a result, our current formalization constitutes a midpoint, in terms of the set of programs accepted, between Nipkow's formalization and our implementation. It is trivial to implement the requirement on the stack size as a model-checked property in our implementation; however, it would be more interesting to lift this requirement in our Isabelle model, for example, by further generalizing the notion of a state type to be a set of types.

References

1. D. Basin, S. Friedrich, M.J. Gawkowski, and J. Posegga. Bytecode Model Checking: An Experimental Analysis. In *9th International SPIN Workshop on Model Checking of Software*, 2002, volume 2318 of *LNCS*, pages 42–59, Grenoble. Springer-Verlag, 2002.
2. Y. Bertot. A Coq formalization of a type checker for object initialization in the Java Virtual Machine. Technical Report RR-4047, INRIA, Nov. 2000.
3. A. Coglio, A. Goldberg, and Z. Qian. Toward a provably-correct implementation of the JVM bytecode verifier. In *Proc. DARPA Information Survivability Conference and Exposition (DISCEX'00), Vol. 2*, pages 403–410. IEEE Computer Society Press, 2000.
4. S. Freund and J. Mitchell. A type system for object initialisation in the Java bytecode language. In *ACM Conf. Object-Oriented Programming: Systems, Languages and Applications*, 1998.
5. S.N. Freund and J. C. Mitchell. A formal framework for the java bytecode language and verifier. In *ACM Conf. Object-Oriented Programming: Systems, Languages and Applications*, 1999.
6. A. Goldberg. A specification of Java loading and bytecode verification. In *Proc. 5th ACM Conf. Computer and Communications Security*, 1998.
7. M. Hagiya and A. Tozawa. On a new method for dataflow analysis of Java virtual machine subroutines. In G. Levi, editor. *Static Analysis (SAS'98)*, volume 1503 of *LNCS*, pages 17-32. Springer-Verlag, 1998.
8. G.A. Kildall. A unified approach to global program optimization. In *Proc. ACM Symp. Principles of Programming Languages*, pages 194-206, 1973.

9. X. Leroy. Java Bytecode Verification: An Overview. In G. Berry, H. Comon, and A. Finkel, editors. *CAV 2001, LNCS*, pages 265-285. Springer-Verlag, 2001.
10. T. Lindholm and F. Yellin. *The Java Virtual Machine Specification.* Addison-Wesley, 1996.
11. T. Nipkow. Verified Bytecode Verifiers. In *Foundations of Software Science and Computation Structure (FOSSACS'01)*, pages 347–363, Springer-Verlag, 2001.
12. T. Nipkow and D. v. Oheimb. Java$_{ligth}$ is type-safe – definitely. In *Proc. 25th ACM Symp. Principles of Programming Languages*, pages 161-170, 1998.
13. T. Nipkow, D.v. Oheimb, and C. Pusch. μJava: Embedding a programming language in a theorem prover. In F. Bauer and R. Steinbrüggen, editors, *Foundations of Secure Computation*, pages 117-144. IOS Press, 2000.
14. J. Posegga and H. Vogt. Java bytecode verification using model checking. In *Workshop Fundamental Underspinnings of Java*, 1998.
15. C. Pusch. Proving the soundness of a Java bytecode verifier specification in Isabelle/HOL. In W. Cleaveland, editor, *Tools and Algorithms for the Construction and Analysis of Systems (TACAS'99)*, volume 1597 of *LNCS*, pages 89-103. Springer-Verlag, 1999.
16. D.A. Schmidt. Data flow analysis is model checking of abstract interpretations. In *POPL'98*, pages 38-48. ACM Press 1998.
17. Z. Qian. A formal specification of Java Virtual Machine instructions for objects, methods and subroutines. In J. Alves-Foss, editor, *Formal Syntax and Semantics of Java*, volume 1523 of *LNCS*, pages 271-311. Springer-Verlag, 1999.
18. Z. Qian. Standard fixpoint iteration for Java bytecode verification. *ACM Trans. Programming Languages and Systems*, 22(4):638-672, 2000.
19. R. Stata and M. Abadi. A type system for Java bytecode subroutines. In *Proc 25th ACM Symp. Principles of Programming Languages*, pages 149-161. ACM Press, 1998.

The 5 Colour Theorem in Isabelle/Isar

Gertrud Bauer and Tobias Nipkow

Technische Universität München
Institut für Informatik, Arcisstrasse 21, 80290 München, Germany
http://www4.in.tum.de/~bauerg/
http://www.in.tum.de/~nipkow/

Abstract. Based on an inductive definition of triangulations, a theory of undirected planar graphs is developed in Isabelle/HOL. The proof of the 5 colour theorem is discussed in some detail, emphasizing the readability of the computer assisted proofs.

1 Introduction

It is well known that traditional mathematics is in principle reducible to logic. There are two main motivations for carrying out computer-based formalizations of mathematics in practice: to demonstrate the actual feasibility (in a particular domain), and to develop the formal underpinnings for some applications. Our formalization of graph theory is driven by both motivations.

First of all we would like to explore how feasible computer proofs are in an area that is (over)loaded with graphic intuition and does not have a strong algebraic theory, namely the realm of planar graphs. A first experiment in that direction was already reported by Yamamoto *et al.* [18]. Their main result was Euler's formula. This is just one of many stepping stones towards our main result, the 5 colour theorem (5CT). At the same time we try to lay the foundations for a project one or two orders of magnitude more ambitious: a computer-assisted proof of the 4 colour theorem (4CT). We come back to this point in the conclusion. Finally we have an orthogonal aim, namely of demonstrating the advantage of Wenzel's extension of Isabelle, called Isar [14], namely readable proofs. Hence some of the formal proofs are included, together with informal comments. It is not essential that the reader understands every last detail of the formal proofs. Isar is as generic as Isabelle but our work is based on the logic HOL.

There is surprisingly little work on formalizing graph theory in the literature. Apart from the article by Yamamoto *et al.* on planar graphs [18], from which we inherit the inductive approach, we are only aware of three other publications [17, 6, 19], none of which deal with planarity.

Before we embark on the technical details, let us briefly recall the history and main proof idea for the 4 and 5 colour theorem. The 4CT was first conjectured in 1853 by Francis Guthrie as a map colouring problem. The first incorrect proof attempts were given 1879 by Kempe and 1880 by Tait. Kempe already used an argument today known as Kempe chains. The incorrectness of Kempe's proof was

V.A. Carreño, C. Muñoz, S. Tahar (Eds.): TPHOLs 2002, LNCS 2410, pp. 67–82, 2002.

pointed out by Heawood in 1890, but he could weaken Kempe's argument to the 5CT. In 1969 Heesch [9] introduced the methods of reducibility and discharging which were used in the first proof by Appel and Haken in 1976 [1, 2, 3]. The proof was simplified and improved 1995 by Robertson, Sanders, Seymour and Thomas [11, 12]. Both proofs rely heavily on large case analyses performed by (unverified) computer programs, which is contrary to the tradition of mathematical proofs.

The proof of the 5CT is by induction on the size of the graph g (although many textbooks [15, 8] prefer an indirect argument). Let g be non-empty. It can be shown that any non-empty planar graph has a vertex of degree ≤ 5. Let v be such a vertex in g and let g' be g without v. By induction hypothesis g' is 5-colourable because g' is smaller than g. Now we distinguish two cases. If v has degree less than 5, there are fewer neighbours than colours, and thus there is at least one colour left for v. Thus g is also 5-colorable. If v has degree 5, it is more complicated to construct a 5-colouring of g' from one of g. This core of the proof is explained in the body of the paper. In principle, the proof of the 4CT is similar, but immensely more complicated in its details.

We have intentionally not made an effort to find the slickest proof of the 5CT but worked with a standard one (except for using induction rather than a minimal counterexample). This is because we intended the proof to be a benchmark for our formalization of graph theory.

1.1 Notation

We briefly list the most important non-standard notations in Isabelle/Isar.

The notation $[\![P_1; \ldots; P_n]\!] \Longrightarrow P$ is short for the iterated implication $P_1 \Longrightarrow \ldots \Longrightarrow P_n \Longrightarrow P$. The symbol $\bigwedge$ denotes Isabelle's universal quantifier from the meta-logic. Set comprehension is written $\{x.P\}$ instead of $\{x \mid P\}$. Function *card* maps a finite set to its cardinality, a natural number. The notation c ‘ A is used for the image of a set A under a function c.

The keyword **constdefs** starts the declaration and definition of new constants; "≡"is definitional equality. Propositions in definitions, lemmas, and proofs can be preceded by a name, written "*name*: *proposition*"; We use the technique of suppressing subproofs, denoted by the symbol "⟨*proof*⟩", to illustrate the structure of a proof. The subproofs may be trivial subproofs, by simplification or predicate calculus or involve some irrelevant Isabelle specific steps. If a proposition can be proved directly, its proof starts with the keyword **by** followed by some builtin proof method like *simp*, *arith*, *rule*, *blast* or *auto* [10]; The name *?thesis* in an Isar proof refers to the proposition we are proving at that moment.

1.2 Overview

In §2 we introduce graphs and their basic operations. In §3 we give an inductive definition of near triangulations and define planarity using triangulations. In §4 we discuss two fundamental topological properties of planar graphs and their representations in our formalization. In §5 we introduce colourings and Kempe chains and present the Isar proof of the 5CT.

2 Formalization of Graphs

2.1 Definition of Graphs

In this section we define the type of graphs and introduce some basic graph operations (insertion and deletion of edges and deletion of vertices) and some classes of graphs.

A (finite) graph g consists of a finite set of vertices $\mathcal{V}\ g$ and a set of edges $\mathcal{E}\ g$, where each edge is a two-element subset of $\mathcal{V}\ g$. In the field of planar graphs or graph colouring problems, isolated vertices do not matter much, since we can always assign them any colour. Hence we consider only vertices which are connected to another vertex, and leave out an explicit representation of $\mathcal{V}\ g$. This leads to an edge point of view of graphs, which keeps our formalisation as simple as possible. We do not restrict the definition of graphs to simple graphs but most of the graphs we actually work with are simple by construction.

We define a new type α *graph* as an isomorphic copy of the set of finite sets of two-element sets with the morphisms *edges* and *graph-of*:

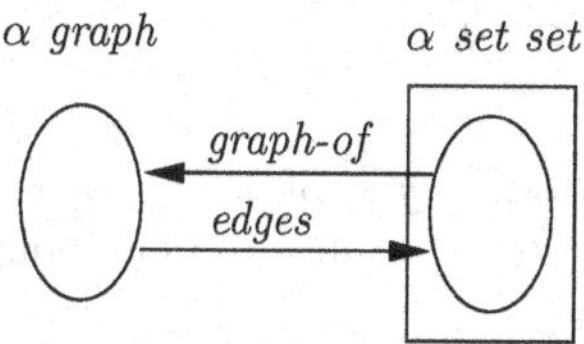

Fig. 1. Definition of α *graph*

typedef α *graph* $= \{g{::}\alpha\ set\ set.\ finite\ g \wedge (\forall e \in g.\ \exists x\ y.\ e = \{x, y\})\}$
morphisms *edges graph-of*

The representing set is not empty as it contains, for example, the empty set.

We introduce the symbol $\mathcal{E}$ for the function *edges* and define $\mathcal{V}\ g$ as the set of all vertices which have an incident edge.

constdefs *vertices* $::\ \alpha\ graph \Rightarrow \alpha\ set \quad (\mathcal{V})$
$\mathcal{V}\ g \equiv \bigcup(\mathcal{E}\ g)$

2.2 Basic Graph Operations.

Size We define the *size* of a graph as the (finite) number of its vertices.

$size\ g \equiv card\ (\mathcal{V}\ g)$

Empty Graph, Insertion, and Deletion of Edges. We introduce the following basic graph operations: the empty graph $\emptyset$, insertion $g \oplus \{x, y\}$ of an edge, deletion $g \ominus \{x, y\}$ of an edge and deletion $g \odot x$ of a vertex x, i.e. deletion of all edges which are incident with x.

constdefs *empty-graph* :: α *graph* $(\emptyset)$
$\emptyset \equiv$ *graph-of* $\{\}$

ins :: α *graph* $\Rightarrow \alpha$ *set* $\Rightarrow \alpha$ *graph* (**infixl** $\oplus$ 60)
$e = \{x, y\} \Longrightarrow g \oplus e \equiv$ *graph-of* (*insert* e $(\mathcal{E}\ g)$)

del :: α *graph* $\Rightarrow \alpha$ *set* $\Rightarrow \alpha$ *graph* (**infixl** $\ominus$ 60)
$g \ominus e \equiv$ *graph-of* $(\mathcal{E}\ g - \{e\})$

del-vertex :: α *graph* $\Rightarrow \alpha \Rightarrow \alpha$ *graph* (**infixl** $\odot$ 60)
$g \odot x \equiv$ *graph-of* $\{e.\ e \in \mathcal{E}\ g \wedge x \notin e\}$

Subgraph. A graph g is a (spanning) subgraph of h, $g \preceq h$, iff g can be extended to h by only inserting edges between existing vertices.

constdefs *subgraph* :: α *graph* $\Rightarrow \alpha$ *graph* $\Rightarrow$ *bool* (**infixl** $\preceq$ 60)
$g \preceq h \equiv \mathcal{V}\,g = \mathcal{V}\,h \wedge \mathcal{E}\,g \subseteq \mathcal{E}\,h$

Degree. The neighbourhood $\Gamma\ g\ x$ of x is the set of incident vertices. The degree of a vertex $x \in \mathcal{V}\ g$ is the number of neighbours of x.

constdefs *neighbours* :: α *graph* $\Rightarrow \alpha \Rightarrow \alpha$ *set* (Γ)
$\Gamma\ g\ x \equiv \{y.\ \{x, y\} \in \mathcal{E}\ g\}$

degree :: α *graph* $\Rightarrow \alpha \Rightarrow$ *nat*
degree $g\ x \equiv$ *card* $(\Gamma\ g\ x)$

2.3 Some Classes of Graphs

Complete Graphs with n vertices are simple graphs where any two different vertices are incident. For example a triangle is a complete graph with 3 vertices.

constdefs K_3 :: $\alpha \Rightarrow \alpha \Rightarrow \alpha \Rightarrow \alpha$ *graph* (K_3)
$K_3\ a\ b\ c \equiv \emptyset \oplus \{a, b\} \oplus \{a, c\} \oplus \{b, c\}$

Cycles are simple connected graphs where any vertex has degree 2. We do not require connectivity, and thus obtain sets of cycles.

constdefs *cycles* :: α *graph* $\Rightarrow$ *bool*
cycles $g \equiv$ *simple* $g \wedge (\forall v \in \mathcal{V}\ g.\ \textit{degree}\ g\ v = 2)$

We introduce abbreviations *cycle-ins-vertex* and *cycle-del-vertex* for insertion and deletion of vertices in cycles (see Fig. 2).

constdefs *cycle-ins-vertex* :: α *graph* $\Rightarrow \alpha \Rightarrow \alpha \Rightarrow \alpha \Rightarrow \alpha$ *graph*
cycle-ins-vertex $g\ x\ y\ z \equiv g \ominus \{x, z\} \oplus \{x, y\} \oplus \{y, z\}$

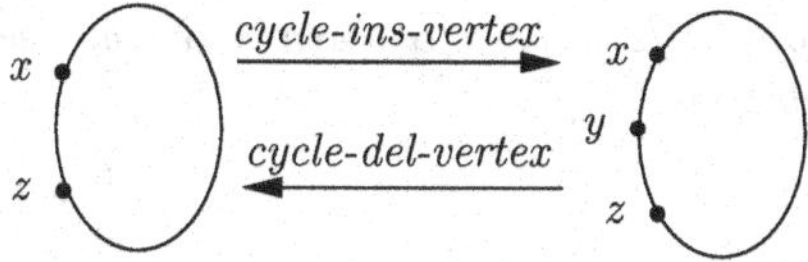

Fig. 2. Insertion and deletion of vertices in cycles

cycle-del-vertex :: α *graph* $\Rightarrow \alpha \Rightarrow \alpha \Rightarrow \alpha \Rightarrow \alpha$ *graph*
cycle-del-vertex g x y z $\equiv$ $g \ominus \{x,\ y\} \ominus \{y,\ z\} \oplus \{x,\ z\}$

3 Triangulations and Planarity

3.1 Inductive Construction of Near Triangulations

To lay the foundations for our definition of planarity based on triangulations, we start in this section with the inductive definition of near triangulations. The idea of defining planar graphs inductively is due to Yamamoto *et al.* [18], but the formulation in terms of near triangulations is due to Wiedijk [16]. We will then derive some fundamental properties of near triangulations such as Euler's theorem and the existence of a vertex of degree at most 5.

Triangulations are plane graphs where any face is a triangle. Near triangulations are plane graphs where any face except the outer face is a triangle. We define near triangulations g with boundary h and inner faces f in an inductive way (see Fig. 3): any triangle K_3 a b c of three distinct vertices is a near triangulation. Near triangulations are extended by adding triangles on the outside in two different ways: in the first case we introduce a new vertex b by inserting two edges joining b with a and c, where a and c are incident on h. In the other case we insert a new edge joining a and c, where a and c are both incident with b on h. The boundary of the new near triangulation is constructed from the original

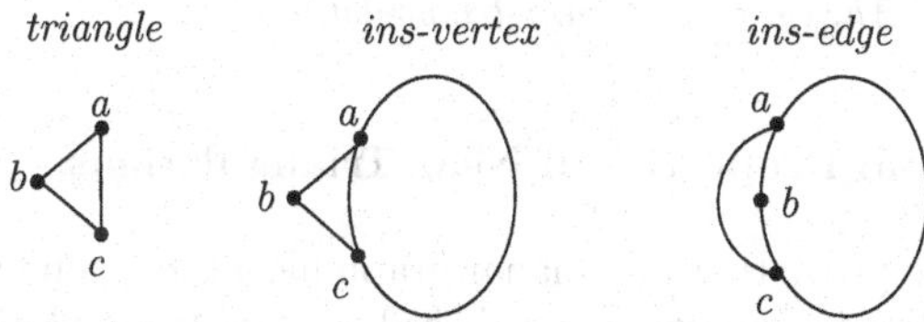

Fig. 3. Inductive definition of near triangulations

one by inserting or deleting the vertex b on h. The set of inner faces f is extended in both cases by the new triangle K_3 a b c.

consts *near-triangulations* :: (α *graph* $\times$ α *graph* $\times$ α *graph set*) *set*
inductive *near-triangulations*
intros
triangle:
$distinct\ [a, b, c] \Longrightarrow (K_3\ a\ b\ c,\ K_3\ a\ b\ c,\ \{K_3\ a\ b\ c\}) \in near\text{-}triangulations$

ins-vertex:
$[\![\ distinct\ [a, b, c];\ \{a, c\} \in \mathcal{E}\ h;\ b \notin \mathcal{V}\ g;$
$(g, h, f) \in near\text{-}triangulations\ ]\!] \Longrightarrow$
$(g \oplus \{a, b\} \oplus \{b, c\},\ cycle\text{-}ins\text{-}vertex\ h\ a\ b\ c,\ f \cup \{K_3\ a\ b\ c\}) \in near\text{-}triangulations$

ins-edge:
$[\![\ distinct\ [a, b, c];\ \{a, b\} \in \mathcal{E}\ h;\ \{b, c\} \in \mathcal{E}\ h;\ \{a, c\} \notin \mathcal{E}\ g;$
$(g, h, f) \in near\text{-}triangulations\ ]\!] \Longrightarrow$
$(g \oplus \{a, c\},\ cycle\text{-}del\text{-}vertex\ h\ a\ b\ c,\ f \cup \{K_3\ a\ b\ c\}) \in near\text{-}triangulations$

A near triangulation is a graph g which can be constructed in this way. A triangulation is a special near triangulation where the boundary consist of 3 vertices. Of course there may be many different ways of constructing a near triangulation. A boundary $\mathcal{H}\ g$ is some boundary in one possible construction of g, $\mathcal{F}\ g$ is some set of inner faces in one possible construction of g. Note that both the inner faces and the boundary (which can be counsidered as the outer face) are not unique, since some planar graphs can be drawn in different ways.

Note that *SOME* is Hilbert's ε-operator, the choice construct in HOL.

constdefs *near-triangulation* :: α *graph* $\Rightarrow$ *bool*
$near\text{-}triangulation\ g\ \equiv\ \exists h\, f.\ (g, h, f) \in near\text{-}triangulations$

triangulation :: α *graph* $\Rightarrow$ *bool*
$triangulation\ g\ \equiv\ \exists h\, f.\ (g, h, f) \in near\text{-}triangulations \wedge card\ (\mathcal{V}\ h) = 3$

boundary :: α *graph* $\Rightarrow$ α *graph* ($\mathcal{H}$)
$\mathcal{H}\ g\ \equiv\ SOME\ h.\ \exists f.\ (g, h, f) \in near\text{-}triangulations$

faces :: α *graph* $\Rightarrow$ α *graph set* ($\mathcal{F}$)
$\mathcal{F}\ g\ \equiv\ SOME\ f.\ \exists h.\ (g, h, f) \in near\text{-}triangulations$

3.2 Fundamental Properties of Near Triangulations

Our next goal is to prove that any planar graph has at least one vertex of degree ≤ 5. Since our definition of planarity is based on near triangulations, we will see that it suffices to show that any near triangulation contains a vertex of degree ≤ 5. To this end we will first establish a series of equations about the number of edges, vertices and faces in near triangulations. Finally we will show that any near triangulation contains a vertex of degree ≤ 5.

The first equation relates the number of inner faces and the number of edges. Any face is adjoined by 3 edges, any edge belongs to 2 faces except the edges on the boundary which belong only to one face. The proof is by induction over the construction of near triangulations.

lemma *nt-edges-faces*:
$(g, h, f) \in \textit{near-triangulations} \Longrightarrow 2 * \textit{card}\ (\mathcal{E}\ g) = 3 * \textit{card}\ f + \textit{card}\ (\mathcal{V}\ h)$ ⟨*proof*⟩

Euler's theorem for near triangulations is also proved by induction. Note that $\mathcal{F}\ g$ contains only the inner faces of g, so the statement contains a 1 instead of the customary 2. We present the top-level structure of the proof, relating the considered quantities in the extended and original near triangulation. Note that we introduce an abbreviation *?P* for the statement we want to prove.

theorem *Euler*:
$\textit{near-triangulation}\ g \Longrightarrow 1 + \textit{card}\ (\mathcal{E}\ g) = \textit{card}\ (\mathcal{V}\ g) + \textit{card}\ (\mathcal{F}\ g)$
proof –
  assume $\textit{near-triangulation}\ g$
  then show $1 + \textit{card}\ (\mathcal{E}\ g) = \textit{card}\ (\mathcal{V}\ g) + \textit{card}\ (\mathcal{F}\ g)$ (**is** $?P\ g\ (\mathcal{F}\ g)$)
  proof (*induct rule*: *nt-f-induct*)
    case (*triangle a b c*)
    show $?P\ (K_3\ a\ b\ c)\ \{K_3\ a\ b\ c\}$ ⟨*proof*⟩
  next
    case (*ins-vertex a b c f g h*)
    have $\textit{card}\ (\mathcal{E}\ (g \oplus \{a, b\} \oplus \{b, c\})) = \textit{card}\ (\mathcal{E}\ g) + 2$ ⟨*proof*⟩
    moreover have $\textit{card}\ (\mathcal{V}\ (g \oplus \{a, b\} \oplus \{b, c\})) = 1 + \textit{card}\ (\mathcal{V}\ g)$ ⟨*proof*⟩
    moreover have $\textit{card}\ (f \cup \{K_3\ a\ b\ c\}) = \textit{card}\ f + 1$ ⟨*proof*⟩
    moreover assume $?P\ g\ f$
    ultimately show $?P\ (g \oplus \{a, b\} \oplus \{b, c\})\ (f \cup \{K_3\ a\ b\ c\})$ **by** *simp*
  next
    case (*ins-edge a b c f g h*)
    have $\textit{card}\ (\mathcal{E}\ (g \oplus \{a, c\})) = 1 + \textit{card}\ (\mathcal{E}\ g)$ ⟨*proof*⟩
    moreover have $\textit{card}\ (\mathcal{V}\ (g \oplus \{a, c\})) = \textit{card}\ (\mathcal{V}\ g)$ ⟨*proof*⟩
    moreover have $\textit{card}\ (f \cup \{K_3\ a\ b\ c\}) = 1 + \textit{card}\ f$ ⟨*proof*⟩
    moreover assume $?P\ g\ f$
    ultimately show $?P\ (g \oplus \{a, c\})\ (f \cup \{K_3\ a\ b\ c\})$ **by** *simp*
  qed
qed

The following result is easily achieved by combining the previous ones.

lemma *edges-vertices*: $\textit{near-triangulation}\ g \Longrightarrow$
$\textit{card}\ (\mathcal{E}\ g) = 3 * \textit{card}\ (\mathcal{V}\ g) - 3 - \textit{card}\ (\mathcal{V}\ (\mathcal{H}\ g))$ ⟨*proof*⟩

The sum of the degrees of all vertices counts the number of neighbours over all vertices, so every edge is counted twice. The proof is again by induction.

theorem *degree-sum*: $\textit{near-triangulation}\ g \Longrightarrow$
$(\sum v \in \mathcal{V}\ g.\ \textit{degree}\ g\ v) = 2 * \textit{card}\ (\mathcal{E}\ g)$ ⟨*proof*⟩

Now we prove our claim by contradiction. We can complete the proof with a short calculation combining the two preceding equations.

theorem *degree5*: *near-triangulation* $g \Longrightarrow \exists v \in \mathcal{V}\, g.\ \textit{degree}\ g\ v \leq 5$
proof –
  assume (*): *near-triangulation* g **then have** (**): $3 \leq \textit{card}\ (\mathcal{V}\, g)$ **by** (*rule card3*)
  show *?thesis*
  proof (*rule classical*)
    assume $\neg\ (\exists v \in \mathcal{V}\, g.\ \textit{degree}\ g\ v \leq 5)$
    then have $\bigwedge v.\ v \in \mathcal{V}\, g \Longrightarrow 6 \leq \textit{degree}\ g\ v$ **by** *auto*
    then have $6 * \textit{card}\ (\mathcal{V}\, g) \leq (\sum v \in \mathcal{V}\, g.\ \textit{degree}\ g\ v)$ **by** *simp*
    also have $\ldots = 2 * \textit{card}\ (\mathcal{E}\, g)$ **by** (*rule degree-sum*)
    also from (**) **and** (*) **have** $\ldots < 6 * \textit{card}\ (\mathcal{V}\, g)$
      by (*simp add*: *edges-vertices*) *arith*
    finally show *?thesis* **by** *arith*
  qed
qed

3.3 Planar Graphs

In text books, planar graphs are usually defined as those graphs which have an embedding in the two-dimensional Euclidean plane with no crossing edges. However, this approach requires a significant amount of topology. We start from the well-known and easy observation that planar graphs with at least 3 vertices are exactly those which can be extended to a triangulation by adding only edges. We do not consider graphs with less than 3 vertices, since a graph with at most n vertices can always be n-coloured. This leads to the following definition.

constdefs *planar* :: α *graph* $\Rightarrow$ *bool*
  $\textit{planar}\ g \equiv \exists g'.\ \textit{triangulation}\ g' \wedge g \preceq g'$

Obviously triangulations are planar. But we also need to know that any near triangulation g is planar. Of course this is also true, we only have to insert edges until there are only 3 vertices on the boundary h of g. We can prove this by induction on the number of vertices in h.

lemma *near-triangulation-planar*: *near-triangulation* $g \Longrightarrow$ *planar* g $\langle$*proof*$\rangle$

We have glossed over one issue so far: is the usual topological definition of triangulation the same as our inductive one? And thus, is our notion of planarity the same as the topological one? It is clear that all of our inductively generated near triangulations are near triangulations in the topological sense, and hence that any planar graph in our sense is also planar in the topological sense. In the other direction let g be a planar graph in the topological sense such that g has at least 3 vertices. Then $g \preceq g'$ for some 2-connected planar graph g'. By Proposition 4 of [18] g' can be generated by the inductive definition in [18], where arbitrary polygons rather than just triangles are added in each step. Now there is a triangulation g'' in our sense such that $g' \preceq g''$ — just add edges to the polygons to subdivide them into triangles. Thus *planar* g because $g \preceq g''$. This finishes the informal proof that our inductive and the usual topological notion of planarity coincide.

4 Topological Properties of Near Triangulations

4.1 Crossing Paths on Near Triangulations

This section is dedicated to an obvious topological property of near triangulations shown in Fig. 4 on the left, where the circle is the boundary: any two paths between v_0 and v_2 and between v_1 and v_3 must intersect, i.e. have a vertex in common. In fact, this is true for all planar graphs. We give an inductive proof for near triangulations: given two paths in the graph extended by a new triangle, this induces two sub-paths in the old graph, which must intersect by induction hypothesis. The cases where the two paths lie completely within the old graph are trivial by induction hypothesis. Then there are numerous symmetric and degenerate cases where one of the two paths include (some of) the newly added edges. In Fig. 4 on the right you see the two main cases, one for inserting a vertex and one for inserting an edge. In both cases the sub-paths between v_0 and a and between v_1 and v_3 lie within the old graph and therefore intersect. In the first case one can also replace the sub-path abc by ac to obtain a path within the old graph, thus cutting down on the number of case distinctions, which is what we did in the formal proof.

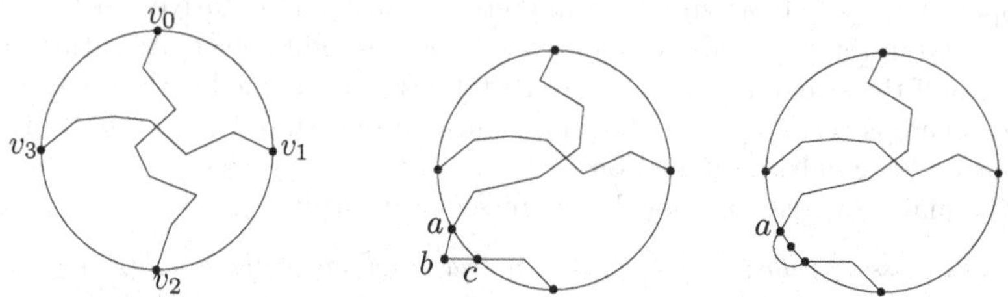

Fig. 4. Crossing paths

Although it took us a moment to find the pictorial proof, the shock came when the formal proof grew to 2000 lines. We will not discuss any details but merely the concepts involved.

A path could either be a graph or a list of vertices. We opted for lists because of their familiarity and large library. This may have been a mistake because lists are better suited for directed paths and may require additional case distinctions in proofs in an undirected setting. Given a graph g, $(x,p,y) \in paths\ g$ means that p is a list of distinct vertices leading from x to y. The definition is inductive:

consts $paths :: \alpha\ graph \Rightarrow (\alpha \times \alpha\ list \times \alpha)set$
inductive $paths\ g$
intros
basis: $(x,\ [x],\ x) \in paths\ g$
step: $[\![\ (y,\ p,\ z) \in paths\ g;\ \{x,\ y\} \in \mathcal{E}\ g;\ x \notin set\ p\ ]\!] \Longrightarrow (x,\ x\#p,\ z) \in paths\ g$

Function *set* converts a list into a set and the infix # separates the head from the tail of a list.

From a path p it is easy to recover a set of edges *pedges p*. Keyword **recdef** starts a recursive function definition and *measure size* is the hint for the automatic termination prover:

consts *pedges* :: α *list* $\Rightarrow$ α *set set*
recdef *measure length*
  pedges $(x\#y\#zs) = \{\{x,\ y\}\} \cup$ *pedges* $(y\#zs)$
  pedges $xs = \{\}$

Now we come to the key assumption of our main lemma shown in Fig. 4: the vertices v_0, v_1, v_2 and v_3 must appear in the right order on the boundary. We express this with the predicate *ortho h x y u v* where h is the boundary, x must be opposite y and u opposite v:

constdefs *ortho* :: α *graph* $\Rightarrow \alpha \Rightarrow \alpha \Rightarrow \alpha \Rightarrow \alpha \Rightarrow$ *bool*
  ortho $h\ x\ y\ u\ v \equiv x \neq y \wedge u \neq v \wedge$
    $(\exists p\ q.\ (x,\ p,\ y) \in \mathit{paths}\ h \wedge (y,\ q,\ x) \in \mathit{paths}\ h \wedge u \in \mathit{set}\ p \wedge v \in \mathit{set}\ q \wedge$
      $\mathcal{E}\ h = \mathit{pedges}\ p \cup \mathit{pedges}\ q \wedge \mathit{set}\ p \cap \mathit{set}\ q \subseteq \{x,\ y\})$

This definition is attractive because it follows very easily that x can be swapped with y and u with v. But there are many alternatives to this definition, and in retrospect we wonder if we made the right choice. As it turns out, almost half the size of this theory, i.e. 1000 lines, is concerned with proofs about *ortho*. There is certainly room for more automation. All of theses proofs involve first-order logic only, no induction.

The main theorem can now be expressed very easily:

theorem *crossing-paths*: $[\![\ (g,\ h,\ f) \in \mathit{near\text{-}triangulations};\ \mathit{ortho}\ h\ v_0\ v_2\ v_1\ v_3;$
  $(v_0,\ p,\ v_2) \in \mathit{paths}\ g;\ (v_1,\ q,\ v_3) \in \mathit{paths}\ g\ ]\!] \Longrightarrow \mathit{set}\ p \cap \mathit{set}\ q \neq \{\}$ ⟨*proof*⟩

We also need to know that we can find 4 orthogonal vertices in a large enough graph:

lemma *cycles-ortho-ex*: $[\![\ \mathit{cycles}\ h;\ 4 \leq \mathit{card}\ (\mathcal{V}\ h)\ ]\!] \Longrightarrow$
  $\exists v_0\ v_1\ v_2\ v_3.\ \mathit{ortho}\ h\ v_0\ v_2\ v_1\ v_3 \wedge \mathit{distinct}\ [v_0,\ v_1,\ v_2,\ v_3]$ ⟨*proof*⟩

4.2 Inversions of Triangulations

For the proof of the 5CT we need the following topological property, which is illustrated in Fig. 5: if we delete a vertex v in a triangulation g, we obtain (in two steps) a near triangulation with boundary h where the vertices of h are exactly the neighbours v_0, v_1, v_2, v_3, v_4 of v.

theorem *del-vertex-near-triangulation*: $[\![\ \mathit{triangulation}\ g;\ 4 \leq \mathit{card}\ (\mathcal{V}\ g)\ ]\!] \Longrightarrow$
  $\exists h\ f.(g \odot v,\ h,\ f) \in \mathit{near\text{-}triangulations} \wedge \mathcal{V}\ h = \Gamma\ g\ v$

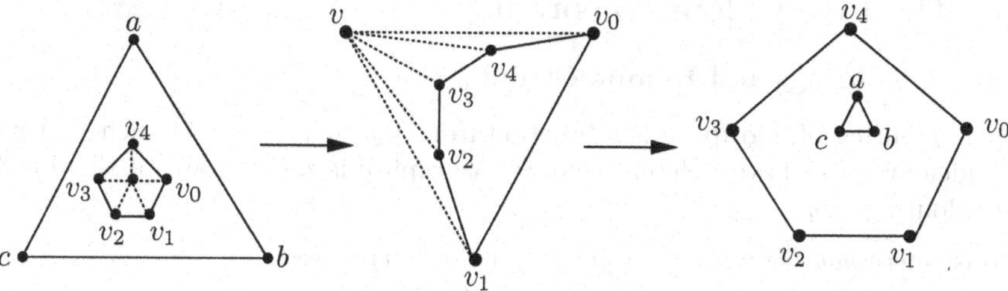

Fig. 5. Deletion of a vertex in a triangulation

This property is based on another fundamental but non-trivial property of triangulations (see Fig. 6). Given a triangulation g, we can consider any face of g as the boundary of g, i.e. we may invert the given inductive construction of g such that the boundary is one of the inner faces.

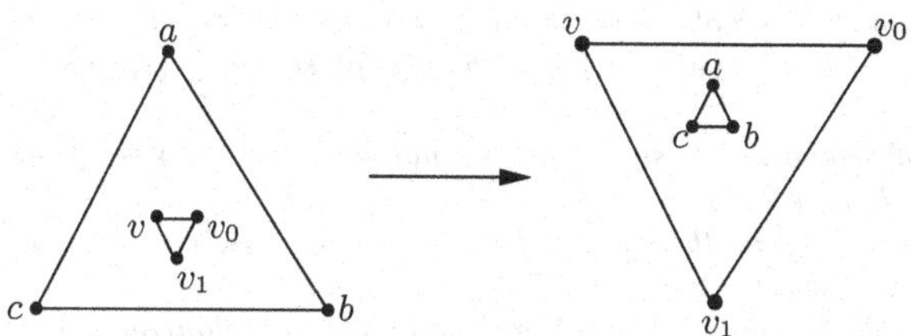

Fig. 6. Inversion of a triangulation

Now we can complete the proof of our previous claim. First we invert the triangulation such that one of the faces adjoining v is the outer one (first step in Fig. 5). Now we cut off all edges to neighbours of v and obtain a near triangulation where the vertices on h are exactly the neighbours of v. This is the only theorem we have not formally proved so far, but it is a topological property and does not concern colouring. It also is a special case of the following topological property, which may be seen as a formulation of the Jordan Curve Theorem: if we delete the inner faces of a near triangulation from a triangulation we again obtain a near triangulation.

5 The Five Colour Theorem

5.1 Colourings and Kempe Chains

A n-colouring of a graph g is a function from $\mathcal{V}\ g$ to $\{k.\ k < n\}$ such that any incident vertices have different colours. A graph g is n-colourable if there is an n-colouring of g.

constdefs $colouring :: nat \Rightarrow \alpha\ graph \Rightarrow (\alpha \Rightarrow nat) \Rightarrow bool$
$colouring\ n\ g\ c\ \equiv$
$(\forall v.\ v \in \mathcal{V}\ g \longrightarrow c\ v < n) \wedge (\forall v\ w.\ \{v,\ w\} \in \mathcal{E}\ g \longrightarrow c\ v \neq c\ w)$

constdefs $colourable :: nat \Rightarrow \alpha\ graph \Rightarrow bool$
$colourable\ n\ g\ \equiv\ \exists c.\ colouring\ n\ g\ c$

A Kempe chain of colours i and j is the subset of all vertices of g which are connected with v by a path of vertices of colours i and j.

We are able to prove the following major result: we can obtain a new n-colouring c' from a n-colouring c by transposing the colours of all vertices in a Kempe chain.

constdefs $kempe :: \alpha\ graph \Rightarrow (\alpha \Rightarrow nat) \Rightarrow \alpha \Rightarrow nat \Rightarrow nat \Rightarrow \alpha\ set$ $(\mathcal{K})$
$\mathcal{K}\ g\ c\ v\ i\ j\ \equiv\ \{w.\ \exists p.\ (v,\ p,\ w) \in paths\ g \wedge c\text{`}(set\ p) \subseteq \{i,\ j\}\}$

constdefs $swap\text{-}colours :: \alpha\ set \Rightarrow nat \Rightarrow nat \Rightarrow (\alpha \Rightarrow nat) \Rightarrow (\alpha \Rightarrow nat)$
$swap\text{-}colours\ K\ i\ j\ c\ v\ \equiv$
$if\ v \in K\ then\ if\ c\ v = i\ then\ j\ else\ if\ c\ v = j\ then\ i\ else\ c\ v\ else\ c\ v$

lemma $swap\text{-}colours\text{-}kempe$: $[\![\ v \in \mathcal{V}\ g;\ i < n;\ j < n;\ colouring\ n\ g\ c\ ]\!] \Longrightarrow$
$colouring\ n\ g\ (swap\text{-}colours\ (\mathcal{K}\ g\ c\ v\ i\ j)\ i\ j\ c)$ $\langle proof \rangle$

5.2 Reductions

A important proof principle in the proof of the 5CT is reduction of the problem of colouring g to the problem of colouring a smaller graph. For the proof of the 5CT we only need two different reductions. One for the case that g has a vertex v of degree less than 4 and one for the case that g has a vertex v of degree 5. Both reductions are based on the following property: if a colouring c on $g \odot v$ uses less than 5 colours for the neighbours $\Gamma\ g\ v$ of v, then c can be extended to a colouring on g. We only have to assign to v one of the colours which are not used in $\Gamma\ g\ v$.

lemma *free-colour-reduction*:
$[\![\ card\ (c\ \text{`}\ \Gamma\ g\ v) < 5;\ colouring\ 5\ (g \odot v)\ c;\ triangulation\ g\ ]\!] \Longrightarrow$
$colourable\ 5\ g$ $\langle proof \rangle$

The reduction for the case that v has degree less than 5 is quite obvious: there are only 4 neighbours so they cannot use more than 4 colours.

lemma *degree4-reduction*:
$[\![\ colourable\ 5\ (g \odot v);\ degree\ g\ v < 5;\ triangulation\ g\ ]\!] \Longrightarrow colourable\ 5\ g$ $\langle proof \rangle$

The reduction for the cases that v has degree 5 is a bit more complicated [8, 7]: we construct a colouring of g from a colouring of $g \odot v$ where g is a triangulation and *degree* $g\ v = 5$. We assume $g \odot v$ is colourable, so we can choose an (arbitrary) colouring c of $g \odot v$. In the case that c uses less than 5 colours in the neighbourhood $\Gamma\ g\ v$ of v, we obviously can extend c on g.

lemma *degree5-reduction*:
  ⟦ *colourable* 5 ($g \odot v$); *degree* $g\ v = 5$; *triangulation* g ⟧ $\Longrightarrow$ *colourable* 5 g
proof –
  assume (∗): *triangulation* g **and** *degree* $g\ v = 5$
  assume *colourable* 5 ($g \odot v$) **then obtain** c **where** *colouring* 5 ($g \odot v$) c ..

  have *card* (c ‘ $\Gamma\ g\ v$) ≤ 5 ⟨*proof*⟩ **then show** *colourable* 5 g
  proof *cases*
    assume *card* (c ‘ $\Gamma\ g\ v$) < 5 **show** *colourable* 5 g **by** (*rule free-colour-reduction*)
  next

In the case that c uses all 5 colours on the neighbours $\Gamma\ g\ v$ of v, we consider the near triangulation $g \odot v$ with boundary h on $\Gamma\ g\ v$. We can find four distinct orthogonal vertices $v_0\ v_1\ v_2\ v_3$ on h.

    assume *card* (c ‘ $\Gamma\ g\ v$) $= 5$ **also have** (∗∗): *card* ($\Gamma\ g\ v$) $= 5$ ⟨*proof*⟩
    finally have *inj-on* c ($\Gamma\ g\ v$) ⟨*proof*⟩

    have $4 \leq$ *card* ($\mathcal{V}\ g$) ⟨*proof*⟩
    with (∗) **obtain** $h\ f$ **where** ($g \odot v$, h, f) $\in$ *near-triangulations* **and**
      $\mathcal{V}\ h = \Gamma\ g\ v$ **by** (*rules dest*: *del-vertex-near-triangulation*)
    with (∗∗) **have** $4 \leq$ *card* ($\mathcal{V}\ h$) **and** *cycles* h ⟨*proof*⟩
    then obtain $v_0\ v_1\ v_2\ v_3$ **where** *ortho* $h\ v_0\ v_2\ v_1\ v_3$ **and** *distinct* $[v_0, v_1, v_2, v_3]$
      by (*auto dest*: *cycles-ortho-ex*)

There cannot be a path in $g \odot v$ in the colours of v_0 and v_2 between v_0 and v_2 and at the same time one in the colours of v_1 and v_3 between v_1 and v_3. The reason is that the paths would intersect according to the *crossing-paths* lemma (see Fig. 7 on the left). This is a contradiction to the fact that all neighbours have different colours. Thus we can obtain v_0' and v_2' such that there is no path in the colours of v_0' and v_2' between v_0' and v_2'.

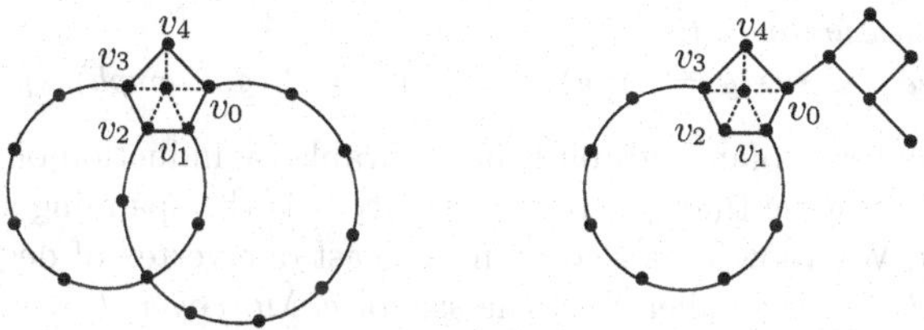

Fig. 7. Reduction for the case *degree* $g\ v = 5$

have $\neg$ $((\exists p.\ (v_0,\ p,\ v_2) \in \mathit{paths}\ (g \odot v) \wedge c`(\mathit{set}\ p) \subseteq \{c\ v_0,\ c\ v_2\}) \wedge$
$(\exists p.\ (v_1,\ p,\ v_3) \in \mathit{paths}\ (g \odot v) \wedge c`(\mathit{set}\ p) \subseteq \{c\ v_1,\ c\ v_3\}))$
proof –
{ **fix** p **assume** $(v_0,\ p,\ v_2) \in \mathit{paths}\ (g \odot v)$ **and** $c`(\mathit{set}\ p) \subseteq \{c\ v_0,\ c\ v_2\}$
fix q **assume** $(v_1,\ q,\ v_3) \in \mathit{paths}\ (g \odot v)$ **and** $c`(\mathit{set}\ q) \subseteq \{c\ v_1,\ c\ v_3\}$
have $\mathit{set}\ p \cap \mathit{set}\ q \neq \{\}$ **by** (*rule crossing-paths*)
then obtain s **where** $s \in \mathit{set}\ p$ **and** $s \in \mathit{set}\ q$ **by** *auto*
have $c\ s \in \{c\ v_0,\ c\ v_2\}$ **and** $c\ s \in \{c\ v_1,\ c\ v_3\}$ ⟨*proof*⟩
moreover have $c\ v_0 \neq c\ v_1$ **and** $c\ v_0 \neq c\ v_3$ **and**
$c\ v_1 \neq c\ v_2$ **and** $c\ v_2 \neq c\ v_3$ ⟨*proof*⟩
ultimately have *False* **by** *auto*
} **then show** *?thesis* **by** *auto*
qed
then obtain $v_0{}'\ v_2{}'$ **where** $v_0{}' = v_0 \wedge v_2{}' = v_2 \vee v_0{}' = v_1 \wedge v_2{}' = v_3$ **and**
$\neg\ (\exists p.\ p \in \mathit{paths}\ (g \odot v)\ v_0{}'\ v_2{}' \wedge c`(\mathit{set}\ p) \subseteq \{c\ v_0{}',\ c\ v_2{}'\})$ **by** *auto*

We consider the Kempe chain K of all vertices which are connected with $v_0{}'$ by a path in the colours of $v_0{}'$ and $v_2{}'$. We define a new colouring c' on $g \odot v$ by transposing the colours of $v_0{}'$ and $v_2{}'$ in K. According to our previous considerations $v_2{}'$ is separated from $v_0{}'$, so $v_2{}'$ does not lie in K. (See Fig. 7, on the right.) We can extend the colouring c' to g by assigning the colour of $v_0{}'$ to v.

def $K \equiv \mathcal{K}\ (g \odot v)\ c\ v_0{}'\ (c\ v_0{}')\ (c\ v_2{}')$
def $c' \equiv$ *swap-colours* $K\ (c\ v_0{}')\ (c\ v_2{}')\ c$
obtain *colouring* $5\ (g \odot v)\ c'$ ⟨*proof*⟩

have $v_2{}' \notin K$ ⟨*proof*⟩ **then have** $c'\ v_2{}' = c\ v_2{}'$ ⟨*proof*⟩
also have $v_0{}' \in K$ ⟨*proof*⟩ **then have** $c'\ v_0{}' = c\ v_2{}'$ ⟨*proof*⟩
finally have $c'\ v_0{}' = c'\ v_2{}'$.
then have $\neg$ *inj-on* $c'\ (\mathcal{V}\ h)$ ⟨*proof*⟩ **also have** $\mathcal{V}\ h = \Gamma\ g\ v$.
finally have *card* $(c'\ `\ \Gamma\ g\ v) < 5$ ⟨*proof*⟩
then show *colourable* $5\ g$ **by** (*rule free-colour-reduction*)
qed
qed

5.3 The Proof of the Five Colour Theorem

We prove the 5CT using the following rule for induction on *size g*.

lemma *graph-measure-induct*:
$(\bigwedge g.\ (\bigwedge h.\ \mathit{size}\ h < \mathit{size}\ g \Longrightarrow P\ h) \Longrightarrow P\ g) \Longrightarrow P\ (g{::}\alpha\ \mathit{graph})$ ⟨*proof*⟩

We show that any planar graph is 5-colourable by induction on *size g* (compare [8, 7]). We assume that g is planar. Hence it is the spanning subgraph of a triangulation t. We know that t contains at least one vertex of degree less than 5. The size of $t \odot v$ is smaller than the size of g. Moreover, $t \odot v$ is a near triangulation and therefore planar. It follows from the induction hypothesis (*IH*) that $t \odot v$ must be 5-colourable. Now we construct a colouring of t depending on a case distinction on the degree of v by applying an appropriate reduction.

theorem *5CT*: *planar g* $\Longrightarrow$ *colourable* 5 *g*
proof –
  assume *planar g* **then show** *colourable* 5 *g*
  proof (*induct rule*: *graph-measure-induct*)
    fix g::α *graph*
    assume *IH*: $\bigwedge g'$::α *graph*. *size* g' < *size* $g \Longrightarrow$ *planar* $g' \Longrightarrow$ *colourable* 5 g'
    assume *planar g* **then obtain** *t* **where** *triangulation t* **and** $g \preceq t$ ..
    then obtain *v* **where** $v \in \mathcal{V}\, t$ **and** *d*: *degree t v* ≤ 5 **by** (*blast dest*: *degree5*)

    have *size* $(t \odot v)$ < *size t* .. **also have** *size t* = *size g* ..
    also have *planar* $(t \odot v)$ ..
    ultimately obtain *colourable* 5 $(t \odot v)$ **by** (*rules dest*: *IH*)

    from *d* **have** *colourable* 5 *t*
    proof *cases*
      assume *degree t v* < 5 **show** *colourable* 5 *t* **by** (*rule degree4-reduction*)
    next
      assume *degree t v* = 5 **show** *colourable* 5 *t* **by** (*rule degree5-reduction*)
    qed
    then show *colourable* 5 *g* ..
  qed
qed

6 Conclusion

Although we have proved what we set out to, the resulting formalization and proofs are still only a beginning. For a start, the size of the proofs for the *crossing-paths* lemma in §4.1 are too large for comfort, and the proof of theorem *del-vertex-neartriangulation* still needs to be formalized in Isabelle. It also appears that both propositions could be obtained as special cases of a discrete version of the Jordan Curve Theorem. However, in other parts, e.g. Euler's theorem and the reduction lemmas, the proofs are roughly in line with what one would expect from the informal proofs. The complete proof, without the topological properties, consists of about 20 pages basic graph theory, 15 pages of triangulations, and 7 pages for the 5CT including Kempe chains and reductions.

So what about a full formalization of one of the existing proofs of the 4CT [4, 13]? The top-level proof by Robertson *et al.* consists of two computer-validated theorems plus a lemma by Birkhoff [5]. The latter is more complicated than the 5CT, but still reasonably simple. The crucial part are the two computer-validated theorems. One would either have to verify the correctness of the algorithms, or let the algorithms produce a trace that can be turned into a proof. Although the second appears more tractable, in either case one needs a considerable amount of additional graph theory. Such an undertaking appears within reach of a dedicated group of experts in theorem proving and graph theory, and its benefits could be significant.

References

1. K. Appel and W. Haken. Every planar map is four colourable. *Bulletin of the American mathematical Society*, 82:711–112, 1977.
2. K. Appel and W. Haken. Every planar map is four colourable. I: Discharging. *Illinois J. math.*, 21:429–490, 1977.
3. K. Appel and W. Haken. Every planar map is four colourable. II: Reducibility. *Illinois J. math.*, 21:491–567, 1977.
4. K. Appel and W. Haken. *Every planar map is four colourable*, volume 98. Contemporary Mathematics, 1989.
5. G. D. Birkhoff. The reducibility of maps. *Amer. J. Math.*, 35:114–128, 1913.
6. C.-T. Chou. A Formal Theory of Undirected Graphs in Higher-Order Logic. In T. Melham and J. Camilleri, editors, *Higher Order Logic Theorem Proving and Its Applications*, volume 859 of *LNCS*, pages 158–176, 1994.
7. R. Diestel. *Graph Theory.* Graduate Texts in Mathematics. Springer, 2000. Electronic Version: `http://www.math.uni-hamburg.de/home/diestel/books/graph.theory/`.
8. T. Emden-Weinert, S. Hougardy, B. Kreuter, H. Prömel, and A. Steger. Einführung in Graphen und Algorithmen, 1996. `http://www.informatik.hu-berlin.de/Institut/struktur/algorithmen/ga/`.
9. H. Heesch. *Untersuchungen zum Vierfarbenproblem.* Number 810/a/b in Hochschulskriptum. Bibliographisches Institut, Mannheim, 1969.
10. T. Nipkow, L. Paulson, and M. Wenzel. *Isabelle/HOL — A Proof Assistant for Higher-Order Logic*, volume 2283 of *LNCS.* Springer, 2002. To appear.
11. N. Robertson, D. Sanders, P. Seymour, and R. Thomas. The four colour theorem. `http://www.math.gatech.edu/~thomas/FC/fourcolor.html`, 1995.
12. N. Robertson, D. Sanders, P. Seymour, and R. Thomas. A new proof of the four colour theorem. *Electron. Res. Announce. Amer. Math Soc.*, 2(1):17–25, 1996.
13. N. Robertson, D. Sanders, P. Seymour, and R. Thomas. The four colour theorem. *J. Combin. Theory Ser. B*, 70:2–44, 1997.
14. M. Wenzel. *Isabelle/Isar — A Versatile Environment for Human-Readable Formal Proof Documents.* PhD thesis, Institut für Informatik, Technische Universität München, 2002. `http://tumb1.biblio.tu-muenchen.de/publ/diss/in/2002/wenzel.htm`.
15. D. West. *Introduction to Graph Theory.* Prentice Hall, New York, 1996.
16. F. Wiedijk. The Four Color Theorem Project. `http://www.cs.kun.nl/~freek/4ct/`, 2000.
17. W. Wong. A simple graph theory and its application in railway signalling. In M. Archer, J. Joyce, K. Levitt, and P. Windley, editors, *Proc. 1991 International Workshop on the HOL Theorem Proving System and its Applications*, pages 395–409. IEEE Computer Society Press, 1992.
18. M. Yamamoto, S.-y. Nishizaha, M. Hagiya, and Y. Toda. Formalization of Planar Graphs. In E. Schubert, P. Windley, and J. Alves-Foss, editors, *Higher Order Logic Theorem Proving and Its Applications*, volume 971 of *LNCS*, pages 369–384, 1995.
19. M. Yamamoto, K. Takahashi, M. Hagiya, S.-y. Nishizaki, and T. Tamai. Formalization of Graph Search Algorithms and Its Applications. In J. Grundy and M. Newey, editors, *Theorem Proving in Higher Order Logics*, volume 1479 of *LNCS*, pages 479–496, 1998.

Type-Theoretic Functional Semantics

Yves Bertot, Venanzio Capretta, and Kuntal Das Barman

Project LEMME, INRIA Sophia Antipolis
{Yves.Bertot,Venanzio.Capretta,Kuntal.Das_Barman}@sophia.inria.fr

Abstract. We describe the operational and denotational semantics of a small imperative language in type theory with inductive and recursive definitions. The operational semantics is given by natural inference rules, implemented as an inductive relation. The realization of the denotational semantics is more delicate: The nature of the language imposes a few difficulties on us. First, the language is Turing-complete, and therefore the interpretation function we consider is necessarily partial. Second, the language contains strict sequential operators, and therefore the function necessarily exhibits nested recursion. Our solution combines and extends recent work by the authors and others on the treatment of general recursive functions and partial and nested recursive functions. The first new result is a technique to encode the approach of Bove and Capretta for partial and nested recursive functions in type theories that do not provide simultaneous induction-recursion. A second result is a clear understanding of the characterization of the definition domain for general recursive functions, a key aspect in the approach by iteration of Balaa and Bertot. In this respect, the work on operational semantics is a meaningful example, but the applicability of the technique should extend to other circumstances where complex recursive functions need to be described formally.

1 Introduction

There are two main kinds of semantics for programming languages.

Operational semantics consists in describing the steps of the computation of a program by giving formal rules to derive judgments of the form $\langle p, a\rangle \rightsquigarrow r$, to be read as "the program p, when applied to the input a, terminates and produces the output r".

Denotational semantics consists in giving a mathematical meaning to data and programs, specifically interpreting data (input and output) as elements of certain domains and programs as functions on those domains; then the fact that the program p applied to the input a gives r as result is expressed by the equality $[\![p]\!]([\![a]\!]) = [\![r]\!]$, where $[\![-]\!]$ is the interpretation.

Our main goal is to develop operational and denotational semantics inside type theory, to implement them in the proof-assistant **Coq** [12], and to prove their main properties formally. The most important result in this respect is a soundness and completeness theorem stating that operational and denotational semantics agree.

V.A. Carreño, C. Muñoz, S. Tahar (Eds.): TPHOLs 2002, LNCS 2410, pp. 83–97, 2002.

The implementation of operational semantics is straightforward: The derivation system is formalized as an inductive relation whose constructors are direct rewording of the derivation rules.

The implementation of denotational semantics is much more delicate. Traditionally, programs are interpreted as partial functions, since they may diverge on certain inputs. However, all function of type theory are total. The problem of representing partial functions in a total setting has been the topic of recent work by several authors [7, 5, 13, 4, 14]. A standard way of solving it is to restrict the domain to those elements that are interpretations of inputs on which the program terminates and then interpret the program as a total function on the restricted domain. There are different approaches to the characterization of the restricted domain. Another approach is to lift the co-domain by adding a bottom element, this approach is not applicable here because the expressive power of the programming language imposes a limit to computable functions.

Since the domain depends on the definition of the function, a direct formalization needs to define domain and function simultaneously. This is not possible in standard type theory, but can be achieved if we extend it with Dybjer's simultaneous induction-recursion [6]. This is the approach adopted in [4].

An alternative way, adopted by Balaa and Bertot in [1], sees the partial function as a fixed point of an operator F that maps total functions to total functions. It can be approximated by a finite number of iterations of F on an arbitrary base function. The domain can be defined as the set of those elements for which the iteration of F stabilizes after a finite number of steps independently of the base function.

The drawback of the approach of [4] is that it is not viable in standard type theories (that is, without Dybjer's schema). The drawback of the approach of [1] is that the defined domain is the domain of a fixed point of F that is not in general the least fixed point. This maybe correct for lazy functional languages (call by name), but is incorrect for strict functional languages (call by value), where we need the least fixed point. The interpretation of an imperative programming language is essentially strict and therefore the domain is too large: The function is defined for values on which the program does not terminate.

Here we combine the two approaches of [4] and [1] by defining the domain in a way similar to that of [4], but disentangling the mutual dependence of domain and function by using the iteration of the functional F with a variable index in place of the yet undefined function.

We claim two main results. First, we develop denotational semantics in type theory. Second, we model the accessibility method in a weaker system, that is, without using simultaneous induction-recursion.

Here is the structure of the paper.

In Section 2 we define the simple imperative programming language IMP. We give an informal description of its operational and denotational semantics. We formalize the operational semantics by an inductive relation. We explain the difficulties related to the implementation of the denotational semantics.

In Section 3 we describe the iteration method. We point out the difficulty in characterizing the domain of the interpretation function by the convergence of the iterations.

In Section 4 we give the denotational semantics using the accessibility method. We combine it with the iteration technique to formalize nested recursion without the use of simultaneous induction-recursion.

All the definitions have been implemented in **Coq** and all the results proved formally in it. We use here an informal mathematical notation, rather than giving **Coq** code. There is a direct correspondence between this notation and the **Coq** formalization. Using the **PCoq** graphical interface (available on the web at the location `http://www-sop.inria.fr/lemme/pcoq/index.html`), we also implemented some of this more intuitive notation. The **Coq** files of the development are on the web at `http://www-sop.inria.fr/lemme/Kuntal.Das_Barman/imp/`.

2 IMP and Its Semantics

Winskel [15] presents a small programming language IMP with *while* loops. IMP is a simple imperative language with integers, truth values true and false, memory locations to store the integers, arithmetic expressions, boolean expressions and commands. The formation rules are

arithmetic expressions: $a ::= n \mid X \mid a_0 + a_1 \mid a_0 - a_1 \mid a_0 * a_1;$
boolean expressions: $b ::= \mathsf{true} \mid \mathsf{false} \mid a_0 = a_1 \mid a_0 \leq a_1 \mid \neg b \mid b_0 \vee b_1 \mid b_0 \wedge b_1;$
commands: $c ::= \mathsf{skip} \mid X \leftarrow a \mid c_0; c_1 \mid \mathsf{if}\ b\ \mathsf{then}\ c_0\ \mathsf{else}\ c_1 \mid \mathsf{while}\ b\ \mathsf{do}\ c$

where n ranges over integers, X ranges over locations, a ranges over arithmetic expressions, b ranges over boolean expressions and c ranges over commands.

We formalize it in **Coq** by three inductive types AExp, BExp, and Command.

For simplicity, we work with natural numbers instead of integers. We do so, as it has no significant importance in the semantics of IMP. Locations are also represented by natural numbers. One should not confuse the natural number denoting a location with the natural number contained in the location. Therefore, in the definition of AExp, we denote the constant value n by $\mathsf{Num}(n)$ and the memory location with address v by $\mathsf{Loc}(v)$

We see commands as state transformers, where a state is a map from memory locations to natural numbers. The map is in general partial, indeed it is defined only on a finite number of locations. Therefore, we can represent a state as a list of bindings between memory locations and values. If the same memory location is bound twice in the same state, the most recent binding, that is, the leftmost one, is the valid one.

$$\begin{array}{l}\mathsf{State}\colon \mathbf{Set}\\ []\colon \mathsf{State}\\ [\cdot \mapsto \cdot, \cdot]\colon \mathbb{N} \to \mathbb{N} \to \mathsf{State} \to \mathsf{State}\end{array}$$

The state $[v \mapsto n, s]$ is the state s with the content of the location v replaced by n.

Operational semantics consists in three relations giving meaning to arithmetic expressions, boolean expressions, and commands. Each relation has three arguments: The expression or command, the state in which the expression is evaluated or the command executed, and the result of the evaluation or execution.

$$\begin{aligned}(\langle\cdot,\cdot\rangle_{\mathsf{A}} \rightsquigarrow \cdot)&: \mathsf{AExp} \to \mathsf{State} \to \mathbb{N} \to \mathbf{Prop}\\(\langle\cdot,\cdot\rangle_{\mathsf{B}} \rightsquigarrow \cdot)&: \mathsf{BExp} \to \mathsf{State} \to \mathbb{B} \to \mathbf{Prop}\\(\langle\cdot,\cdot\rangle_{\mathsf{C}} \rightsquigarrow \cdot)&: \mathsf{Command} \to \mathsf{State} \to \mathsf{State} \to \mathbf{Prop}\end{aligned}$$

For arithmetic expressions we have that constants are interpreted in themselves, that is, we have axioms of the form

$$\langle \mathsf{Num}(n), \sigma\rangle_{\mathsf{A}} \rightsquigarrow n$$

for every $n\colon \mathbb{N}$ and $\sigma\colon \mathsf{State}$. Memory locations are interpreted by looking up their values in the state. Consistently with the spirit of operational semantics, we define the lookup operation by derivation rules rather than by a function.

$$\frac{(\mathsf{value\_ind}\ \sigma\ v\ n)}{\langle \mathsf{Loc}(v), \sigma\rangle_{\mathsf{A}} \rightsquigarrow n}$$

where

$$\begin{aligned}&\mathsf{value\_ind}\colon \mathsf{State} \to \mathbb{N} \to \mathbb{N} \to \mathbf{Prop}\\&\mathsf{no\_such\_location}\colon (v\colon \mathbb{N})(\mathsf{value\_ind}\ []\ v\ 0)\\&\mathsf{first\_location}\colon (v, n\colon \mathbb{N}; \sigma\colon \mathsf{State})(\mathsf{value\_ind}\ [v \mapsto n, \sigma]\ v\ n)\\&\mathsf{rest\_locations}\colon (v, v', n, n'\colon \mathbb{N}; \sigma\colon \mathsf{State})\\&\qquad v \neq v' \to (\mathsf{value\_ind}\ \sigma\ v\ n) \to (\mathsf{value\_ind}\ [v' \mapsto n', \sigma]\ v\ n)\end{aligned}$$

Notice that we assign the value 0 to empty locations, rather that leaving them undefined. This corresponds to giving a default value to uninitialized variables rather than raising an exception.

The operations are interpreted in the obvious way, for example,

$$\frac{\langle a_0, \sigma\rangle_{\mathsf{A}} \rightsquigarrow n_0 \quad \langle a_1, \sigma\rangle_{\mathsf{A}} \rightsquigarrow n_1}{\langle a_0 + a_1, \sigma\rangle_{\mathsf{A}} \rightsquigarrow n_0 + n_1}$$

where the symbol $+$ is overloaded: $a_0 + a_1$ denotes the arithmetic expression obtained by applying the symbol $+$ to the expressions a_0 and a_1, $n_0 + n_1$ denotes the sum of the natural numbers n_0 and n_1.

In short, the operational semantics of arithmetic expressions is defined by the inductive relation

$$\begin{aligned}&(\langle\cdot,\cdot\rangle_{\mathsf{A}} \rightsquigarrow \cdot)\colon \mathsf{AExp} \to \mathsf{State} \to \mathbb{N} \to \mathbf{Prop}\\&\mathsf{eval\_Num}\colon (n\colon \mathbb{N}; \sigma\colon \mathsf{State})(\langle \mathsf{Num}(n), \sigma\rangle_{\mathsf{A}} \rightsquigarrow n)\\&\mathsf{eval\_Loc}\colon (v, n\colon \mathbb{N}; \sigma\colon \mathsf{State})(\mathsf{value\_ind}\ \sigma\ v\ n) \to (\langle \mathsf{Loc}(v), \sigma\rangle_{\mathsf{A}} \rightsquigarrow n)\\&\mathsf{eval\_Plus}\colon (a_0, a_1\colon \mathsf{AExp}; n_0, n_1\colon \mathbb{N}; \sigma\colon \mathsf{State})\\&\qquad (\langle a_0, \sigma\rangle_{\mathsf{A}} \rightsquigarrow n_0) \to (\langle a_1, \sigma\rangle_{\mathsf{A}} \rightsquigarrow n_0) \to\\&\qquad (\langle a_0 + a_1, \sigma\rangle_{\mathsf{A}} \rightsquigarrow n_0 + n_1)\\&\mathsf{eval\_Minus}\colon \cdots\\&\mathsf{eval\_Mult}\colon \cdots\end{aligned}$$

For the subtraction case the cutoff difference is used, that is, $n - m = 0$ if $n \le m$.

The definition of the operational semantics of boolean expressions is similar and we omit it.

The operational semantics of commands specifies how a command maps states to states. skip is the command that does nothing, therefore it leaves the state unchanged.

$$\frac{}{\langle \mathsf{skip}, \sigma\rangle_{\mathsf{C}} \rightsquigarrow \sigma}$$

The assignment $X \leftarrow a$ evaluates the expression a and then updates the contents of the location X to the value of a.

$$\frac{\langle a, \sigma\rangle_{\mathsf{A}} \rightsquigarrow n \qquad \sigma_{[X \mapsto n]} \rightsquigarrow \sigma'}{\langle X \leftarrow a, \sigma\rangle_{\mathsf{C}} \rightsquigarrow \sigma'}$$

where $\sigma_{[X \mapsto n]} \rightsquigarrow \sigma'$ asserts that σ' is the state obtained by changing the contents of the location X to n in σ. It could be realized by simply $\sigma' = [X \mapsto n, \sigma]$. This solution is not efficient, since it duplicates assignments of existing locations and it would produce huge states during computation. A better solution is to look for the value of X in σ and change it.

$(\cdot_{[\cdot \mapsto \cdot]} \rightsquigarrow \cdot)\colon \mathsf{State} \to \mathbb{N} \to \mathbb{N} \to \mathsf{State} \to \mathbf{Prop}$
$\mathsf{update\_no\_location}\colon (v, n\colon \mathbb{N})([]_{[v \mapsto n]} \rightsquigarrow [])$
$\mathsf{update\_first}\colon (v, n_1, n_2\colon \mathbb{N}; \sigma\colon \mathsf{State})([v \mapsto n_1, \sigma]_{[v \mapsto n_2]} \rightsquigarrow [v \mapsto n_2, \sigma])$
$\mathsf{update\_rest}\colon (v_1, v_2, n_1, n_2\colon \mathbb{N}; \sigma_1, \sigma_2\colon \mathbb{N}) v_1 \neq v_2 \to$
$\quad (\sigma_{1[v_2 \mapsto n_2]} \rightsquigarrow \sigma_2) \to ([v_1 \mapsto n_1, \sigma_1]_{[v_2 \mapsto n_2]} \rightsquigarrow [v_1 \mapsto n_1, \sigma_2])$

Notice that we require a location to be already defined in the state to update it. If we try to update a location not present in the state, we leave the state unchanged. This corresponds to requiring that all variables are explicitly initialized before the execution of the program. If we use an uninitialized variable in the program, we do not get an error message, but an anomalous behavior: The value of the variable is always zero.

Evaluating a sequential composition $c_1; c_2$ on a state σ consists in evaluating c_1 on σ, obtaining a new state σ_1, and then evaluating c_2 on σ_1 to obtain the final state σ_2.

$$\frac{\langle c_1, \sigma\rangle_{\mathsf{C}} \rightsquigarrow \sigma_1 \qquad \langle c_2, \sigma_1\rangle_{\mathsf{C}} \rightsquigarrow \sigma_2}{\langle c_1; c_2, \sigma\rangle_{\mathsf{C}} \rightsquigarrow \sigma_2}$$

Evaluating conditionals uses two rules. In both rules, we evaluate the boolean expression b, but they differ on the value returned by this step and the sub-instruction that is executed.

$$\frac{\langle b, \sigma\rangle_{\mathsf{B}} \rightsquigarrow \mathsf{true} \quad \langle c_1, \sigma\rangle_{\mathsf{C}} \rightsquigarrow \sigma_1}{\langle \mathsf{if}\ b\ \mathsf{then}\ c_1\ \mathsf{else}\ c_2, \sigma\rangle_{\mathsf{C}} \rightsquigarrow \sigma_1} \qquad \frac{\langle b, \sigma\rangle_{\mathsf{B}} \rightsquigarrow \mathsf{false} \quad \langle c_2, \sigma\rangle_{\mathsf{C}} \rightsquigarrow \sigma_2}{\langle \mathsf{if}\ b\ \mathsf{then}\ c_1\ \mathsf{else}\ c_2, \sigma\rangle_{\mathsf{C}} \rightsquigarrow \sigma_2}$$

As for conditionals, we have two rules for *while* loops. If b evaluates to true, c is evaluated on σ to produce a new state σ', on which the loop is evaluated recursively. If b evaluates to false, we exit the loop leaving the state unchanged.

$$\frac{\langle b, \sigma\rangle_{\mathsf{B}} \rightsquigarrow \mathsf{true} \quad \langle c, \sigma\rangle_{\mathsf{C}} \rightsquigarrow \sigma' \quad \langle \mathsf{while}\ b\ \mathsf{do}\ c, \sigma'\rangle_{\mathsf{C}} \rightsquigarrow \sigma''}{\langle \mathsf{while}\ b\ \mathsf{do}\ c, \sigma\rangle_{\mathsf{C}} \rightsquigarrow \sigma''} \qquad \frac{\langle b, \sigma\rangle_{\mathsf{B}} \rightsquigarrow \mathsf{false}}{\langle \mathsf{while}\ b\ \mathsf{do}\ c, \sigma\rangle_{\mathsf{C}} \rightsquigarrow \sigma}$$

The above rules can be formalized in **Coq** in a straightforward way by an inductive relation.

$$
\begin{array}{l}
\langle\cdot,\cdot\rangle_{\mathsf{C}} \leadsto \cdot : \mathsf{Command} \to \mathsf{State} \to \mathsf{State} \to \mathbf{Prop} \\
\mathsf{eval\_skip}\colon (\sigma\colon \mathsf{State})(\langle \mathsf{skip}, \sigma\rangle_{\mathsf{C}} \leadsto \sigma) \\
\mathsf{eval\_assign}\colon (\sigma, \sigma'\colon \mathsf{State}; v, n\colon \mathbb{N}; a\colon \mathsf{AExp}) \\
\qquad (\langle a, \sigma\rangle_{\mathsf{A}} \leadsto n) \to (\sigma_{[v \mapsto n]} \leadsto \sigma') \to (\langle v \leftarrow a, \sigma\rangle_{\mathsf{C}} \leadsto \sigma') \\
\mathsf{eval\_scolon}\colon (\sigma, \sigma_1, \sigma_2\colon \mathsf{State}; c_1, c_2\colon \mathsf{Command}) \\
\qquad (\langle c_1, \sigma\rangle_{\mathsf{C}} \leadsto \sigma_1) \to (\langle c_2, \sigma_1\rangle_{\mathsf{C}} \leadsto \sigma_2) \to (\langle c_1; c_2, \sigma\rangle_{\mathsf{C}} \leadsto \sigma_2) \\
\mathsf{eval\_if\_true}\colon (b\colon \mathsf{BExp}; \sigma, \sigma_1\colon \mathsf{State}; c_1, c_2\colon \mathsf{Command}) \\
\qquad (\langle b, \sigma\rangle_{\mathsf{B}} \leadsto \mathsf{true}) \to (\langle c_1, \sigma\rangle_{\mathsf{C}} \leadsto \sigma_1) \to \\
\qquad (\langle \mathsf{if}\ b\ \mathsf{then}\ c_1\ \mathsf{else}\ c_2, \sigma\rangle_{\mathsf{C}} \leadsto \sigma_1) \\
\mathsf{eval\_if\_false}\colon (b\colon \mathsf{BExp}; \sigma, \sigma_2\colon \mathsf{State}; c_1, c_2\colon \mathsf{Command}) \\
\qquad (\langle b, \sigma\rangle_{\mathsf{B}} \leadsto \mathsf{false}) \to (\langle c_2, \sigma\rangle_{\mathsf{C}} \leadsto \sigma_2) \to \\
\qquad (\langle \mathsf{if}\ b\ \mathsf{then}\ c_1\ \mathsf{else}\ c_2, \sigma\rangle_{\mathsf{C}} \leadsto \sigma_2) \\
\mathsf{eval\_while\_true}\colon (b\colon \mathsf{BExp}; c\colon \mathsf{Command}; \sigma, \sigma', \sigma''\colon \mathsf{State}) \\
\qquad (\langle b, \sigma\rangle_{\mathsf{B}} \leadsto \mathsf{true}) \to (\langle c, \sigma\rangle_{\mathsf{C}} \leadsto \sigma') \to \\
\qquad (\langle \mathsf{while}\ b\ \mathsf{do}\ c, \sigma'\rangle_{\mathsf{C}} \leadsto \sigma'') \to (\langle \mathsf{while}\ b\ \mathsf{do}\ c, \sigma\rangle_{\mathsf{C}} \leadsto \sigma'') \\
\mathsf{eval\_while\_false}\colon (b\colon \mathsf{BExp}; c\colon \mathsf{Command}; \sigma\colon \mathsf{State}) \\
\qquad (\langle b, \sigma\rangle_{\mathsf{B}} \leadsto \mathsf{false}) \to (\langle \mathsf{while}\ b\ \mathsf{do}\ c, \sigma\rangle_{\mathsf{C}} \leadsto \sigma)
\end{array}
$$

For the rest of the paper we leave out the subscripts A, B, and C in $\langle\cdot,\cdot\rangle \leadsto \cdot$.

3 Functional Interpretation

Denotational semantics consists in interpreting program evaluation as a function rather than as a relation. We start by giving a functional interpretation to expression evaluation and state update. This is quite straightforward, since we can use structural recursion on expressions and states. For example, the interpretation function on arithmetic expressions is defined as

$$
\begin{array}{l}
[\![\cdot]\!]\colon \mathsf{AExp} \to \mathsf{State} \to \mathbb{N} \\
[\![\mathsf{Num}(n)]\!]_\sigma := n \\
[\![\mathsf{Loc}(v)]\!]_\sigma := \mathsf{value\_rec}(\sigma, v) \\
[\![a_0 + a_1]\!]_\sigma := [\![a_0]\!]_\sigma + [\![a_1]\!]_\sigma \\
[\![a_0 - a_1]\!]_\sigma := [\![a_0]\!]_\sigma - [\![a_1]\!]_\sigma \\
[\![a_0 * a_1]\!]_\sigma := [\![a_0]\!]_\sigma \cdot [\![a_1]\!]_\sigma
\end{array}
$$

where $\mathsf{value\_rec}(\cdot,\cdot)$ is the function giving the contents of a location in a state, defined by recursion on the structure of the state. It differs from value_ind because it is a function, not a relation; value_ind is its graph. We can now prove that this interpretation function agrees with the operational semantics given by the inductive relation $\langle\cdot,\cdot\rangle \leadsto \cdot$ (all the lemmas and theorems given below have been checked in a computer-assisted proof).

Lemma 1. $\forall\sigma\colon \mathsf{State}.\forall a\colon \mathsf{AExp}.\forall n\colon \mathbb{N}.\langle\sigma, a\rangle \leadsto n \Leftrightarrow [\![a]\!]_\sigma = n$.

In the same way, we define the interpretation of boolean expressions

$$[\![\cdot]\!]\colon \mathsf{BExp} \to \mathsf{State} \to \mathbb{B}$$

and prove that it agrees with the operational semantics.

Lemma 2. $\forall \sigma\colon \mathsf{State}.\forall b\colon \mathsf{BExp}.\forall t\colon \mathbb{B}.\langle \sigma, b\rangle \rightsquigarrow t \Leftrightarrow [\![a]\!]_\sigma = t.$

We overload the Scott brackets $[\![\cdot]\!]$ to denote the interpretation function both on arithmetic and boolean expressions (and later on commands).

Similarly, we define the update function

$$\cdot[\cdot/\cdot]\colon \mathsf{State} \to \mathbb{N} \to \mathbb{N} \to \mathsf{State}$$

and prove that it agrees with the update relation

Lemma 3. $\forall \sigma, \sigma'\colon \mathsf{State}.\forall v, n\colon \mathbb{N}.\sigma_{[v\mapsto n]} \rightsquigarrow \sigma' \Leftrightarrow \sigma[n/v] = \sigma'.$

The next step is to define the interpretation function $[\![\cdot]\!]$ on commands. Unfortunately, this cannot be done by structural recursion, as for the cases of arithmetic and boolean expressions. Indeed we should have

$$\begin{array}{l}
[\![\cdot]\!]\colon \mathsf{Command} \to \mathsf{State} \to \mathsf{State}\\
[\![\mathsf{skip}]\!]_\sigma := \sigma\\
[\![X \leftarrow a]\!]_\sigma := \sigma[[\![a]\!]_\sigma/X]\\
[\![c_1; c_2]\!]_\sigma := [\![c_1]\!]_{[\![c_2]\!]_\sigma}\\
[\![\mathsf{if}\ b\ \mathsf{then}\ c_1\ \mathsf{else}\ c_2]\!]_\sigma := \begin{cases} [\![c_1]\!]_\sigma \ \mathsf{if}\ [\![b]\!]_\sigma = \mathsf{true}\\ [\![c_2]\!]_\sigma \ \mathsf{if}\ [\![b]\!]_\sigma = \mathsf{false}\end{cases}\\
[\![\mathsf{while}\ b\ \mathsf{do}\ c]\!]_\sigma := \begin{cases} [\![\mathsf{while}\ b\ \mathsf{do}\ c]\!]_{[\![c]\!]_\sigma} & \mathsf{if}\ [\![b]\!]_\sigma = \mathsf{true}\\ \sigma & \mathsf{if}\ [\![b]\!]_\sigma = \mathsf{false}\end{cases}
\end{array}$$

but in the clause for *while* loops the interpretation function is called on the same argument if the boolean expression evaluates to true. Therefore, the argument of the recursive call is not structurally smaller than the original argument.

So, it is not possible to associate a structural recursive function to the instruction execution relation as we did for the lookup, update, and expression evaluation relations. The execution of *while* loops does not respect the pattern of structural recursion and termination cannot be ensured: for good reasons too, since the language is Turing complete. However, we describe now a way to work around this problem.

3.1 The Iteration Technique

A function representation of the computation can be provided in a way that respects typing and termination if we don't try to describe the execution function itself but the *second order function of which the execution function is the least fixed point.* This function can be defined in type theory by cases on the structure of the command.

$$\begin{array}{l}
\mathsf{F}\colon (\mathsf{Command} \to \mathsf{State} \to \mathsf{State}) \to \mathsf{Command} \to \mathsf{State} \to \mathsf{State}\\
(\mathsf{F}\ f\ \mathsf{skip}\ \sigma) := \sigma\\
(\mathsf{F}\ f\ (X \leftarrow a)\ \sigma) := \sigma[[\![a]\!]_\sigma/X]\\
(\mathsf{F}\ f\ (c_1; c_2)\ \sigma) := (f\ c_2\ (f\ c_1\ \sigma))\\
(\mathsf{F}\ f\ (\mathsf{if}\ b\ \mathsf{then}\ c_1\ \mathsf{else}\ c_2)\ \sigma) := \begin{cases} (f\ c_1\ \sigma)\ \mathsf{if}\ [\![b]\!]_\sigma = \mathsf{true}\\ (f\ c_2\ \sigma)\ \mathsf{if}\ [\![b]\!]_\sigma = \mathsf{false}\end{cases}\\
(\mathsf{F}\ f\ (\mathsf{while}\ b\ \mathsf{do}\ c)\ \sigma) := \begin{cases} (f\ (\mathsf{while}\ b\ \mathsf{do}\ c)\ (f\ c\ \sigma)) & \mathsf{if}\ [\![b]\!]_\sigma = \mathsf{true}\\ \sigma & \mathsf{if}\ [\![b]\!]_\sigma = \mathsf{false}\end{cases}
\end{array}$$

Intuitively, writing the function F is exactly the same as writing the recursive execution function, except that the function being defined is simply replaced by a bound variable (here f). In other words, we replace recursive calls with calls to the function given in the bound variable f.

The function F describes the computations that are performed at each iteration of the execution function and the execution function performs the same computation as the function F when the latter is repeated *as many times as needed.* We can express this with the following theorem.

Theorem 1 (eval_com_ind_to_rec).

$$\forall c\colon \mathsf{Command}.\forall \sigma_1, \sigma_2\colon \mathsf{State}.$$
$$\langle c, \sigma_1\rangle \rightsquigarrow \sigma_2 \Rightarrow \exists k\colon \mathbb{N}.\forall g\colon \mathsf{Command} \rightarrow \mathsf{State} \rightarrow \mathsf{State}.(\mathsf{F}^k\ g\ c\ \sigma_1) = \sigma_2$$

where we used the following notation

$$\mathsf{F}^k = (\mathsf{iter}\ (\mathsf{Command} \rightarrow \mathsf{State} \rightarrow \mathsf{State})\ \mathsf{F}\ k) = \lambda g.\underbrace{(\mathsf{F}\ (\mathsf{F}\ \cdots\ (\mathsf{F}}_{k\ times}\ g)\ \cdots))$$

definable by recursion on k,

$$\mathsf{iter}\colon (A\colon \mathbf{Set})(A \rightarrow A) \rightarrow \mathbb{N} \rightarrow A \rightarrow A$$
$$(\mathsf{iter}\ A\ f\ 0\ a) := a$$
$$(\mathsf{iter}\ A\ f\ (\mathsf{S}\ k)\ a) := (f\ (\mathsf{iter}\ A\ f\ k\ a)).$$

Proof. Easily proved using the theorems described in the previous section and an induction on the derivation of $\langle c, \sigma_1\rangle \rightsquigarrow \sigma_2$: This kind of induction is also called *rule induction* in [15]. □

3.2 Extracting an Interpreter

The **Coq** system provides an *extraction* facility [10], which makes it possible to produce a version of any function defined in type theory that is written in a functional programming language's syntax, usually the **OCaml** implementation of ML. In general, the extraction facility performs some complicated program manipulations, to ensure that arguments of functions that have only a logical content are not present anymore in the extracted code. For instance, a division function is a 3-argument function inside type theory: The first argument is the number to be divided, the second is the divisor, and the third is a proof that the second is non-zero. In the extracted code, the function takes only two arguments: The extra argument does not interfere with the computation and its presence cannot help ensuring typing, since the programming language's type system is too weak to express this kind of details.

The second order function F and the other recursive functions can also be extracted to ML programs using this facility. However, the extraction process is a simple translation process in this case, because none of the various function actually takes proof arguments.

To perform complete execution of programs, using the ML translation of F, we have the possibility to compute using the extracted version of the iter function. However, we need to guess the right value for the k argument. One way to cope with this is to create an artificial "infinite" natural number, that will always appear to be big enough, using the following recursive data definition:

$$\mathsf{letrec}\ \omega = (\mathsf{S}\ \omega).$$

This definition does not correspond to any natural number that can be manipulated inside type theory: It is an infinite tree composed only of S constructors. In memory, it corresponds to an S construct whose only field points to the whole construct: It is a loop.

Using the extracted iter with ω is not very productive. Since ML evaluates expressions with a call-by-value strategy, evaluating

$$(\mathsf{iter}\ \mathsf{F}\ g\ \omega\ c\ \sigma)$$

imposes that one evaluates

$$(\mathsf{F}\ (\mathsf{iter}\ \mathsf{F}\ g\ \omega)\ c\ \sigma)$$

which in turn imposes that one evaluates

$$(\mathsf{F}\ (\mathsf{F}\ (\mathsf{iter}\ \mathsf{F}\ g\ \omega))\ c\ \sigma)$$

and so on. Recursion unravels unchecked and this inevitably ends with a stack overflow error. However, it is possible to use a variant of the iteration function that avoids this infinite looping, even for a call-by-value evaluation strategy. The trick is to η-expand the expression that provokes the infinite loop, to force the evaluator to stop until an extra value is provided, before continuing to evaluate the iterator. The expression to define this variant is as follows:

$$\begin{array}{l}
\mathsf{iter}'\colon (A, B\colon \mathbf{Set})((A \to B) \to A \to B) \to \mathbb{N} \to (A \to B) \to A \to B \\
(\mathsf{iter}'\ A\ B\ G\ 0\ f) := f \\
(\mathsf{iter}'\ A\ B\ G\ (\mathsf{S}\ k)\ f) := (G\ \lambda a\colon A.(\mathsf{iter}'\ A\ B\ G\ k\ f\ a))
\end{array}$$

Obviously, the expression $\lambda a\colon A.(\mathsf{iter}'\ A\ B\ G\ k\ f\ a)$ is η-equivalent to the expression $(\mathsf{iter}'\ A\ B\ G\ k\ f)$. However, for call-by-value evaluation the two expression are not equivalent, since the λ-expression in the former stops the evaluation process that would lead to unchecked recursion in the latter.

With the combination of iter′ and ω we can now execute any terminating program without needing to compute in advance the number of iterations of F that will be needed. In fact, ω simply acts as a *natural number that is big enough.* We obtain a functional interpreter for the language we are studying, that is (almost) proved correct with respect to the inductive definition $\langle\cdot,\cdot\rangle \leadsto \cdot$.

Still, the use of ω as a natural number looks rather like a dirty trick: This piece of data cannot be represented in type theory, and we are taking advantage of important differences between type theory and ML's memory and computation models: How can we be sure that what we proved in type theory is valid for what we execute in ML? A first important difference is that, while executions of iter or iter′ are sure to terminate in type theory, $(\mathsf{iter}'\ \mathsf{F}\ \omega\ g)$ will loop if the program passed as argument is a looping program.

The purpose of using ω and iter′ is to make sure that F will be called as many times as needed when executing an arbitrary program, with the risk of non-termination when the studied program does not terminate. This can be done more easily by using a *fixpoint* function that simply returns the fixpoint of F. This fixpoint function is defined in ML by

$$\mathsf{letrec}\ (\mathsf{fix}\ f) = f(\lambda x.\mathsf{fix}\ f\ x).$$

Obviously, we have again used the trick of η-expansion to avoid looping in the presence of a call-by-value strategy. With this fix function, the interpreter function is

$$\begin{array}{l}
\mathsf{interp}\colon \mathsf{Command} \to \mathsf{State} \to \mathsf{State} \\
\mathsf{interp} := \mathsf{fix}\ \mathsf{F}.
\end{array}$$

To obtain a usable interpreter, it is then only required to provide a parser and printing functions to display the results of evaluation. This shows how we can build an interpreter for IMP in ML. But we realized it by using some tricks of functional programming that are not available in type theory. If we want to define an interpreter for IMP in type theory, we have to find a better solution to the problem of partiality.

3.3 Characterizing Terminating Programs

Theorem 1 gives one direction of the correspondence between operational semantics and functional interpretation through the iteration method. To complete the task of formalizing denotational semantics, we need to define a function in type theory that interprets each command. As we already remarked, this function cannot be total, therefore we must first restrict its domain to the terminating commands. This is done by defining a predicate D over commands and states, and then defining the interpretation function $[\![\cdot]\!]$ on the domain restricted by this predicate. Theorem 1 suggests the following definition:

$$\begin{array}{l} D\colon \mathsf{Command} \to \mathsf{State} \to \mathbf{Prop} \\ (D\ c\ \sigma) := \exists k\colon \mathbb{N}.\forall g_1, g_2\colon \mathsf{Command} \to \mathsf{State} \to \mathsf{State}. \\ \qquad\qquad (\mathsf{F}^k\ g_1\ c\ \sigma) = (\mathsf{F}^k\ g_2\ c\ \sigma). \end{array}$$

Unfortunately, this definition is too weak. In general, such an approach cannot be used to characterize terminating "nested" iteration. This is hard to see in the case of the IMP language, but it would appear plainly if one added an exception instruction with the following semantics:

$$\langle \mathsf{exception}, \sigma \rangle \rightsquigarrow [].$$

Intuitively, the programmer could use this instruction to express that an exceptional situation has been detected, but all information about the execution state would be destroyed when this instruction is executed.

With this new instruction, there are some commands and states for which the predicate D is satisfied, but whose computation does not terminate.

$$c := \mathsf{while\ true\ do\ skip; exception}.$$

It is easy to see that for any state σ the computation of c on σ does not terminate. In terms of operational semantics, for no state σ' is the judgment $\langle c, \sigma \rangle \rightsquigarrow \sigma'$ derivable.

However, $(D\ c\ \sigma)$ is provable, because $(\mathsf{F}^k\ g\ c\ \sigma) = []$ for any $k > 1$.

In the next section we work out a stronger characterization of the domain of commands, that turn out to be the correct one in which to interpret the operational semantics.

4 The Accessibility Predicate

A common way to represent partial functions in type theory is to restrict their domain to those arguments on which they terminate. A partial function $f\colon A \rightharpoonup B$ is then represented by first defining a predicate $D_f\colon A \to \mathbf{Prop}$ that characterizes the domain of f, that is, the elements of A on which f is defined; and then formalizing the function itself as $f\colon (\Sigma x\colon A.(D_f\ x)) \to B$, where $\Sigma x\colon A.(D_f\ x)$ is the type of pairs $\langle x, h \rangle$ with $x\colon A$ and $h\colon (D_f\ x)$.

The predicate D_f cannot be defined simply by saying that it is the domain of definition of f, since, in type theory, we need to define it before we can define f. Therefore, D_f must be given before and independently from f. One way to do it is to characterize D_f as the predicate satisfied by those elements of A for which the iteration technique converges to the same value for every initial function. This is a good definition when we try to model lazy functional programming languages, but, when interpreting strict programming languages or imperative languages, we find that this predicate would be too weak, being satisfied by elements for which the associated program diverges, as we have seen at the end of the previous section.

Sometimes the domain of definition of a function can be characterized independently of the function by an inductive predicate called *accessibility* [11, 7, 5, 3]. This simply states that an element of a can be proved to be in the domain if the application of f on a calls f recursively on elements that have already been proved to be in the domain. For example, if in the recursive definition of f there is a clause of the form

$$f(e) := \cdots f(e_1) \cdots f(e_2) \cdots$$

and a matches e, that is, there is a substitution of variables ρ such that $a = \rho(e)$; then we add a clause to the inductive definition of Acc of type

$$\mathsf{Acc}(e_1) \to \mathsf{Acc}(e_2) \to \mathsf{Acc}(e).$$

This means that to prove that a is in the domain of f, we must first prove that $\rho(e_1)$ and $\rho(e_2)$ are in the domain.

This definition does not always work. In the case of nested recursive calls of the function, we cannot eliminate the reference to f in the clauses of the inductive definition Acc. If, for example, the recursive definition of f contains a clause of the form

$$f(e) := \cdots f(f(e')) \cdots$$

then the corresponding clause in the definition of Acc should be

$$\mathsf{Acc}(e') \to \mathsf{Acc}(f(e')) \to \mathsf{Acc}(e)$$

because we must require that all arguments of the recursive calls of f satisfy Acc to deduce that also e does. But this definition is incorrect because we haven't defined the function f yet and so we cannot use it in the definition of Acc. Besides, we need Acc to define f, therefore we are locked in a vicious circle.

In our case, we have two instances of nested recursive clauses, for the sequential composition and *while* loops. When trying to give a semantics of the commands, we come to the definition

$$[\![c_1; c_2]\!]_\sigma := [\![c_2]\!]_{[\![c_1]\!]_\sigma}$$

for sequential composition and

$$[\![\mathsf{while}\ b\ \mathsf{do}\ c]\!]_\sigma := [\![\mathsf{while}\ b\ \mathsf{do}\ c]\!]_{[\![c]\!]_\sigma}$$

for a *while* loop, if the interpretation of b in state σ is true.

Both cases contain a nested occurrence of the interpretation function $[\![-]\!]$.

An alternative solution, presented in [4], exploits the extension of type theory with simultaneous induction-recursion [6]. In this extension, an inductive type or inductive family can be defined simultaneously with a function on it. For the example above we would have

$$\begin{array}{l}\mathsf{Acc}\colon A \to \mathbf{Prop}\\ f\colon (x\colon A)(\mathsf{Acc}\ x) \to B\\ \vdots\\ \mathsf{acc}_n\colon (h'\colon(\mathsf{Acc}\ e'))(\mathsf{Acc}\ (f\ e'\ h')) \to (\mathsf{Acc}\ e)\\ \vdots\\ (f\ e\ (\mathsf{acc}_n\ h'\ h)) := \cdots (f\ (f\ e'\ h)\ h) \cdots\\ \vdots\end{array}$$

This method leads to the following definition of the accessibility predicate and interpretation function for the imperative programming language IMP:

$$\begin{array}{l}\mathsf{comAcc}\colon \mathsf{Command} \to \mathsf{State} \to \mathbf{Prop}\\ [\![]\!]\colon (c\colon \mathsf{Command}; \sigma\colon \mathsf{State})(\mathsf{comAcc}\ c\ \sigma) \to \mathsf{State}\end{array}$$

$$\begin{array}{l}\mathsf{accSkip}\colon (\sigma\colon \mathsf{State})(\mathsf{comAcc}\ \mathsf{skip}\ \sigma)\\ \mathsf{accAssign}\colon (v\colon \mathbb{N}; a\colon \mathsf{AExp}; \sigma\colon \mathsf{State})(\mathsf{comAcc}\ (v \leftarrow a)\ \sigma)\\ \mathsf{accScolon}\colon (c_1, c_2\colon \mathsf{Command}; \sigma\colon \mathsf{State}; h_1\colon (\mathsf{comAcc}\ c_1\ \sigma))(\mathsf{comAcc}\ c_2\ [\![c_1]\!]_\sigma^{h_1})\\ \qquad \to (\mathsf{comAcc}\ (c_1; c_2)\ \sigma)\\ \mathsf{accIf\_true}\colon (b\colon \mathsf{BExp}; c_1, c_2\colon \mathsf{Command}; \sigma\colon \mathsf{State})[\![b]\!]_\sigma = \mathsf{true} \to (\mathsf{comAcc}\ c_1\ \sigma)\\ \qquad \to (\mathsf{comAcc}\ (\mathsf{if}\ b\ \mathsf{then}\ c_1\ \mathsf{else}\ c_2)\ \sigma)\\ \mathsf{accIf\_false}\colon (b\colon \mathsf{BExp}; c_1, c_2\colon \mathsf{Command}; \sigma\colon \mathsf{State})[\![b]\!]_\sigma = \mathsf{false} \to (\mathsf{comAcc}\ c_2\ \sigma)\\ \qquad \to (\mathsf{comAcc}\ (\mathsf{if}\ b\ \mathsf{then}\ c_1\ \mathsf{else}\ c_2)\ \sigma)\\ \mathsf{accWhile\_true}\colon (b\colon \mathsf{BExp}; c\colon \mathsf{Command}; \sigma\colon \mathsf{State})[\![b]\!] = \mathsf{true}\\ \qquad \to (h\colon (\mathsf{comAcc}\ c\ \sigma))(\mathsf{comAcc}\ (\mathsf{while}\ b\ \mathsf{do}\ c)\ [\![c]\!]_\sigma^h)\\ \qquad \to (\mathsf{comAcc}(\mathsf{while}\ b\ \mathsf{do}\ c)\ \sigma)\\ \mathsf{accWhile\_false}\colon (b\colon \mathsf{BExp}; c\colon \mathsf{Command}; \sigma\colon \mathsf{State})[\![b]\!] = \mathsf{false}\\ \qquad \to (\mathsf{comAcc}\ (\mathsf{while}\ b\ \mathsf{do}\ c)\ \sigma)\end{array}$$

$$\begin{array}{l}[\![\mathsf{skip}]\!]_\sigma^{(\mathsf{accSkip}\ \sigma)} := \sigma\\ [\![(v := a)]\!]_\sigma^{(\mathsf{accAssign}\ v\ a\ \sigma)} := \sigma[a/v]\\ [\![(c_1; c_2)]\!]_\sigma^{(\mathsf{accScolon}\ c_1\ c_2\ \sigma\ h_1\ h_2)} := [\![c_2]\!]^{h_2}_{[\![c_1]\!]_\sigma^{h_1}}\\ [\![\mathsf{if}\ b\ \mathsf{then}\ c_1\ \mathsf{else}\ c_2]\!]_\sigma^{(\mathsf{accIf\_true}\ b\ c_1\ c_2\ \sigma\ p\ h_1)} := [\![c_1]\!]_\sigma^{h_1}\\ [\![\mathsf{if}\ b\ \mathsf{then}\ c_1\ \mathsf{else}\ c_2]\!]_\sigma^{(\mathsf{accIf\_false}\ b\ c_1\ c_2\ \sigma\ q\ h_2)} := [\![c_2]\!]_\sigma^{h_2}\\ [\![\mathsf{while}\ b\ \mathsf{do}\ c]\!]_\sigma^{(\mathsf{accWhile\_true}\ b\ c\ \sigma\ p\ h\ h')} := [\![\mathsf{while}\ b\ \mathsf{do}\ c]\!]^{h'}_{[\![c]\!]_\sigma^h}\\ [\![\mathsf{while}\ b\ \mathsf{do}\ c]\!]_\sigma^{(\mathsf{accWhile\_false}\ b\ c\ \sigma\ q)} := \sigma\end{array}$$

This definition is admissible in systems that implement Dybjer's schema for simultaneous induction-recursion. But on systems that do not provide such schema, for example **Coq**, this definition is not valid.

We must disentangle the definition of the accessibility predicate from the definition of the evaluation function. As we have seen before, the evaluation function can be seen as the limit of the iteration of the functional F on an arbitrary base function $f\colon \mathsf{Command} \to \mathsf{State} \to \mathsf{State}$. Whenever the evaluation of a command c is defined on a state σ, we have that $[\![c]\!]_\sigma$ is equal to $(F_f^k\ c\ \sigma)$ for a sufficiently large number of iterations k. Therefore, we consider the functions F_f^k as approximations to the interpretation function being defined. We can formulate the accessibility predicate by using such approximations in place of the explicit occurrences of the evaluation function. Since the iteration approximation has two extra parameters, the number of iterations k and the base function f, we must also add them as new arguments of comAcc. The resulting inductive definition is

$$
\begin{array}{l}
\mathsf{comAcc}\colon \mathsf{Command} \to \mathsf{State} \to \mathbb{N} \to (\mathsf{Command} \to \mathsf{State} \to \mathsf{State}) \to \mathbf{Prop}\\
\mathsf{accSkip}\colon (\sigma\colon \mathsf{State}; k\colon \mathbb{N}; f\colon \mathsf{Command} \to \mathsf{State} \to \mathsf{State})(\mathsf{comAcc}\ \mathsf{skip}\ \sigma\ k+1\ f)\\
\mathsf{accAssign}\colon (v\colon \mathbb{N}; a\colon \mathsf{AExp}; \sigma\colon \mathsf{State}; k\colon \mathbb{N}; f\colon \mathsf{Command} \to \mathsf{State} \to \mathsf{State})\\
\qquad (\mathsf{comAcc}\ (v \leftarrow a)\ \sigma\ k+1\ f)\\
\mathsf{accScolon}\colon (c_1, c_2\colon \mathsf{Command}; \sigma\colon \mathsf{State}; k\colon \mathbb{N}; f\colon (\mathsf{Command} \to \mathsf{State} \to \mathsf{State}))\\
\qquad (\mathsf{comAcc}\ c_1\ \sigma\ k\ f) \to (\mathsf{comAcc}\ c_2\ (F_f^k\ c_1\ \sigma)\ k\ f)\\
\qquad \to (\mathsf{comAcc}\ (c_1; c_2)\ \sigma\ k+1\ f)\\
\mathsf{accIf\_true}\colon (b\colon \mathsf{BExp}; c_1, c_2\colon \mathsf{Command}; \sigma\colon \mathsf{State};\\
\qquad k\colon \mathbb{N}; f\colon \mathsf{Command} \to \mathsf{State} \to \mathsf{State})(\langle b, \sigma\rangle \rightsquigarrow \mathsf{true})\\
\qquad \to (\mathsf{comAcc}\ c_1\ \sigma\ k\ f) \to (\mathsf{comAcc}\ (\mathsf{if}\ b\ \mathsf{then}\ c_1\ \mathsf{else}\ c_2)\ \sigma\ k+1\ f)\\
\mathsf{accIf\_false}\colon (b\colon \mathsf{BExp}; c_1, c_2\colon \mathsf{Command}; \sigma\colon \mathsf{State};\\
\qquad k\colon \mathbb{N}; f\colon \mathsf{Command} \to \mathsf{State} \to \mathsf{State})(\langle b, \sigma\rangle \rightsquigarrow \mathsf{false})\\
\qquad \to (\mathsf{comAcc}\ c_2\ \sigma\ k\ f) \to (\mathsf{comAcc}\ (\mathsf{if}\ b\ \mathsf{then}\ c_1\ \mathsf{else}\ c_2)\ \sigma\ k+1\ f)\\
\mathsf{accWhile\_true}\colon (b\colon \mathsf{BExp}; c\colon \mathsf{Command}; \sigma\colon \mathsf{State};\\
\qquad k\colon \mathbb{N}; f\colon \mathsf{Command} \to \mathsf{State} \to \mathsf{State})(\langle b, \sigma\rangle \rightsquigarrow \mathsf{true})\\
\qquad \to (\mathsf{comAcc}\ c\ \sigma\ k\ f) \to (\mathsf{comAcc}\ (\mathsf{while}\ b\ \mathsf{do}\ c)\ (F_f^k\ c\ \sigma))\\
\qquad \to (\mathsf{comAcc}(\mathsf{while}\ b\ \mathsf{do}\ c)\ \sigma\ k+1\ f)\\
\mathsf{accWhile\_false}\colon (b\colon \mathsf{BExp}; c\colon \mathsf{Command}; \sigma\colon \mathsf{State};\\
\qquad k\colon \mathbb{N}; f\colon \mathsf{Command} \to \mathsf{State} \to \mathsf{State})(\langle b, \sigma\rangle \rightsquigarrow \mathsf{false})\\
\qquad \to (\mathsf{comAcc}\ (\mathsf{while}\ b\ \mathsf{do}\ c)\ \sigma\ k+1\ f).
\end{array}
$$

This accessibility predicate characterizes the points in the domain of the program parametrically on the arguments k and f. To obtain an independent definition of the domain of the evaluation function we need to quantify on them. We quantify existentially on k, because if a command c and a state σ are accessible in k steps, then they will still be accessible in a higher number of steps. We quantify universally on f because we do not want the result of the computation to depend on the choice of the base function.

$$
\begin{array}{l}
\mathsf{comDom}\colon \mathsf{Command} \to \mathsf{State} \to \mathbf{Set}\\
(\mathsf{comDom}\ c\ \sigma) = \Sigma k\colon \mathbb{N}.\forall f\colon \mathsf{Command} \to \mathsf{State} \to \mathsf{State}.(\mathsf{comAcc}\ c\ \sigma\ k\ f)
\end{array}
$$

The reason why the sort of the predicate comDom is **Set** and not **Prop** is that we need to extract the natural number k from the proof to be able to compute the following evaluation function:

$$
\begin{array}{l}
[\![]\!]\colon (c\colon \mathsf{Command}; \sigma\colon \mathsf{State}; f\colon \mathsf{Command} \to \mathsf{State} \to \mathsf{State})\\
\qquad (\mathsf{comDom}\ c\ \sigma) \to \mathsf{State}\\
[\![c]\!]_{\sigma,f}^{\langle k,h\rangle} = (F_f^k\ c\ \sigma)
\end{array}
$$

To illustrate the meaning of these definitions, let us see how the interpretation of a sequential composition of two commands is defined. The interpretation of the command $(c_1; c_2)$ on the state σ is $[\![c_1; c_2]\!]_\sigma^H$, where H is a proof of (comDom $(c_1; c_2)$ σ). Therefore H must be in the form $\langle k, h\rangle$, where $k\colon \mathbb{N}$ and $h\colon \forall f\colon \mathsf{Command} \to \mathsf{State} \to \mathsf{State}.(\mathsf{comAcc}\ (c_1; c_2)\ \sigma\ k\ f)$. To see how h can be constructed, let us assume that $f\colon \mathsf{Command} \to \mathsf{State} \to \mathsf{State}$ and prove (comAcc $(c_1; c_2)$ σ k f). This can be done only by using the constructor accScolon. We see that it must be $k = k' + 1$ for some k' and we must have proofs h_1: (comAcc c_1 σ k' f) and h_2: (comAcc c_2 ($F_f^{k'}$ c_1 σ) k' f). Notice that in h_2 we don't need to refer to the evaluation function $[\![]\!]$ anymore, and therefore the definitions of comAcc does not depend on the evaluation function anymore. We have now that $(h\ f) :=$ (accScolon c_1 c_2 σ k' f h_1 h_2). The definition of $[\![c_1; c_2]\!]_\sigma^H$ is also not recursive anymore, but consists just in iterating F k times, where k is obtained from the proof H.

We can now prove an exact correspondence between operational semantics and denotational semantics given by the interpretation operator $[\![\cdot]\!]$.

Theorem 2.

$$\forall c\colon \mathsf{Command}.\forall \sigma, \sigma'\colon \mathsf{State}.$$
$$\langle c, \sigma\rangle \rightsquigarrow \sigma' \Leftrightarrow \exists H\colon (\mathsf{comDom}\ c\ \sigma).\forall f\colon \mathsf{Command} \to \mathsf{State} \to \mathsf{State}.[\![c]\!]_{\sigma,f}^H = \sigma'.$$

Proof. From left to right, it is proved by rule induction on the derivation of $\langle c, \sigma\rangle \rightsquigarrow \sigma'$. The number of iterations k is the depth of the proof and the proof of the comAcc predicate is a translation step by step of it. From right to left, it is proved by induction on the proof of comAcc.

5 Conclusions

The combination of the iteration technique and the accessibility predicate has, in our opinion, a vast potential that goes beyond its application to denotational semantics. Not only does it provide a path to the implementation and reasoning about partial and nested recursive functions that does not require simultaneous induction-recursion; but it gives a finer analysis of convergence of recursive operators. As we pointed out in Section 3, it supplies not just any fixed point of an operator, but the least fixed point.

We were not the first to formalize parts of Winskel's book in a proof system. Nipkow [9] formalized the first 100 pages of it in Isabelle/HOL. The main difference between our work and his, is that he does not represent the denotation as a function but as a subset of State × State that happens to be the graph of a function. Working on a well developed library on sets, he has no problem in using a least-fixpoint operator to define the subset associated to a *while* loop: But this approach stays further removed from functional programming than an approach based directly on the functions provided by the prover. In this respect, our work is the first to reconcile a theorem proving framework with total functions with denotational semantics. One of the gains is directly executable code (through extraction or ι-reduction). The specifications provided by Nipkow are only executable in the sense that they all belong to the subset of inductive properties that can be translated to Prolog programs. In fact, the reverse process has been used and those specifications had all been obtained by a translation from a variant of Prolog to a theorem prover [2]. However, the prover's function had not been used to represent the semantics.

Our method tries to maximize the potential for automation: Given a recursive definition, the functional operator F, the iterator, the accessibility predicate, the domain, and the evaluation function can all be generated automatically. Moreover, it is possible to automate the proof of the accessibility predicate, since there is only one possible proof step for any given argument; and the obtained evaluation function is computable inside type theory.

We expect this method to be widely used in the future in several areas of formalization of mathematics in type theory.

References

1. Antonia Balaa and Yves Bertot. Fix-point equations for well-founded recursion in type theory. In Harrison and Aagaard [8], pages 1–16.
2. Yves Bertot and Ranan Fraer. Reasoning with executable specifications. In *International Joint Conference of Theory and Practice of Software Development (TAPSOFT/FASE'95)*, volume 915 of *LNCS*. Springer-Verlag, 1995.
3. A. Bove. Simple general recursion in type theory. *Nordic Journal of Computing*, 8(1):22–42, Spring 2001.
4. Ana Bove and Venanzio Capretta. Nested general recursion and partiality in type theory. In Richard J. Boulton and Paul B. Jackson, editors, *Theorem Proving in Higher Order Logics: 14th International Conference, TPHOLs 2001*, volume 2152 of *Lecture Notes in Computer Science*, pages 121–135. Springer-Verlag, 2001.
5. Catherine Dubois and Véronique Viguié Donzeau-Gouge. A step towards the mechanization of partial functions: Domains as inductive predicates. Presented at CADE-15, Workshop on Mechanization of Partial Functions, 1998.
6. Peter Dybjer. A general formulation of simultaneous inductive-recursive definitions in type theory. *Journal of Symbolic Logic*, 65(2), June 2000.
7. Simon Finn, Michael Fourman, and John Longley. Partial functions in a total setting. *Journal of Automated Reasoning*, 18(1):85–104, February 1997.
8. J. Harrison and M. Aagaard, editors. *Theorem Proving in Higher Order Logics: 13th International Conference, TPHOLs 2000*, volume 1869 of *Lecture Notes in Computer Science*. Springer-Verlag, 2000.
9. Tobias Nipkow. Winskel is (almost) right: Towards a mechanized semantics textbook. In V. Chandru and V. Vinay, editors, *Foundations of Software Technology and Theoretical Computer Science*, volume 1180 of *LNCS*, pages 180–192. Springer, 1996.
10. Christine Paulin-Mohring and Benjamin Werner. Synthesis of ML programs in the system Coq. *Journal of Symbolic Computation*, 15:607–640, 1993.
11. Lawrence C. Paulson. Proving termination of normalization functions for conditional expressions. *Journal of Automated Reasoning*, 2:63–74, 1986.
12. The Coq Development Team. LogiCal Project. *The Coq Proof Assistant. Reference Manual. Version 7.2.* INRIA, 2001.
13. K. Slind. Another look at nested recursion. In Harrison and Aagaard [8], pages 498–518.
14. Freek Wiedijk and Jan Zwanenburg. First order logic with domain conditions. Available at `http://www.cs.kun.nl/~freek/notes/partial.ps.gz`, 2002.
15. Glynn Winskel. *The Formal Semantics of Programming Languages, an introduction.* Foundations of Computing. The MIT Press, 1993.

A Proposal for a Formal OCL Semantics in Isabelle/HOL

Achim D. Brucker and Burkhart Wolff

Institut für Informatik, Albert-Ludwigs-Universität Freiburg
Georges-Köhler-Allee 52, D-79110 Freiburg, Germany
{brucker,wolff}@informatik.uni-freiburg.de
http://www.informatik.uni-freiburg.de/~{brucker,wolff}

Abstract. We present a formal semantics as a conservative shallow embedding of the Object Constraint Language (OCL). OCL is currently under development within an open standardization process within the OMG; our work is an attempt to accompany this process by a proposal solving open questions in a consistent way and exploring alternatives of the language design. Moreover, our encoding gives the foundation for tool supported reasoning over OCL specifications, for example as basis for test case generation.
Keywords: Isabelle, OCL, UML, shallow embedding, testing.

1 Introduction

The Unified Modeling Language (UML) [13] has been widely accepted throughout the software industry and is successfully applied to diverse domains [6]. UML is supported by major CASE tools and integrated into a software development process model that stood the test of time. The Object Constraint Language (OCL) [12,17,18] is a textual extension of the UML. OCL is in the tradition of data-oriented formal specification languages like Z [16] or VDM [5]. For short, OCL is a three-valued Kleene-Logic with equality that allows for specifying constraints on graphs of object instances whose structure is described by UML class diagrams.

In order to achieve a maximum of acceptance in industry, OCL is currently developed within an open standardization process by the OMG. Although the OCL is part of the UML standard since version 1.3, at present, the official OCL standard 1.4 concentrates on the concrete syntax, covers only in parts the well-formedness of OCL and handles nearly no formal semantics. So far, the description of the OCL is merely an informal requirement analysis document with many examples, which are sometimes even contradictory.

Consequently, there is a need[1] for both software engineers and CASE tool developers to clarify the concepts of OCL formally and to put them into perspective of more standard semantic terminology. In order to meet this need, we started to provide a conservative embedding of OCL into Isabelle/HOL. As far

[1] This work was partially funded by the OMG member Interactive Objects Software GmbH (http://www.io-software.com).

V.A. Carreño, C. Muñoz, S. Tahar (Eds.): TPHOLs 2002, LNCS 2410, pp. 99–114, 2002.

as this was possible, we tried to follow the design decisions of OCL 1.4 in order to provide insight into the possible design choices to be made in the current standardization process of version 2.0.

Attempting to be a "practical formalism" [17], OCL addresses software developers who do not have a strong mathematical background. Thus, OCL deliberately avoids mathematical notation; rather, it uses a quite verbose, programming language oriented syntax and attempts to hide concepts such as logical quantifiers. This extends also to a design rationale behind the semantics: OCL is still viewed as an object-oriented assertion language and has thus more similarities with an object-oriented programming language than a conventional specification language. For example, standard library operators such as "concat" on sequences are defined as *strict operations* (i.e. they yield an explicit value *undefined* as result whenever an argument is undefined), only bounded quantifiers are admitted, and there is a tendency to define infinite sets away wherever they occur. As a result, OCL has a particularly executable flavor which comes handy when generating code for assertions or when animating specifications.

Object-oriented languages represent a particular challenge for the "art of embedding languages in theorem provers" [10]. This holds even more for a shallow embedding, which we chose since we aim at reasoning in OCL specifications and not at meta-theoretic properties of our OCL representation. In a shallow embedding, the types of OCL language constructs have to be represented by types of HOL and concepts such as undefinedness, mutual recursion between object instances, dynamic types, and extensible class hierarchies have to be handled.

In this paper, we present a new formal model of OCL in form of a conservative embedding into Isabelle/HOL that can cope with the challenges discussed above. Its modular organization has been used to investigate the interdependence of certain language features (method recursion, executability, strictness, smashing, flattening etc.) in order to provide insight into the possible design choices for the current design process [2]. We extend known techniques for the shallow representation of object orientation and automated proof techniques to lift lemmas from the HOL-library to the OCL level. As a result, we provide a first calculus to formally reason over OCL specifications and provide some foundation for automated reasoning in OCL.

This paper proceeds as follows: After a introduction into our running example using UML/OCL, we will guide through the layers of our OCL semantics, namely the object model and resulting semantic universes, the states and state relations, the OCL logic and the OCL library. It follows a description of automated deduction techniques based on derived rules, let it be on modified tableaux-deduction techniques or congruence rewriting. We will apply these techniques in a paradigmatic example for test-case generation in a black-box test setting.

2 A Guided Tour through UML/OCL

The UML provides a variety of diagram types for describing dynamic (e.g. state charts, activity diagrams, etc.) and static (class diagrams, object diagrams, etc.) system properties. One of the more prominent diagram types of the UML is the

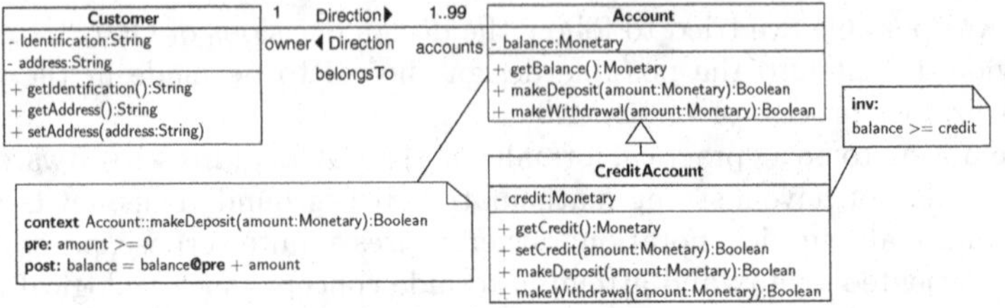

Fig. 1. Modeling a simple banking scenario with UML

class diagram for modeling the underlying data model of a system in an object oriented manner. The class diagram in our running example in Fig. 1 illustrates a simple banking scenario describing the relations between the classes **Customer**, **Account** and its specialization **CreditAccount**. To be more precise, the relation between data of the classes **Account** and **CreditAccount** is called *subtyping*. A class does not only describe a set of record-like data consisting of *attributes* such as balance but also functions (*methods*) defined over the classes data model.

It is characteristic for the object oriented paradigm, that the functional behavior of a class and all its methods are also accessible for all subtypes; this is called *inheritance*. A class is allowed to redefine an inherited method, as long as the method interface does not change; this is called *overwriting*, as it is done in the example for the method makeWithdrawal().

It is possible to model relations between classes (*association*), possibly constrained by *multiplicities*. In Fig. 1, the association belongsTo requires, that every instance of **Account** is associated with exactly one instance of **Customer**. Associations were represented by introducing implicit set-valued attributes into the objects, while multiplicity were mapped to suitable data invariants. In the following, we assume that associations have already been "parsed away".

Understanding OCL as a data-oriented specification formalism, it seems natural to refine class diagrams using OCL for specifying invariants, pre- and postconditions of methods. For example, see Fig. 1, where the specification of the method makeWithdrawal() is given by its pair of pre- and postcondition.

In UML class members can contain attributes of the type of the defining class. Thus, UML can represent (mutually) recursive data types. Moreover, OCL introduces also recursively specified methods [12]; however, at present, a dynamic semantics of a method call is missing (see [2] for a short discussion of the resulting problems).

3 Representing OCL in Isabelle/HOL

OCL formulae are built over expressions that access the underlying state. In postconditions, path-expressions can access the *current* and the *previous state*, e.g. balance = balance**@pre** + amount. Accesses on both states may be arbitrarily mixed, e.g. **self**.x**@pre**.y denotes an object, that was constructed by dereferencing in the previous state and selecting attribute x in it, while the next dereferencing step via y is done in the current state. Thus, method specifications represent state transition relations built from the conjunction of pre and post condition, where the state consists of a graph of object instances whose type is defined by the underlying class diagram. Since the fundamental OCL-operator allInstances allows for the selection of all instances (objects) of a certain class, there must be a means to reconstruct their dynamic types in the object graph.

In a shallow embedding, the key question arises how to represent the static type structure of the objects *uniformly* within a state or as argument of a dynamic type test is_T. Constructions like "the universe of all class diagram interpretations" are too large to be represented in the simple type system underlying higher-order logic.

Our solution is based on the observation that we need not represent *all* class diagram interpretations inside the logic; it suffices if we can provide an extra-logical mechanism that represents any concrete class hierarchy and that allows for extensions of any given class hierarchy. For practical reasons, we require that such an *extension* mechanism is logically conservative both in the sense that only definitional axioms are used and that all existing proofs on data of the former class hierarchy remain valid for an extended one.

Based on these considerations, *HOL-OCL* is organized in several layers:

- a *semantic coding scheme* for the *object model* layer along the class-hierarchy,
- the *system state* and *relations* over it, forming the denotational domain for the semantics of methods,
- the *OCL logic* for specifying state relations or class invariants,
- the *OCL library* describing predefined basic types.

3.1 The Encoding of Extensible Object Models

The main goals of our encoding scheme is to provide typed constructors and accessor functions for a given set of classes or enumeration types to be inserted in a previous class hierarchy. The coding scheme will be represented in two steps: in this section, we will describe *raw* constructors and accessors, while in Sec. 3.2 a refined scheme for accessors is presented.

The basic configuration of any class hierarchy is given by the OCL standard; the library for this basic configuration is described in Sec. 3.5.

Handling Undefinedness. In the OCL standard 1.4, the notion of explicit undefinedness is part of the language, both for the logic and the basic values.

> Object Constraint Language Specification [12] (version 1.4), page 6–58
>
> Whenever an OCL-expression is being evaluated, there is the possibility that one or more queries in the expression are undefined. If this is the case, then the complete expression will be undefined.

This requirement postulates the strictness of all operations (the logical operators are explicit exceptions) and rules out a modeling of undefinedness via underspecification. Thus, the language has a similar flavor than LCF or SPECTRUM [1] and represents a particular challenge for automated reasoning.

In order to handle undefinedness, we introduce for each type τ a *lifted type* $\tau_\bot$, i.e. we introduce a special type constructor. It adds to each given type an additional value $\bot$. The function $\lfloor\_\rfloor : \alpha \Rightarrow \alpha_\bot$ denotes the injection, the function $\lceil\_\rceil : \alpha_\bot \Rightarrow \alpha$ its inverse. Moreover, we have the case distinction function case_up $c\ f\ x$ that returns c if $x = \bot$ and $f\ k$ if $x = \lfloor k \rfloor$. We will also write case x of $\lfloor k \rfloor \Rightarrow f\ k \mid \bot \Rightarrow c$.

Note that the definition of lifted types leads to the usual construction of *flat cpo's* well known from the theory of complete partial orders (*cpo*) and denotational semantics [19]. For the sake of simplification, we avoid a full-blown cpo-structure here (while maintaining our semantics "cpo-ready") and define only a tiny fragment of it that provides concepts such as *definedness* $\mathrm{DEF}(x) \equiv (x \neq \bot)$ or *strictness of a function* is_strict $f \equiv (f \bot = \bot)$.

Managing Holes in Universes. Since objects can be viewed as records consisting of *attributes*, and since the object alternatives can be viewed as variants, it is natural to construct the "type of all objects", i.e. the semantic universe $\mathcal{U}_x$ corresponding to a certain class hierarchy x, by Cartesian products and by type sums (based on the constructors $\mathrm{Inl} : \alpha \Rightarrow \alpha + \beta$ and $\mathrm{Inr} : \beta \Rightarrow \alpha + \beta$ from the Isabelle/HOL library). In order to enable extensibility, we provide systematically polymorphic variables — the "holes" — into a universe that were filled when extending a class hierarchy.

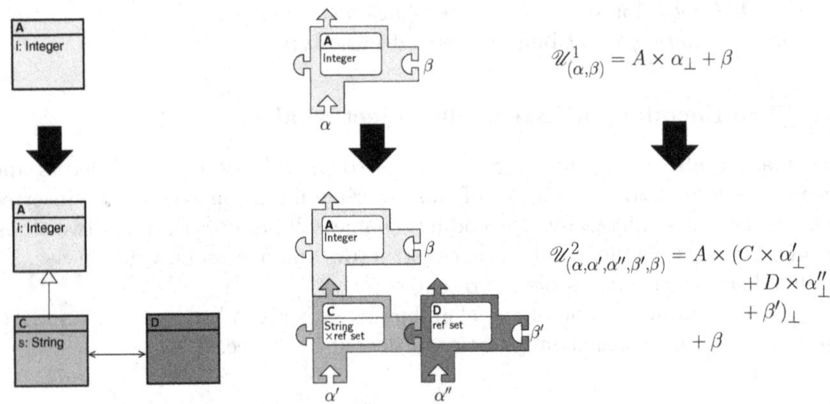

Fig. 2. Extending Class Hierarchies and Universes with Holes

In our scheme, a class can be extended in two ways: either, an alternative to the class is added *at the same level* which corresponds to the creation of an alternative subtype of the supertype (the *β-instance*), or a class is added *below* which corresponds to the creation of a subtype (the *α-instance*). The insertion of a class corresponds to *filling a hole* ν *by a record* T is implemented by the particular type instance:

$$\nu \mapsto ((T \times \alpha_\bot) + \beta)$$

As a consequence, the universe $\mathcal{U}^2$ in Fig. 2 is just a type instance of $\mathcal{U}^1$, in particular: an α-extension with C β-extended by D. Thus, properties proven over $\mathcal{U}^1$ also holds for $\mathcal{U}^2$.

The *initial* universe corresponding to the minimal class hierarchy of the OCL library consists of the real numbers, strings and bool. It is defined as follows:

$$\text{Real} = \text{real}_\bot \quad \text{Boolean} = \text{bool}_\bot \quad \text{String} = \text{string}_\bot \quad \mathsf{OclAny}_\alpha = \alpha_\bot$$
$$\mathcal{U}_\alpha = \text{Real} + \text{Boolean} + \text{String} + \mathsf{OclAny}_\alpha$$

Note that the α-extensions were all lifted, i.e. additional $\bot$ elements were added. Thus, there is a uniform way to denote "closed" objects, i.e. objects whose potential extension is not used. Consequently, it is possible to determine the dynamic type by testing for closing $\bot$'s. For example, the OclAny type has exactly one object represented by $\text{Inr}(\text{Inr}(\text{Inr}\,\bot))$ in any universe $\mathcal{U}_\alpha$.

Note, moreover, that all types of attributes occurring in the records A,C or D may contain basic types, and sets, sequences or bags over them, but not references to the types induced by class declarations. These references were replaced the abstract type ref to be defined later; thus, recursion is not represented at the level of the universe construction, that just provides "raw data".

Outlining the Coding Scheme. Now we provide raw constructors, raw accessors, and tests for the dynamic types over the data universe.

The idea is to provide for each class T with attributes $t_1 : \tau_1, \ldots, t_n : \tau_n$ a type $T = \tau_1 \times \cdots \times \tau_n$ and a constructor $\text{mk\_T} : T \Rightarrow \mathcal{U}_x$, which embeds a record of type T into the actual version of the universe (for example mk_Boolean with type $\text{Boolean} \Rightarrow \mathcal{U}_\alpha$ is defined by $\text{mk\_Boolean} \equiv \text{Inr} \circ \text{Inl}$). Accordingly, there is a test $\text{is\_T} : \mathcal{U}_x \Rightarrow \text{bool}$ that checks if an object in the universe is embedded as T, and an accessor $\text{get\_T} : \mathcal{U}_x \Rightarrow T$ that represents the corresponding projection. Finally, a constant $T : \mathcal{U}_x$ set is provided that contains the *characteric set* of T in the sense of a set of all objects of class T.

Data invariants I are represented by making the constructor partial w.r.t. I, i.e. the constructor will be defined only for input tuples T that fulfill I; correspondingly, the test for the dynamic type is also based on I.

At present, we encode our examples by hand. The task of implementing a compiler that converts representations of UML-diagrams (for example, formats produced by ArgoUML) is desirable but not in the focus of this research.

3.2 System State

Basic Definitions. The task of defining the state or state transitions is now straight-forward: We define an abstract type ref for "references" or "locations", and a state that is a partial mapping from references to objects in a universe:

$$\begin{aligned} \textbf{types}\quad & \text{ref} \\ & \sigma\,\text{state} = \text{ref} \Rightarrow \alpha\,\text{option} \\ & \sigma\,\text{st} \quad = \sigma\,\text{state} \times \sigma\,\text{state} \end{aligned}$$

Based on state, we define the only form of universal quantification of OCL: the operator allInstances extracts all objects of a "type" — represented by its characteristic set — from the current state. The standard specifies allInstances as being undefined for Integer or Real or String. Thus, infinite sets are avoided in an ad-hoc manner. However, nothing prevents from having infinite states; and we did not enforce finiteness by additional postulates.

$$\begin{aligned}
&\mathsf{allInstances} : [\mathcal{U}_\alpha\,\mathrm{set}, \mathcal{U}_\alpha\,\mathrm{st}] \Rightarrow \beta\,\mathrm{Set}\\
&\mathsf{allInstances}\,\mathit{type} \equiv \lambda(s, s') \bullet \mathbf{if}\,(\mathit{type} = \mathrm{Integer} \vee \mathit{type} = \mathrm{Real} \vee \mathit{type} = \mathrm{String})\\
&\qquad\qquad \mathbf{then}\,\bot\ \mathbf{else\,if}\,(\mathit{type} = \mathrm{Boolean})\\
&\qquad\qquad\qquad \mathbf{then}\lfloor \mathrm{Boolean} \rfloor\ \mathbf{else} \lfloor \mathit{type} \cap (\mathrm{range}\, s') \rfloor
\end{aligned}$$

Defining OCL operators like oclIsNew : $\mathcal{U}_\alpha \Rightarrow \mathcal{U}_\alpha\,\mathrm{st} \Rightarrow \mathrm{Boolean}$ or oclIsTypeOf is now routine; the former checks if an object is defined in the current but not in the previous state, while the latter redefines the test is_T of Sec. 3.1.

The HOL Type of OCL Expressions. Functions from state transition pairs to lifted OCL values will be the denotational domain of our OCL semantics. From a transition pair, all values will be extracted (via path expressions), that can be passed as arguments to to basic operations or user-defined methods. More precisely, all expressions with OCL type τ will be represented by an *HOL-OCL* expression of type $V_\alpha(\tau_\bot)$ defined by:

$$\mathbf{types}\quad V_\alpha(\theta) = \mathcal{U}_\alpha\,\mathrm{st} \Rightarrow \theta$$

where α will represent the type of the state transition. For example, all *logical HOL-OCL* expressions have the type $V_\gamma(\mathrm{Boolean})$ (recall that Boolean is the lifted type bool from HOL).

As a consequence, all operations and methods embedded into *HOL-OCL* will have to pass the context state transition pair to its argument expressions, collect the values according their type, and compute a result value. For example, let a function f have the OCL type $\tau_1 \times \cdots \times \tau_n \Rightarrow \tau_{n+1}$, then our representation f' will have the type $V_\alpha(\tau_1) \times \cdots \times V_\alpha(\tau_n) \Rightarrow V_\alpha(\tau_{n+1})$. Now, when defining f', we proceed by $f'(e_1, \ldots, e_n)(c) = E(e_1 c, \ldots, e_n c)$ for some E and some context transition c. We call the structure of definitions *state passing*. Functions with one argument of this form are characterized semantically:

$$\mathrm{pass}\, f = \exists E \bullet \forall X\ c \bullet f\ X\ c = E(X\ c)$$

Being state passing will turn out to be an important invariant of our shallow semantics and will be essential for our OCL calculus. The conversion of a function f into f' in state passing style will be called the *lifting* of f (which should not be confused with the lifting on types of Sec. 3.1).

A Coding Scheme for State-Based Accessors. Now we define the accessor functions with the "real OCL signature", that can be used to build up path expressions, on the basis of raw accessors. Two problems have to be solved: first, references to class names occurring in types of attributes must be handled (i.e. ref set in raw accessors types must be mapped to Set(A)), and second, raw accessors must be lifted to state passing style. In a simple example, an accessor x of type τ in a class C must have the HOL type $x : \mathcal{U}_x \Rightarrow V_x(\tau)$ which is normally achieved by wrapping the raw accessor x_0 in an additional abstraction: $x\ u = \lambda\, c \bullet x_0\ u$. If the class has a class invariant, a test for the invariant must be added (violations are considered as undefinedness and therefore treated like to undefined references into the state). If the accessor yields an OCL-type with references to other classes (e.g. in Set(A)), these references must be accessed and inserted into the surrounding collection; this may involve smashing (see the discussion of collection types in Sec. 3.5).

Following this extended code scheme, we can define conservatively new accessors over some extended universe whenever we extend a class hierarchy; this allows for modeling mutual data recursion that is introduced by extension while maintaining static typechecking.

3.3 Encoding Our Example

In order to encode our UML model in Fig. 1, we declare the type for the class **Account** (we skip **CreditAccount** and **Customer**). An account is a tuple describing the balance (of type Monetary) and the encoded association end '*owner*' (of type ref set). For our first universe, with the two "holes" α (for extending "below") and β (for extending on "the same level"), we define:

$$\begin{aligned}\textbf{types}\quad \text{Account\_type} &= \text{Monetary} \times \text{ref set}\\ \text{Account} &= \text{Account\_type}_\bot\\ \mathcal{U}^1_{(\alpha,\beta)} &= \mathcal{U}_{(\text{Account\_type}\times\alpha_\bot+\beta)}\end{aligned}$$

We need the raw constructor for an account object. Note that this function "lives" in the universe $\mathcal{U}^3_{(\alpha',\alpha'',\beta',\beta)}$ which contains *all* classes from Fig. 1.

$$\begin{aligned}\text{mk\_Account} &: \text{Account\_type} \Rightarrow \mathcal{U}^3_{(\alpha',\alpha'',\beta',\beta)}\\ \text{mk\_Account} &\equiv \text{mk\_OclAny} \circ \text{lift} \circ \text{Inl} \circ \lambda\, x \bullet (x, \bot)\end{aligned}$$

In the next step, we need to define the the accessors for the attribute. As an example, we present the definition for accessing the association end *owner* (of type Customer) in the current state s':

$$\begin{aligned}.\text{owner} &: \mathcal{U}^3_{(\alpha',\alpha'',\beta',\beta)} \Rightarrow V_{(\alpha',\alpha'',\beta',\beta)}(\text{Customer})\\ (obj.\text{owner}) &\equiv (\lambda(s, s') \bullet \text{up\_case}(\text{lift} \circ ((\text{op}\ ``)(\lambda\, x \bullet \text{option\_Case}\ \bot\\ &\qquad \text{get\_Customer}(s'\ x))) \circ \text{snd})(\bot)(\text{get\_Account}\ obj))\end{aligned}$$

Note, that accessor functions are lifted, e.g. they operate over a previous and current state pair (s, s').

3.4 OCL Logic

We turn now to a key chapter of the OCL-semantics: the logics. According to the OCL standard (which follows SPECTRUM here), the logic operators have to be defined as Kleene-Logic, requiring that any logical operator reduces to a defined logical value whenever possible.

In itself, the logic will turn out to be completely independent from an underlying state transition pair or universe and is therefore valid in all universes. An OCL formula is a function that is either true, false or undefined depending on its underlying state transition. Logical expressions are just special cases of OCL expressions and must produce Boolean values. Consequently, the general type of logical formulae must be:

$$\textbf{types} \quad \mathrm{BOOL}_\alpha = V_\alpha(\mathrm{Boolean})$$

The logical constants true resp. false can be defined as constant functions, that yield the lifted value for meta-logical undefinedness, truth or falsehood, i.e. the HOL values of the HOL type bool. Moreover, the predicate is_def decides for any OCL expression X that its value (evaluated in the context c) is defined:

$$\begin{array}{llr} \textbf{constdefs} & \mathsf{is\_def} : V_\alpha(\beta_\bot) \Rightarrow \mathrm{BOOL}_\alpha & \mathsf{is\_def}\, X \equiv \lambda\, c \bullet \lfloor \mathrm{DEF}(X\ c) \rfloor \\ & \bot_{\mathcal{L}} : \mathrm{BOOL}_\alpha & \bot_{\mathcal{L}} \equiv \lambda\, x \bullet \lfloor \bot \rfloor \\ & \mathsf{true} : \mathrm{BOOL}_\alpha & \mathsf{true} \equiv \lambda\, x \bullet \lfloor \mathrm{True} \rfloor \\ & \mathsf{false} : \mathrm{BOOL}_\alpha & \mathsf{false} \equiv \lambda\, x \bullet \lfloor \mathrm{False} \rfloor \end{array}$$

The definition of the strict not and and, or, and implies are straight-forward:

$$\begin{array}{l} \mathsf{not} : \mathrm{BOOL}_\alpha \Rightarrow \mathrm{BOOL}_\alpha \quad \mathsf{not}\, S \equiv \lambda\, c \bullet \textbf{if}\ \mathrm{DEF}(S\ c)\ \textbf{then} \lfloor \neg \lceil S\ c \rceil \rfloor\ \textbf{else} \bot \\ \mathsf{and} : [\mathrm{BOOL}_\alpha, \mathrm{BOOL}_\alpha] \Rightarrow \mathrm{BOOL}_\alpha \\ S\, \mathsf{and}\, T \equiv \lambda\, c \bullet \textbf{if}\ \mathrm{DEF}(S\ c)\ \textbf{then}\ \textbf{if}\ \mathrm{DEF}(T\ c)\ \textbf{then} \lfloor \lceil S\ c \rceil \wedge \lceil T\ c \rceil \rfloor\ \textbf{else} \\ \qquad\qquad \textbf{if}(S\ c = \lfloor \mathrm{False} \rfloor)\ \textbf{then} \lfloor \mathrm{False} \rfloor\ \textbf{else} \bot\ \textbf{else} \\ \qquad\qquad\qquad \textbf{if}(T\ c = \lfloor \mathrm{False} \rfloor)\ \textbf{then} \lfloor \mathrm{False} \rfloor\ \textbf{else} \bot \end{array}$$

From these definitions, the following rules of the truth table were derived:

a	not a
true	false
false	true
$\bot_{\mathcal{L}}$	$\bot_{\mathcal{L}}$

a	b	a and b
false	false	false
false	true	false
false	$\bot_{\mathcal{L}}$	false

a	b	a and b
true	false	false
true	true	true
true	$\bot_{\mathcal{L}}$	$\bot_{\mathcal{L}}$

a	b	a and b
$\bot_{\mathcal{L}}$	false	false
$\bot_{\mathcal{L}}$	true	$\bot_{\mathcal{L}}$
$\bot_{\mathcal{L}}$	$\bot_{\mathcal{L}}$	$\bot_{\mathcal{L}}$

Based on these basic equalities, it is not difficult to derive with Isabelle the laws of the perhaps surprisingly rich algebraic structure of Kleene-Logics: Both and and or enjoy not only the usual associativity, commutativity and idempotency laws, but also both distributivity and de Morgan laws. It is essentially this richness and algebraic simplicity that we will exploit in the example in Sec. 5.

OCL needs own equalities, a logical one called *strong equality* ($\triangleq$) and a strictified version of it called *weak equality* ($\doteq$) that is executable. They have the type $[V_\alpha(\beta), V_\alpha(\beta)] \Rightarrow \mathrm{BOOL}_\alpha$ and are defined similarly to and above based on the standard HOL equality.

3.5 The Library: OCL Basic Data Types

The library provides operations for Integer, Real (not supported at present) and Strings. Moreover, the parametric data types Set, Sequence and Bag with their functions were also provided; these were types were grouped into a class "Collection" in the standard. At present, the standard prescribes only ad-hoc polymorphism for the operations of the library and not *late binding*.

Since the standard suggests a uniform semantic structure of all functions in the library, we decided to make the uniformity explicit and to exploit it in the proof support deriving rules over them.

In the library, all operations are *lifted*, *strictified* and (as far as collection functions are concerned; see below) *smashed* versions of functions from the HOL library. However, *methodification* (i.e. introduction of late binding), is not needed here due to the standards preference of ad-hoc polymorphism. However, we also consider general recursion based on fixed-point semantics and a shallow representation for methodification, which is an essential feature of an object-oriented specification language in our view. The interested reader is referred to the extended version of this paper.

The generic functions for lift_2 and strictify are defined as follows:

$$\text{strictify}\ f\ x \equiv \text{case}\, x\, \text{of}\, \lfloor v \rfloor \Rightarrow (f\ v) | \bot \Rightarrow \bot$$
$$\text{lift}_2\, f \equiv (\lambda X\ Y\ st \bullet f(X\ st)(Y\ st))$$

According to this definition, lift_2 converts a function of type $[\alpha, \beta] \Rightarrow \gamma$ to a function of type $[V_\sigma(\alpha), V_\sigma(\beta)] \Rightarrow V_\sigma(\gamma)$.

A Standard Class: Integer. Based on combinators like strictify and lift_2, the definitions of the bulk of operators follow the same pattern exemplified by:

types $\quad \text{INTEGER}_\alpha = V_\alpha(\text{Integer})$
defs $\quad \text{op} + \equiv \text{lift}_2(\text{strictify}(\lambda\, x : \text{int} \bullet \text{strictify}(\lambda\, y \bullet \lfloor x + y \rfloor)))$

A Collection Class: Set. For collections, the requirement of having strict functions must consequently be extended to the constructors of sets. Since it is desirable to have in data types only denotations that were "generated" by the constructors, this leads to the concept of *smashing* of all collections. For example, smashed sets are identified with $\bot_{\mathscr{L}}$ provided one of their elements is $\bot$: $\{a, \bot_{\mathscr{L}}\} = \bot_{\mathscr{L}}$; Analogously, smashed versions of *Bags*, *Seq* or *Pairs* can be defined. Smashed sets directly represent the execution behavior in usual programming languages such as Java. We omit the details of the construction of smashed sets here for space reasons; apart from smashing arguments, the definitions of set operations such as includes, excludes, union or intersection follow the usual pattern.

The OCL standard prescribes also a particular concept called *flattening*. This means for example that a set $\{a, \{b\}, c\}$ is identified as $\{a, b, c\}$. We consider flattening as a syntactic issue and require that a front-end "parses away" such situations and generates conversions.

4 Towards Automated Theorem Proving in *HOL-OCL*

Based on derived rules, we will provide several calculi and proof techniques for OCL that are oriented towards Isabelle's powerful proof-procedures like *fast_tac* and *simp_tac*. While the former is geared towards natural deduction calculi, the latter is based on rewriting and built for reasoning in equational theories.

4.1 A Natural Deduction-Calculus for OCL

As a foundation, we introduce two notions of validity: a formula may be *valid for all transitions* or just *valid* (written $\models P$) or be *valid for a transition* t (written $t \models P$). We can define these notions by $\models P \equiv P = \mathsf{true}$ or $t \models P \equiv P\ t = \mathsf{true}\ t$ respectively. Recall that a formula may neither be valid nor invalid in a state, it can be undefined:

$$\frac{\begin{array}{c}[st \models not(\text{is\_def}(A))]\\ \vdots\\ R\end{array} \qquad \begin{array}{c}[st \models not(A)]\\ \vdots\\ R\end{array} \qquad \begin{array}{c}[st \models A]\\ \vdots\\ R\end{array}}{R}$$

This rule replaces in a Kleene-Logic the usual *classical* rule. Note that R may be an arbitrary judgment.

The core of the calculus consists of the more conventional rules like:

$$\frac{\models A \quad \models B}{\models A\,\mathsf{and}\,B} \qquad \frac{\models A\,\mathsf{and}\,B}{\models A} \qquad \frac{\models A\,\mathsf{and}\,B}{\models A} \qquad \frac{\models A\,\mathsf{and}\,B \qquad \begin{array}{c}[\models A, \models B]\\ \vdots\\ \models R\end{array}}{\models R}$$

and their counterparts and-valid-in-transition, or-validity, or-valid-in-transition and the suitable not-elimination rules.

Unfortunately, the rules handling the implication are only in parts elegantly:

$$\frac{\models A \quad \models A\,\mathsf{implies}\,B}{\models B} \qquad \frac{\models B}{\models A\,\mathsf{implies}\,B} \qquad \frac{\forall st \bullet \begin{array}{c}[st \models \mathsf{not}(\text{is\_def}(A))]\\ \vdots\\ st \models B\end{array} \qquad \forall st \bullet \begin{array}{c}[st \models A]\\ \vdots\\ st \models B\end{array}}{\models A\,\mathsf{implies}\,B}$$

The problem is the implication introduction rule to the right that combines the two validity levels of the natural deduction calculus.

Undefinedness leads to an own side-calculus in OCL: Since from $\models A\,\mathsf{and}\,B$ we can conclude definedness both for A and B in all contexts, and since from $\models E \triangleq E'$ we can conclude that any subexpression in E and E' is defined (due to strictness to all operations in the expression language of OCL), a lot of definedness information is usually hidden in an OCL formula, let it be a method precondition or an invariant. In order to *efficiently* reason over OCL specification, it may be necessary precompute this information.

At present, we have only developed a simple setup for *fast_tac* according to the rules described above, which is already quite powerful, but not complete.

4.2 Rewriting

Rewriting OCL-formulae seems to have a number of advantages; mainly, it allows for remaining on the level of absolute validity which is easier to interpret, and it allows to hide the definedness concerns miraculously inside the equational calculus. The nastiness of the implication introduction can be shifted a bit further inside in an equational calculus: The two rules

$$A \,\mathsf{implies}(B \,\mathsf{implies}\, C) = (A \,\mathsf{and}\, B) \,\mathsf{implies}\, C$$
$$A \,\mathsf{implies}(B \,\mathsf{or}\, C) = (A \,\mathsf{implies}\, B) \,\mathsf{or}(A \,\mathsf{implies}\, C)$$

hold for all cases which — together with the other lattice rules for the logic — motivates a Hilbert-Style calculus for OCL; unfortunately, the assumption rule $A \,\mathsf{implies}\, A$ holds only if A is defined in all contexts. At least, this gives rise to proof procedures that defer definedness reasoning to local places in a formula.

A useful mechanism to transport definedness information throughout an OCL formula can be based on Isabelle's simplifier that can cope with a particular type of rules. Derived congruence rewriting rules for *HOL-OCL* look like:

$$\frac{\begin{array}{c}[st \models A \vee st \models \mathsf{not}\,\mathsf{is\_def}(A)] \\ \vdots \\ B\ st = B\ st'\end{array}}{(A \,\mathsf{and}\, B)\ st = (A \,\mathsf{and}\, B')\ st} \qquad \frac{A\ st = A'\ st \quad B\ st = B'\ st}{(A \,\mathsf{or}\, B)\ st = (A' \,\mathsf{or}\, B')\ st}$$

allow for replacing, for example, variables occurrences by $\bot_{\mathcal{L}}$ or true if their undefinedness or validity follows somewhere in the context.

We discovered a further interesting technique for proving the equality of two formulae $P\ X$ and $Q\ X$ based on the idea to perform a case split by substituting X by $\bot_{\mathcal{L}}$, true or false. Unfortunately, it turns out that the rule:

$$\frac{P\ \bot_{\mathcal{L}} = P'\ \bot_{\mathcal{L}} \qquad P\ \mathsf{true} = P'\ \mathsf{true} \qquad P\ \mathsf{false} = P'\ \mathsf{false}}{P\ X = P'\ X}$$

is simply unsound due to the fact that it does not hold for *all* functions $P\ X\ c$, only if P and P' are state passing, which represents an invariant of our embedding (see Sec. 3.2). Fortunately, for all logical operators, state passing rules such as:

$$\frac{\mathrm{pass}\, P}{\mathrm{pass}(\lambda X \bullet \mathsf{not}(P\ X))} \qquad \frac{\mathrm{pass}\, P \quad \mathrm{pass}\, P'}{\mathrm{pass}(\lambda X \bullet (P\ X) \,\mathsf{and}(P'\ X))}$$

hold. Moreover, any function constructed by a lifting (and these are all library function definitions) are state passing. This allows for proof procedures built on systematic case distinctions which turn out to be efficient and useful.

The situation is similar for reasoning over strong and weak equality. For strong equality, we have *nearly* the usual rules of an equational theory with reflexivity, symmetry and transitivity. For Leibniz rule (substitutivity), however, we need again that the context P is state passing:

$$\frac{\models a \triangleq b \quad \models P\ a \quad \mathrm{pass}\, P}{\models P\ b}$$

This is similar for strict equality, except for additional definedness constraints.

4.3 Lifting Theorems from HOL to the *HOL-OCL* Level

Since all operations in the library are defined extremely canonically by a combination of (optional) smashing, strictification and lifting operators, it is possible to derive automatically from generic theorems such as strictness rules, definedness propagation etc.:

$$\begin{aligned}
\mathrm{lift}_1(\mathrm{strictify}\, f)\, \bot_{\mathscr{L}} &= \bot_{\mathscr{L}}\\
\mathrm{lift}_2(\mathrm{strictify}(\lambda\, x \bullet \mathrm{strictify}(f\ x)))\, \bot_{\mathscr{L}}\ X &= \bot_{\mathscr{L}}\\
\mathrm{lift}_2(\mathrm{strictify}(\lambda\, x \bullet \mathrm{strictify}(f\ x)))\ X\, \bot_{\mathscr{L}} &= \bot_{\mathscr{L}}\\
\mathrm{is_def}(\mathrm{lift}_1(\mathrm{strictify}(\lambda\, x \bullet \mathrm{lift}(f\ x)))\ X) &= \mathrm{is_def}(X)\\
(\forall x\ y \bullet f\ x\ y = f\ y\ x) \Rightarrow \mathrm{lift}_2\, f\ X\ Y &= \mathrm{lift}_2\, f\ Y\ X
\end{aligned}$$

The last rule is used to lift a commutativity property from the HOL level to the *HOL-OCL*-level. With such lifting theorems, many standard properties were proven automatically in the library.

5 Application: Test Case Generation

A prominent example for automatic test case generation is the triangle problem [11]: Given three integers representing the lengths of the sides of a triangle, a small algorithm has to check, whether these integers describe invalid input or an equilateral, isosceles, or scalene triangle. Assuming a class **Triangle** with the operations isTriangle() (test if the input describes a triangle) and triangle() (classify the valid triangle) leads to the following OCL specification:

```
context Triangle :: isTriangle (s0, s1, s2: Integer ): Boolean
pre:  (s0 > 0) and (s1 > 0) and (s2 > 0)
post: result = (s2 < (s0 + s1)) and (s0 < (s1 + s2)) and (s1 < (s0 + s2))

context Triangle :: triangle (s0, s1, s2: Integer ): TriangType
pre:  (s0 > 0) and (s1 > 0) and (s2 > 0)
post: result = if ( isTriangle (s0,s1,s2)) then if (s0 = s1) then if (s1 = s2)
               then Equilateral :: TriangType else Isosceles :: TriangType endif
               else if (s1 = s2) then Isosceles :: TriangType else if (s0 = s2)
               then Isosceles :: TriangType else Scalene :: TriangType endif
               endif endif else Invalid :: TriangType endif
```

Transforming this specification into *HOL-OCL*[2] leads to the following specification *triangle_spec* of the operation triangle():

$$\begin{aligned}
&\mathrm{triangle_spec} \equiv \lambda\, \mathit{result}\ s_1\ s_2\ s_3 \bullet \mathit{result} \triangleq (\text{if isTriangle } s_1\ s_2\ s_3 \text{ then if } s_0 \triangleq s_1\\
&\text{then if } s_1 \triangleq s_2 \text{ then equilateral else isosceles endif else if } s_1 \triangleq s_2 \text{ then isosceles}\\
&\text{else if } s_0 \triangleq s_2 \text{ then isosceles else scalene endif endif endif else invalid endif})
\end{aligned}$$

[2] In the following, we omit the specification of isTriangle().

For the actual test-case generation, we define triangle, which selects via Hilbert's epsilon operator (@) an eligible "implementation" fulfilling our specification:

$$\text{triangle} : [\text{Integer}_\alpha, \text{Integer}_\alpha, \text{Integer}_\alpha] \Rightarrow \text{Triangle}_\bot$$
$$\text{triangle } s_0\ s_1\ s_2 \equiv @\mathit{result}\bullet \models \text{triangle\_spec } \mathit{result}\ s_0\ s_1\ s_2$$

We follow the approach presented in [3] using a disjunctive normal form (DNF) for partition analysis of the specification and as a basis for the test case generation. In our setting this leads to the following main steps:

1. Eliminate logical operators except and, or, and not.
2. Convert the formula into DNF.
3. Eliminate unsatisfiable disjoints by using concurrence rewriting.
4. Select the actual set of test-cases.

Intermediate results are formulae with over 50 disjoints. The logical simplification can only eliminate simple logical falsifications, but this representation can tremendously be simplified by using *congruence rewriting*. Based on its deeper knowledge of the used data types (taking advantage of e.g. $\not\models$ isosceles $\triangleq$ invalid) this step eliminates many unsatisfiable disjoints caused by conflicting constraints. After the congruence rewriting, only six cases are left, respectively one for invalid inputs and one for equilateral triangles, and three cases describing the possibilities for isosceles triangles.

$$\begin{aligned}
&\text{triangle } s_0\ s_1\ s_2 = @\mathit{result}\bullet \models \mathit{result} \triangleq \text{invalid and not isTriangle s0 s1 s2}\\
&\text{or } \mathit{result} \triangleq \text{equilateral and isTriangle } s_0\ s_1\ s_2 \text{ and } s_0 \triangleq s_1 \text{ and } s_1 \triangleq s_2\\
&\text{or } \mathit{result} \triangleq \text{isosceles and isTriangle } s_0\ s_1\ s_2 \text{ and } s_0 \triangleq s_1 \text{ and } s_1 \not\triangleq s_2\\
&\text{or } \mathit{result} \triangleq \text{isosceles and isTriangle } s_0\ s_1\ s_2 \text{ and } s_0 \triangleq s_2 \text{ and } s_0 \not\triangleq s_1\\
&\text{or } \mathit{result} \triangleq \text{isosceles and isTriangle } s_0\ s_1\ s_2 \text{ and } s_1 \triangleq s_2 \text{ and } s_0 \not\triangleq s_1\\
&\text{or } \mathit{result} \triangleq \text{scalene and isTriangle } s_0\ s_1\ s_2 \text{ and } s_0 \not\triangleq s_1 \text{ and } s_0 \not\triangleq s_2 \text{ and } s_1 \not\triangleq s_2
\end{aligned}$$

These six disjoints represent the partitions, from which test cases can be selected, possible exploiting boundary cases like minimal or maximum Integers of the underlying implementation.

6 Conclusion

6.1 Achievements

We have presented a new formal semantic model of OCL in form of a conservative embedding into Isabelle/HOL that can cope with the requirements and the examples of the OCL standard 1.4. On the basis of the embedding, we derived several calculi and proof techniques for OCL. Since "deriving" means that we proved all rules with Isabelle, we can guarantee both the consistency of the semantics as well as the soundness of the calculi. Our semantics is organized in a modular way such that it can be used to study the interdependence of certain language features (method recursion, executability, strictness, smashing, flattening etc.) which might be useful in the current standardization process of OCL. We have shown the potential for semantic based tools for OCL using automated reasoning by an exemplary test-case generation.

6.2 Related Work

Previous semantic definitions of OCL [8,14,18] are based on "mathematical notation" in the style of "naive set theory", which is in our view quite inadequate to cover so subtle subjects such as inheritance. Moreover, the development of proof calculi and automated deduction for OCL has not been in the focus of interest so far.

In [8], a formal operational semantics together with a formal type system for OCL 1.4 was presented. The authors focus on the issue of subject reduction, but do not define the semantic function for expressions whose evaluation may diverges. In [7], it is claimed that a similar OCL semantics is Turing complete. In contrast, our version of OCL admits an infinite state which turns allInstances into an unbounded universal quantifier; when adding least-fixpoint semantics for recursive methods (as we opt for), we are definitively in the world of non-executable languages.

Using a shallow embedding for an object oriented language is still a challenge. While the basic concepts in our approach of representing subtyping by the subsumption relation on polymorphic types is not new (c.f. for example [15,9]), we have included concepts such as undefinedness, mutual recursion between object instances, dynamic types, recursive method invocation and extensible class hierarchies that pushes the limits of the approach a bit further.

6.3 Future Work

Beyond the usual sigh that the existing library is not developed enough (this type of deficiency is usually resolved after the first larger verification project in an embedding), we see the following extensions of our work:

- While our *fast_tac*-based proof procedure for OCL logic is already quite powerful, it is neither efficient nor complete (but should be for a fragment corresponding to propositional logic extended by definedness). More research is necessary (multivalued logics [4], Decision Diagrams).
- Since *HOL-OCL* is intended to be used over several stages of a software development cycle, a refinement calculus that formally supports this activity may be of particular relevance.
- Combining *HOL-OCL* with a Hoare-Logic such as μJava[10] can pave the way for an integrated formal reasoning over specifications and code.

References

1. Manfred Broy, Christian Facchi, Radu Grosu, Rudi Hettler, Heinrich Hussmann, Dieter Nazareth, Oscar Slotosch, Franz Regensburger, and Ketil Stølen. The requirement and design specification language Spectrum, an informal introduction (V 1.0). Technical Report TUM-I9312, TU München, 1993.
2. Achim D. Brucker and Burkhart Wolff. A note on design decisions of a formalization of the OCL. Technical Report 168, Albert-Ludwigs-Universität Freiburg, 2002.
3. Jeremy Dick and Alain Faivre. Automating the generation and sequencing of test cases from model-based specications. In J.C.P. Woodcock and P.G. Larsen, editors, *FME'93: Industrial-Strength Formal Methods*, volume 670 of *LNCS*, pages 268–284. Springer, 1993.
4. Reiner Hähnle. *Automated Deduction in Multiple-valued Logics.* Oxford University Press, 1994.
5. Cliff B. Jones. *Systematic Software Development Using VDM.* Prentice Hall, 1990.
6. Cris Kobryn. Will UML 2.0 be agile or awkward? *CACM*, 45(1):107–110, 2002.
7. Luis Mandel and Marìa Victoria Cengarle. On the expressive power of OCL. *FM'99*, 1999.
8. Luis Mandel and Marìa Victoria Cengarle. A formal semantics for OCL 1.4. In C. Kobryn M. Gogolla, editor, *UML 2001: The Unified Modeling Language. Modeling Languages, Concepts, and Tools*, volume 2185 of *LNCS*, Toronto, 2001. Springer.
9. Wolfgang Naraschewski and Markus Wenzel. Object-oriented verification based on record subtyping in Higher-Order Logic. In J. Grundy and M. Newey, editors, *Theorem Proving in Higher Order Logics*, volume 1479 of *LNCS*, pages 349–366. Springer, 1998.
10. Tobias Nipkow, David von Oheimb, and Cornelia Pusch. μJava: Embedding a programming language in a theorem prover. In Friedrich L. Bauer and Ralf Steinbrüggen, editors, *Foundations of Secure Computation*, volume 175 of *NATO Science Series F: Computer and Systems Sciences*, pages 117–144. IOS Press, 2000.
11. N. D. North. Automatic test generation for the triangle problem. Technical Report DITC 161/90, National Physical Laboratory, Teddington, 1990.
12. OMG. Object Constraint Language Specification. [13], chapter 6.
13. OMG. *Unified Modeling Language Specification (Version 1.4).* 2001.
14. Mark Richters and Martin Gogolla. On Formalizing the UML Object Constraint Language OCL. In Tok-Wang Ling, Sudha Ram, and Mong Li Lee, editors, *Proc. 17th Int. Conf. Conceptual Modeling (ER'98)*, volume 1507 of *LNCS*, pages 449–464. Springer, 1998.
15. Thomas Santen. *A Mechanized Logical Model of Z and Object-Oriented Specification.* PhD thesis, Technical University Berlin, 1999.
16. J. M. Spivey. *The Z Notation: A Reference Manual.* Prentice Hall, 1992.
17. Jos Warmer and Anneke Kleppe. *The Object Contraint Language: Precise Modelling with UML.* Addison-Wesley Longman, Reading, USA, 1999.
18. Jos Warmer, Anneke Kleppe, Tony Clark, Anders Ivner, Jonas Högström, Martin Gogolla, Mark Richters, Heinrich Hussmann, Steffen Zschaler, Simon Johnston, David S. Frankel, and Conrad Bock. Response to the UML 2.0 OCL RfP. Technical report, 2001.
19. Glynn Winskel. *The Formal Semantics of Programming Languages.* MIT Press, Cambridge, 1993.

Explicit Universes for the Calculus of Constructions

Judicaël Courant

Laboratoire de Recherche en Informatique
Bât 490, Université Paris Sud
91405 Orsay Cedex, France
Judicael.Courant@lri.fr

Abstract. The implicit universe hierarchy implemented in proof assistants such as Coq and Lego, although really needed, is painful, both for the implementer and the user: it interacts badly with modularity features, errors are difficult to report and to understand. Moreover, type-checking is quite complex.
We address these issues with a new calculus, the Explicit Polymorphic Extended Calculus of Constructions. EPECC is a conservative extension of Luo's ECC with universe variables and explicit universe constraints declarations. EPECC behaves better with respect to error reporting and modularity than implicit universes, and also enjoys good metatheoretical properties, notably strong normalization and Church-Rosser properties. Type-inference and type-checking in EPECC are decidable. A prototype implementation is available.

1 Introduction

The type system of the Coq proof-assistant results from a compromise between several requirements:

1. It must be expressive enough to develop significant bodies of mathematics.
2. It must be theoretically sound.
3. It must be as simple as possible.

The original basis of Coq is the Calculus of Constructions (CC), originally introduced by Coquand and Huet [1]. CC is a Pure Type System with an impredicative sort called Prop and an untyped sort Type, the latter being the type of the former. CC fulfills requirements 2 and 3 but is not expressive enough: for instance, adding strong sum would be desirable, but requires to type Type with a new sort since Type : Type leads to Girard's paradox [2] and strong sum is inconsistent with impredicativity [5].

Therefore, Coquand proposed an extension of CC with predicative cumulative universes, called CC^ω [2]. CC^ω replaces Type by Type_0 and has an infinite set of predicative sorts $\{\mathsf{Type}_i \mid i \in \mathbb{N}\}$, called universes, with $\mathsf{Type}_i : \mathsf{Type}_{i+1}$ and Type_i included in Type_{i+1} for all $i \in \mathbb{N}$. CC^ω is sound [2] and more expressive than CC, but using it is tedious: in CC^ω, one can not state a proposition over *all*

V.A. Carreño, C. Muñoz, S. Tahar (Eds.): TPHOLs 2002, LNCS 2410, pp. 115–130, 2002.

universes (such as the existence of a function from Type_i to Type_i for any i) but only on *some* explicitly given universes.

The current version of the Coq proof assistant implements a more flexible system. On the surface, Coq has only one predicative sort, called Type, which has type Type. When one enters a development in Coq, a ***universe inference algorithm*** checks that it can be annotated with universe levels consistently with respect to CC^ω. Coq rejects developments that cannot be annotated consistently.

For instance, if the following development is entered in Coq

```
Definition T := Type.
Definition U := Type.
Definition id := [x:T]x.
```

then Coq checks it can be annotated consistently. Here, any choice of a natural number i for annotating the first occurrence of Type and j for annotating the second is a consistent annotation. However, adding the line

```
Definition u := (id U).
```

is correct if and only if U belongs to the domain of id, that is if and only if $j+1 \leq i$. If one enters instead the line

```
Definition t := (id T).
```

then Coq notices that the definition of t requires T to belong to T, that is $i+1 \leq i$, and as this constraint is unsatisfiable, Coq rejects the definition of t.

This implicit universe hierarchy seems quite convenient. However it has several drawbacks:

1. It prevents separate checking of theories.
2. It prevents the simultaneous loading of separately loadable theories.
3. Errors are non-local.
4. It is not expressive enough.
5. Universe inference is complex.

Our plan is the following: Section 2 details these drawbacks, explains their origin and how we choose to address them. Section 3 formally defines EPECC, our proposal for an extension of the Calculus of Constructions with explicit polymorphic universes. Section 4 is an overview of its metatheoretical properties. Section 5 describes our implementation of EPECC and Section 6 compares EPECC with related works.

2 Drawbacks of Implicit Universes

This section describes several issues related to Coq's implicit universes and analyzes their origins. It explains our design decisions for EPECC, a proposal for explicit polymorphic universes above the Calculus of Constructions.

2.1 Separate Checking

The Coq system V7.2 provides a rudimentary module system: although large developments can be split across several files and compiled into fast-loadable modules, it does not allow for separate checking. One of the reasons for this is the implicit universe mechanism.

Indeed, consider a module A implementing X of type Type and an identity function *id* of type Type $\rightarrow$ Type. The *interface* of A is

```
 X : Type
id : Type -> Type
```

Consider now a module B using A containing only the expression $(id\ X)$. Can we check the implementation of B with the sole knowledge of the *interface* of A? We would like to proceed as if we had two axioms X : Type and id : Type $\rightarrow$ Type, looking for a suitable annotation for all occurrences of Type. For instance, we can annotate all of them by the same natural number n.

Unfortunately, the well-formedness of B under the declarations of the interface of A does not imply that the implementation of A and B are compatible. For instance, consider the following implementation for A:

```
Definition X := Type.
Definition Y := [x:X]x
Definition id := [x:Type](Y x).
```

We can remark the following:

- The implementation of A is correct: if one annotates the first instance of Type by i and the second one by j, the constraint implied by the application (Y x) is $i \geq j$ since x belongs to Type_j and the domain of Y is Type_i *alias* X.
- The implementation of B is correct under the declarations of the interface of A.
- The whole development (A together with B) is incorrect. Indeed, the type of X is Type_{i+1} and the application (id X) generates the constraint $j \geq i+1$ since the domain of id is Type_j. Whence the unsatisfiable constraint $i \geq j \geq i+1$. Therefore Coq rejects this development:

```
Welcome to Coq 7.2 (January 2002)
Coq < Definition X := Type.
X is defined
Coq < Definition Y := [x:X]x.
Y is defined
Coq < Definition id := [x:Type](Y x).
id is defined
Coq < Check (id X).
Error: Universe Inconsistency.
```

In fact the correct modular reading of the checking of the implementation of B using the interface of A is that given some *unknown* parameters i, j and k, we have to check the well-formedness of the implementation of B in a context containing X : Type_i and id : $\mathsf{Type}_j \rightarrow \mathsf{Type}_k$. In other words, we have to check that for any i, j, and k, the implementation of B is well-formed. Unfortunately, although this requirement would ensure separate checking, it is much too strong: if the implementation of A is replaced by

```
Definition X := Type.
Definition id := [x:Type]x.
```

checking B with respect to the interface of A fails, although A plus B as a whole is correct. Moreover, as there is no way to declare relations between i, j and k, there is no way to make the interface of A declare X : Type_i and id : $\mathsf{Type}_j \rightarrow \mathsf{Type}_k$ with $j > i$.

Therefore the occurrences of Type in EPECC are explicitly labeled with universe expressions, possibly containing universe variables, and the typing context contains constraints on universe variables.

CC^ω also poses another problem with respect to separate checking. The natural way to check that some implementation I fulfills some interface J is to infer its most informative interface J' and to check that J' is more informative than J. Unfortunately, although CC^ω may give several types to a given term, it does not define any subtyping relation and there is no notion of principal type in this system. For instance, a term t with type $\mathsf{Type}_0 \rightarrow \mathsf{Type}_0$ may have also the type $\mathsf{Type}_0 \rightarrow \mathsf{Type}_1$ or not, depending on t. In fact, type-checking CC^ω is complex [13,6] and relies on universe inference.

Therefore we choose to replace CC^ω by Zhaohui Luo's Extended Calculus of Constructions [9,10]. ECC extends CC^ω with a real subtyping relation, called the cumulativity relation (thus $\mathsf{Type}_0 \rightarrow \mathsf{Type}_0$ is included in $\mathsf{Type}_0 \rightarrow \mathsf{Type}_1$ in ECC). It enjoys the existence of principal types and has a simple type-checking algorithm.

2.2 Modularity and Locality

Combining together results from different modules is required to develop formal proofs: it is a standard practice in mathematics.

Unfortunately, the implicit universes mechanism prevents this to some extent. In Coq, one can build proof libraries that can be loaded *separately* but not *simultaneously*. Consider for instance a file C containing the following Coq development:

```
Definition T := Type.
Definition U := Type.
```

When checking this development, Coq annotates the first occurrence of Type with a variable i and the second one with a variable j.

Now, consider A containing

```
Require C.
Definition idT := [x:T]x.
Definition u := (idT U).
```

and B containing

```
Require C.
Definition idU := [x:U]x.
Definition t := (idU T).
```

Both files are correct: when checking the first one, Coq loads the compiled version of C and infers the constraint $j + 1 \leq i$; when checking the second one, it infers the constraint $i + 1 \leq j$. However these modules cannot be loaded together: it would lead to the unsatisfiable constraint $j + 1 \leq i \ \wedge \ i + 1 \leq j$.

The reason for this problem is that satisfiability is not modular: type-checking modules generates constraints whose conjunction can be unsatisfiable although each one is satisfiable.

We solve this problem as follows: in EPECC, the user explicitly introduces constraints in the typing environment. The type-checker still generates constraints, but instead of checking that they are *satisfiable*, it checks that they are *enforced* by the user-supplied constraints. This way, type-checking is modular: if the user-supplied constraints imply both c and c', then they imply $c \wedge c'$. In order for EPECC to be sound, we must however ensure that the user-supplied constraints are satisfiable. We do this by allowing only a certain class of constraints.

Another problem is the non-locality of the error reported by Coq: it can be discovered arbitrarily far from the real problem. For instance, consider some modules E and F requiring respectively A and B. Requiring E then F would lead to a message telling that a universe inconsistency has been detected while importing F, without further explanation.

To address this problem, constraints over a universe variable in EPECC can only be given when a universe variable is declared. Let us see how this solves the above example: consider C defining T as Type_i and U as Type_j.

- Assume C gives no constraints on i and j. Then a type error is raised on `(idT U)` for A as we do not have $\forall i \in \mathbb{N}\ \forall j \in \mathbb{N}\ j + 1 \leq i$. Similarly, B is rejected as $i \leq j + 1$ is not valid.
- Assume C gives a constraint implying $i + 1 \leq j$. Then A is rejected and B is accepted.
- Assume C gives a constraint implying $j + 1 \leq i$. Then A is accepted and B is rejected.

In any case, the error can be reported very accurately: either on `(idT U)` or on `(idU T)`.

2.3 Flexibility

The three solutions proposed above have a disadvantage over Coq's implicit universes: one has to choose how C will be used when writing it. In Coq, no choice has to be done when C is written; instead the first usage excludes the other one.

We claim this issue is under the responsibility of some module system, not of EPECC itself: an SML-like module system [11] would let us *parameterize* the

Terms		Sorts	
$t ::= \lambda x{:}t.t$	abstraction	$\sigma ::= \mathsf{Prop}$	
$\mid\ \Pi x{:}t.t$	product	$\mid\ \mathsf{Type}_n$	
$\mid\ (t\ t)$	application	Contexts	
$\mid\ \sigma$	sort	$\Gamma ::= \epsilon$	Empty context
$\mid\ x$	variable	$\mid\ \Gamma; x : t$	Variable declaration

where x ranges over an infinite set of variables and n ranges over $\mathbb{N}$.

Fig. 1. Syntax of ECC^-

module C by a module providing two universe levels i and j. Then, A can introduce two universes i' and j' such that $j' + 1 \leq i'$ and require C with i instantiated by i' and j by j'. Similarly B can introduce two universes i'' and j'' with $i'' + 1 \leq j''$ and require C with i instantiated by i'' and j by j''. Thus, no choice has to be made when C is written and both choices can even be made simultaneously, thanks to the module system.

The prototype presented in Section 5 has such a module system [4]. Therefore, we do not consider further this parameterization issue in this paper.

3 Definition of EPECC

Our proposal for explicit polymorphic universes is based on Luo's ECC. However, because of the lack of space, we do not consider the whole ECC: we remove sigma-types and get a subset of ECC which we call ECC^-. The sequel straightforwardly extends to sigma types. Section 3.1 describes ECC^-, Section 3.2 introduces the syntax of EPECC and Section 3.4 describes its judgments and rules. Note that we do not consider issues related to variable renaming nor variable capture: terms are considered in Barendregt convention and up to bound variable renaming (α-equivalence). Hence, the usual side conditions stating that a variable is fresh are implicit in the presented inference rules.

3.1 ECC^-

Syntax. The syntax of ECC^- is given Figure 1. As in Luo's ECC, the sorts of ECC^- contain an impredicative sort Prop and a family of sorts Type_n, indexed by natural numbers.

Conversion and Cumulativity. The β-reduction is defined as usual, we denote it by $\triangleright_\beta$. The *conversion* relation $\simeq$ is defined as the reflexive, symmetric, transitive closure of the β-reduction. The *cumulativity* relation $\preceq$ is the least partial transitive relation such that:

- $t_1 \preceq t_2$ for any t_1, t_2 such that $t_1 \simeq t_2$;
- $\mathsf{Prop} \preceq \mathsf{Type}_n$ for any $n \in \mathbb{N}$;
- $\mathsf{Type}_n \preceq \mathsf{Type}_m$ for any $(n, m) \in \mathbb{N}^2$ such that $n < m$;
- $\Pi x{:}t_1.t_1' \preceq \Pi x{:}t_2.t_2'$ for any t_1, t_1', t_2, t_2' such that $t_1 \simeq t_2$and $t_1' \preceq t_2'$;

$\preceq$ is a partial order with respect to $\simeq$ (see [10], Lemma 3.1.6, p. 29).

$$\text{T/SORT}\frac{(\sigma_1,\sigma_2)\in\mathcal{A}}{\Gamma\vdash\sigma_1:\sigma_2}\qquad \text{T/APP}\frac{\Gamma\vdash t_1:\Pi x{:}t_3.t_4 \qquad \Gamma\vdash t_2:t_3}{\Gamma\vdash(t_1\ t_2):t_4\{x\leftarrow t_2\}}$$

$$\text{T/PROD}\frac{\Gamma\vdash t_1:\sigma_1 \qquad \Gamma;x:t_1\vdash t_2:\sigma_2 \qquad (\sigma_1,\sigma_2,\sigma_3)\in\mathcal{R}}{\Gamma\vdash \Pi x{:}t_1.t_2:\sigma_3}$$

$$\text{T/LAM}\frac{\Gamma\vdash t_1:\sigma_1 \qquad \Gamma;x:t_1\vdash t_2:t_3}{\Gamma\vdash \lambda x{:}t_1.t_2:\Pi x{:}t_1.t_3}$$

$$\text{T/CUMUL}\frac{\Gamma\vdash t_1:t_3 \qquad \Gamma\vdash t_2:\sigma \qquad t_3\preceq t_2}{\Gamma\vdash t_1:t_2}\qquad \text{T/VAR}\frac{}{\Gamma\vdash x:\Gamma(x)}$$

$$\text{ENV/EMPTY}\frac{}{\vdash \mathsf{ok}}\qquad \text{ENV/DECL}\frac{\Gamma\vdash\mathsf{ok} \qquad \Gamma\vdash t:\sigma}{\Gamma;x:t\vdash\mathsf{ok}}$$

$$\begin{array}{ll}\mathcal{A}= \{\ (\mathsf{Type}_n,\mathsf{Type}_{n+1})\mid n\in\mathbb{N}\ \}\cup\{\ (\mathsf{Prop},\mathsf{Type}_0)\ \} & \\ \mathcal{R}= \quad \{\ (\mathsf{Type}_n,\mathsf{Type}_m,\mathsf{Type}_{\max(n,m)}) & \mid (n,m)\in\mathbb{N}^2\ \} \\ \quad\ \cup\{\ (\mathsf{Type}_n,\mathsf{Prop},\mathsf{Prop}) & \mid n\in\mathbb{N}\ \} \\ \quad\ \cup\{\ (\mathsf{Prop},\mathsf{Type}_n,\mathsf{Type}_n) & \mid n\in\mathbb{N}\ \}\cup\{\ (\mathsf{Prop},\mathsf{Prop},\mathsf{Prop})\}\end{array}$$

Fig. 2. Typing rules of ECC^-

Judgments. In addition to the typing judgment $\Gamma \vdash t_1 : t_2$ from [10], our presentation uses a judgment $\Gamma \vdash \mathsf{ok}$ whose intended meaning is "Γ is a well-formed environment". In [10], there is no such judgment; $\Gamma \vdash \mathsf{Prop} : \mathsf{Type}_0$ is used instead.

Rules. Figure 2 gives the rules of ECC^-. ECC^- can be seen as a PTS up to the following differences:

- As the considered set of products is full, one does not need to check that the product $\Pi x{:}t_1.t_3$ is well-formed in T/LAM (see [13], Section 3).
- The conversion rule is replaced by the cumulativity rule T/CUMUL.

The differences between ECC^- and ECC are the following:

- ECC rules related to sigma types have been removed.
- We do not require Γ to be well-formed in order for a judgment $\Gamma \vdash t_1 : t_2$ to hold in ECC^-. On the contrary, the rules for typing variables and sorts in [10] enforce this. Thus $\Gamma \vdash t_1 : t_2$ holds in [10] if and only if $\Gamma \vdash t_1 : t_2$ and $\Gamma \vdash \mathsf{ok}$ hold in ECC^-.
- When typing products, Luo's presentation requires the domain and range to belong to the same Type_j or the domain to be Prop. The rule T/PROD can seem more permissive as it accepts any sort for the domain and the range, but this is just a derived rule in [10], thanks to the cumulativity rule T/CUMUL.

3.2 Syntax

The syntax of EPECC is obtained from ECC^- as follows:

- We introduce a new non-terminal i, defining the syntactic category of *universe expressions*, with the following rules:

$$\begin{array}{rlll} i ::= & 0 & \\ & | & i+n & \text{lifting a universe by a constant} \\ & | & \mathsf{MAX}(i,i) & \text{maximum of two universes} \\ & | & u & \text{universe variable} \end{array}$$

  where u ranges over an infinite set $\mathfrak{U}$ of universe variables and n ranges over $\mathbb{N}$. We call $\mathfrak{I}$ the set of universe expressions.
- We replace the family of sorts Type_n, for $n \in \mathbb{N}$ by the family of Type_i, for $i \in \mathfrak{I}$.
- In addition to the usual term variables declarations, contexts may also contain universe variable declarations. A universe declaration has the form $u \geq i$ where $u \in \mathfrak{U}$ is a fresh universe variable and $i \in \mathfrak{I}$:

$$\Gamma \mathrel{+}::= \Gamma ; u \geq i$$

  Informally, this declaration means that u denotes a universe level above i. In other words, this means that we have the type inclusion $\mathsf{Type}_i \preceq \mathsf{Type}_u$. We denote the set of universe variables declared in Γ by $UDom(\Gamma)$.

3.3 Conversion and Cumulativity Relation

The cumulativity relation $\preceq$ of ECC^- has to be modified for EPECC. Indeed, whether Type_i is included in Type_j depends not only on i and j but also on the declaration of the variables they contain. Intuitively, Type_i is included in Type_j if and only if for every interpretation of universe variables satisfying the constraints of the context, the value of i is less than or equal to the value of j.

For the same reason, the conversion of Type_i and Type_j is no longer the syntactic equality; instead they are equal if and only if for every interpretation satisfying the constraints, the value of i is equal to the value of j.

In order to define precisely the conversion and cumulativity relations, we now formally define this notion of interpretation.

Definition 1 (Universe Interpretations). *A* universe interpretation *ϕ is a member of $\mathbb{N}^{\mathfrak{U}}$, that is, a function mapping universe variables to natural numbers.*

Given a universe interpretation ϕ, we extend it inductively into a morphism from $\mathfrak{I}$ to $\mathbb{N}$ as follows:

$$\begin{array}{ll} \phi(u) = \phi(u) & \phi(\mathsf{MAX}(i,j)) = \max(\phi(i), \phi(j)) \\ \phi(0) = 0 & \phi(i+n) = \phi(i) + n \end{array}$$

This morphism induces a term interpretation *from EPECC to ECC⁻. The term interpretation* ϕ *associated to the universe interpretation* ϕ *is inductively defined as follows:*

$$\begin{aligned} \phi(\mathsf{Prop}) &= \mathsf{Prop} & \phi(\mathsf{Type}_i) &= \mathsf{Type}_{\phi(i)} \\ \phi(x) &= x & \phi((t_1\ t_2)) &= (\phi(t_1)\ \phi(t_2)) \\ \phi(\lambda x{:}t_1.t_2) &= \lambda x{:}\phi(t_1).\phi(t_2) & \phi(\Pi x{:}t_1.t_2) &= \Pi x{:}\phi(t_1).\phi(t_2) \end{aligned}$$

This term interpretation induces a context interpretation ϕ *from the typing contexts of EPECC to those of ECC⁻ as follows:*

$$\phi(\epsilon) = \epsilon \qquad \phi(\Gamma; x : t) = \phi(\Gamma); x : \phi(t) \qquad \phi(\Gamma; u \geq i) = \phi(\Gamma)$$

We say the interpretations satisfying the constraints of a given typing context Γ are *increasing* on Γ. More formally:

Definition 2 (Increasing Interpretations). *The set of* increasing interpretations *on a context* Γ*, denoted by* $Incr(\Gamma)$*, is defined by induction on* Γ *as follows:*

$$\begin{aligned} Incr(\epsilon) &= \mathbb{N}^{\mathfrak{U}} \\ Incr(\Gamma'; u \geq i) &= \{ \phi \in Incr(\Gamma') \mid \phi(u) \geq \phi(i) \} \\ Incr(\Gamma; x : t) &= Incr(\Gamma) \end{aligned}$$

Given a context Γ, we can now compare universe expressions through the values of their interpretations:

Definition 3 (Γ-Equality and Γ-less-than). *Given a context* Γ *and two universe expressions* i *and* j*, we say* i *and* j *are* Γ*-equal (*resp. i *is* Γ*-less than* j*) and we write* $i =_\Gamma j$ *(*resp. $i \leq_\Gamma j$*) if for any* $\phi \in Incr(\Gamma)$*,* $\phi(i) = \phi(j)$ *(*resp. $\phi(i) \leq \phi(j)$*).*

It is clear that Γ-equality is an equivalence relation, Γ-less-than is a reflexive and transitive relation. Moreover, the former is the symmetric closure of the latter.

Γ-equality induces a congruence on terms, called $\mathfrak{U}$-equivalence:

Definition 4 ($\mathfrak{U}$-Equivalence). *Given a context* Γ*, the* $\mathfrak{U}$*-equivalence over* Γ *is defined as the smallest congruence* $=_\mathfrak{U}$ *over terms such that* $\mathsf{Type}_i =_\mathfrak{U} \mathsf{Type}_j$ *for any* i, j *such that* $i =_\Gamma j$*. We write* $\Gamma \vdash t_1 =_\mathfrak{U} t_2$ *to mean* t_1 *and* t_2 *are* $\mathfrak{U}$*-equivalent over* Γ*.*

In other words, the $\mathfrak{U}$-equivalence over Γ is the smallest reflexive, symmetric, transitive and monotonic relation $=_\mathfrak{U}$ such that $\mathsf{Type}_i =_\mathfrak{U} \mathsf{Type}_j$ for any i, j such that $i =_\Gamma j$ (by "monotonic", we mean that for any t, t_1, t_2 such that $t_1 =_\mathfrak{U} t_2$, we have $t\{x \leftarrow t_1\} =_\mathfrak{U} t\{x \leftarrow t_2\}$).

We now define the conversion relation and the cumulativity relation:

Definition 5 (Conversion). *Given a context* Γ*, the* conversion relation *over* Γ *is defined as the smallest congruence containing the* β*-equivalence and the* $\mathfrak{U}$*-equivalence over* Γ*. "*t_1 *and* t_2 *are convertible over* Γ*" is denoted by* $\Gamma \vdash t_1 \simeq t_2$*.*

Terms:

$$\text{T/VAR}\frac{}{\Gamma \vdash x : \Gamma(x)} \qquad \text{T/SORT}\frac{\Gamma \vdash \sigma_1 : \mathsf{sort} \qquad (\sigma_1, \sigma_2) \in \mathcal{A}}{\Gamma \vdash \sigma_1 : \sigma_2}$$

$$\text{T/PROD}\frac{\Gamma \vdash t_1 : \sigma_1 \qquad \Gamma; x : t_1 \vdash t_2 : \sigma_2 \qquad (\sigma_1, \sigma_2, \sigma_3) \in \mathcal{R}}{\Gamma \vdash \Pi x{:}t_1.t_2 : \sigma_3}$$

$$\text{T/LAM}\frac{\Gamma \vdash t_1 : \sigma_1 \qquad \Gamma; x : t_1 \vdash t_2 : t_3}{\Gamma \vdash \lambda x{:}t_1.t_2 : \Pi x{:}t_1.t_3} \qquad \text{T/APP}\frac{\Gamma \vdash t_1 : \Pi x{:}t_3.t_4 \qquad \Gamma \vdash t_2 : t_3}{\Gamma \vdash (t_1\ t_2) : t_4\{x \leftarrow t_2\}}$$

$$\text{T/CUMUL}\frac{\Gamma \vdash t_1 : t_3 \qquad \Gamma \vdash t_2 : \sigma \qquad \Gamma \vdash t_3 \preceq t_2}{\Gamma \vdash t_1 : t_2}$$

Sorts:

$$\text{SORT/SET}\frac{}{\Gamma \vdash \mathsf{Prop} : \mathsf{sort}} \qquad \text{SORT/TYPE}\frac{\Gamma \vdash i : \mathsf{univ}}{\Gamma \vdash \mathsf{Type}_i : \mathsf{sort}}$$

Universes:

$$\text{UNIV/CONST}\frac{}{\Gamma \vdash 0 : \mathsf{univ}} \qquad \text{UNIV/VAR}\frac{u \in UDom(\Gamma)}{\Gamma \vdash u : \mathsf{univ}}$$

$$\text{UNIV/LIFT}\frac{\Gamma \vdash i : \mathsf{univ}}{\Gamma \vdash i + n : \mathsf{univ}} \qquad \text{UNIV/MAX}\frac{\Gamma \vdash i_1 : \mathsf{univ} \qquad \Gamma \vdash i_2 : \mathsf{univ}}{\Gamma \vdash \mathsf{MAX}(i_1, i_2) : \mathsf{univ}}$$

Environments:

$$\text{ENV/EMPTY}\frac{}{\vdash \mathsf{ok}} \qquad \text{ENV/DECL}\frac{\Gamma \vdash \mathsf{ok} \qquad \Gamma \vdash t : \sigma}{\Gamma; x : t \vdash \mathsf{ok}}$$

$$\text{ENV/UDECL}\frac{\Gamma \vdash \mathsf{ok} \qquad \Gamma \vdash i : \mathsf{univ}}{\Gamma; u \geq i \vdash \mathsf{ok}}$$

$$\begin{aligned}
\mathcal{A} &= \{\ (\mathsf{Type}_i, \mathsf{Type}_{i+1}) \mid i \in \mathfrak{I}\ \} \cup \{\ (\mathsf{Prop}, \mathsf{Type}_0)\ \} \\
\mathcal{R} &= \quad \{(\mathsf{Type}_i, \mathsf{Type}_j, \mathsf{Type}_{\mathsf{MAX}(i,j)}) \mid (i,j) \in \mathfrak{I}^2\ \} \\
&\quad \cup \{(\mathsf{Type}_i, \mathsf{Prop}, \mathsf{Prop}) \qquad \mid i \in \mathfrak{I}\ \} \\
&\quad \cup \{(\mathsf{Prop}, \mathsf{Type}_i, \mathsf{Type}_i) \qquad \mid i \in \mathfrak{I}\ \} \cup \{(\mathsf{Prop}, \mathsf{Prop}, \mathsf{Prop})\}
\end{aligned}$$

Fig. 3. Typing rules of EPECC

Definition 6 (Cumulativity Relation). *Given a context Γ, the* cumulativity relation over Γ *is defined as the smallest transitive relation $\preceq$ over terms such that*

- $t_1 \preceq t_2$ *for all* t_1, t_2 *such that* $\Gamma \vdash t_1 \simeq t_2$;
- $\mathsf{Prop} \preceq \mathsf{Type}_i$ *for all* $i \in \mathfrak{I}$;
- $\mathsf{Type}_i \preceq \mathsf{Type}_j$ *for all* $(i,j) \in \mathfrak{I}^2$ *such that* $i \leq_\Gamma j$;
- $\Pi x{:}t_1.t_1' \preceq \Pi x{:}t_2.t_2'$ *for all* t_1, t_1', t_2, t_2' *such that* $\Gamma \vdash t_1 \simeq t_2$ *and* $t_1' \preceq t_2'$.

We denote "(t_1, t_2) is in the cumulativity relation over Γ" by $\Gamma \vdash t_1 \preceq t_2$".

3.4 Judgments and Rules

The typing rules for EPECC are given in Figure 3. As with ECC$^-$, EPECC defines the judgments $\Gamma \vdash t_1 : t_2$ and $\Gamma \vdash \mathsf{ok}$. It defines two new judgments :

- $\Gamma \vdash i : \mathsf{univ}$, read "$i$ is a well-formed universe expression in Γ". i is a well-formed universe when all the variables it contains are declared in Γ.
- $\Gamma \vdash \sigma : \mathsf{sort}$, read "$\sigma$ is well-formed sort in Γ". The well-formed sort are Prop and the Type_i such that i is well-formed.

The rules dealing with the judgment $\Gamma \vdash t_1 : t_2$ are similar to the ones of ECC$^-$. The typing rule for sorts is close to the standard one, although we check that σ_1 is a valid sort. The set of axioms $\mathcal{A}$ and the set of products are different from the one of ECC$^-$ as integer indices have been replaced by universe expressions. The side condition of the conversion rule T/CUMUL uses the cumulativity relation of EPECC instead of ECC$^-$'s.

4 Metatheory

This section gives the main theoretical properties of EPECC: the elements of $Incr(\phi)$ are morphisms from EPECC to ECC$^-$; the existence of these morphisms let us derive easily the Church-Rosser and strong normalization properties of EPECC. Moreover, EPECC enjoys type-inference and type-checking algorithms.

Proposition 1 (Existence of Increasing Interpretations). *For any context Γ such that $\Gamma \vdash$ ok holds in EPECC, $Incr(\Gamma) \neq \emptyset$.*

Theorem 1 (ϕ Is a Morphism). *For any context Γ and any $\phi \in Incr(\Gamma)$, ϕ is a morphism:*

1. *It preserves β-reduction steps, conversion and cumulativity.*
2. *It transforms derivable typing judgments of EPECC into derivable typing judgments of ECC$^-$.*

Corollary 1. *EPECC is consistent: there is no term t such that $\vdash t : \Pi x{:}\mathsf{Prop}.x$ holds in EPECC.*

Actually, once the adequate embedding of ECC$^-$ in EPECC is defined, Proposition 1 implies that EPECC is a *conservative* extension of ECC$^-$.

Theorem 2 (Strong Normalization). *For any Γ such that $\Gamma \vdash$ ok holds in EPECC, any terms t and t' such that $\Gamma \vdash t : t'$, t is strongly normalizing.*

Proof. As $\Gamma \vdash$ ok holds, there exists $\phi \in Incr(\Gamma)$ by Proposition 1. If there was an infinite reduction starting at t, its image by ϕ would be an infinite reduction starting at $\phi(t)$ (by Theorem 1). Since $\phi(t)$ is well-typed in the well-formed context $\phi(\Gamma)$ (by Theorem 1), this is absurd.

Proposition 2 (No Creation of β-Redexes). *For any Γ, any $\phi \in Incr(\Gamma)$, any t in EPECC, and any t'' in ECC$^-$, if $\phi(t) \triangleright_\beta t''$, there exists t' such that $t \triangleright_\beta t'$ and $\phi(t') = t''$.*

Theorem 3 (Church-Rosser Property Modulo $=_{\mathfrak{U}}$). *For any environment Γ such that $\Gamma \vdash \mathsf{ok}$ holds in EPECC, any terms t_1 and t_2 such that $\Gamma \vdash t_1 \simeq t_2$, there exists t'_1 and t'_2 such that $t_1 \triangleright^{\star}_{\beta} t'_1$, $t_2 \triangleright^{\star}_{\beta} t'_2$ and $\Gamma \vdash t'_1 =_{\mathfrak{U}} t'_2$.*

Proof. This is a consequence of Proposition 1, Theorem 1 and Proposition 2.

The Church-Rosser property can be proved using the usual methods (for an elegant one, see [12]):

Theorem 4 (Church-Rosser Property). *For any terms t_1 and t_2 of EPECC such that t_1 and t_2 are β-equivalent, there exists t' such that $t_1 \triangleright^{\star}_{\beta} t'$ and $t_2 \triangleright^{\star}_{\beta} t'$.*

Subject-reduction can be proved using the same techniques as [10] for ECC:

Theorem 5 (Subject-Reduction). *EPECC enjoys the subject-reduction property. More precisely, if $\Gamma \vdash t_1 : t_2$ holds in EPECC and $t_1 \triangleright_{\beta} t'_1$, we have $\Gamma \vdash t'_1 : t_2$ in EPECC.*

The proof is as usual, the only difficulty is to prove $\Gamma \vdash \Pi x{:}t_1.t'_1 \preceq \Pi x{:}t_2.t'_2$ implies $\Gamma \vdash t_1 \simeq t_2$ and $\Gamma \vdash t'_2 \preceq t'_2$. This involves proving that $\Gamma \vdash t_1 \simeq t_2$ implies the existence of t'_1 and t'_2 such that $t_1 \triangleright^{\star}_{\beta} t'_1$, $t_2 \triangleright^{\star}_{\beta} t'_2$, and $\Gamma \vdash t'_1 =_{\mathfrak{U}} t'_2$.

Our type-checking and type-inference algorithms are similar to Luo's ones for ECC [10]. They rely on an algorithm deciding the cumulativity relation.

Definition 7 (Type-Inference and Type-Checking Algorithms).

- *Given a context Γ and terms t_1 and t_2, we check that t_1 has type t_2 by computing the principal type t of t_1 in Γ, and checking $\Gamma \vdash t \preceq t_2$.*
- *Given a context Γ and a term t, we infer its principal type according to the rules T/VAR, T/SORT, T/PROD, T/LAM and T/APP. As for applications (rule T/APP), we first infer recursively the type of the applied term, then we β-reduce it to a product $\Pi x{:}t_1.t_2$ and we check that the argument of the application has type t_1.*

The conversion test is similar to the one of ECC: in order to decide the conversion of two terms t_1 and t_2, we first reduce them into weak-head normal forms t'_1 and t'_2 and recursively call the conversion test if needed. The only difference is when the normal forms of t_1 and t_2 are Type_{i_1} and Type_{i_2}. In ECC, one checks that i_1 and i_2 are equal natural numbers. In EPECC, we check that i_1 and i_2 are Γ-equal universe expressions.

Similarly, the cumulativity test reduces its arguments into weak-head normal forms and recursively call itself or the conversion test if needed. To compare Type_{i_1} and Type_{i_2}, it decides $i_1 =_{\Gamma} i_2$.

To decide $t_1 =_{\Gamma} t_2$ and $t_1 \leq_{\Gamma} t_2$, we introduce a new judgment $\Gamma \vdash i_1 \leq_n i_2$, where Γ is a typing context, n an integer, and i_1 and i_2 are universe expressions. Its intended meaning is "$\forall \phi \in \mathit{Incr}(\Gamma)\ \phi(i_1) \leq \phi(i_2) + n$". Figure 4 describes inference rules for this judgment.

Theorem 6 (Soundness and Completeness of the Rules). *Given a context Γ such that $\Gamma \vdash \mathsf{ok}$ holds in EPECC, an integer n, and universe expressions i_1*

$$\text{ZERO}\frac{}{\Gamma \vdash 0 \leq_n 0}n \geq 0 \qquad \text{VAR}\frac{}{\Gamma \vdash u \leq_n u}n \geq 0$$

$$\text{LEFT/MAX}\frac{\Gamma \vdash i_1 \leq_n i_3 \qquad \Gamma \vdash i_2 \leq_n i_3}{\Gamma \vdash \text{MAX}(i_1, i_2) \leq_n i_3} \qquad \text{LEFT/LIFT}\frac{\Gamma \vdash i_1 \leq_{n_1 - n_2} i_2}{\Gamma \vdash i_1 + n_2 \leq_{n_1} i_2}$$

$$\text{RIGHT/LIFT}\frac{\Gamma \vdash i_1 \leq_{n+n'} i_2}{\Gamma \vdash i_1 \leq_n i_2 + n'} \qquad \text{RIGHT/VAR}\frac{\Gamma \vdash i \leq_n i'}{\Gamma \vdash i \leq_n u}u \geq i' \in \Gamma$$

$$\text{RIGHT/MAX1}\frac{\Gamma \vdash i \leq_n i_1}{\Gamma \vdash i \leq_n \text{MAX}(i_1, i_2)} \qquad \text{RIGHT/MAX2}\frac{\Gamma \vdash i \leq_n i_2}{\Gamma \vdash i \leq_n \text{MAX}(i_1, i_2)}$$

Fig. 4. Deciding Γ-inequality

and i_2 such that $\Gamma \vdash i_1 : \mathsf{univ}$ and $\Gamma \vdash i_2 : \mathsf{univ}$ hold, $\Gamma \vdash i_1 \leq_n i_2$ can be derived by the rules given in Figure 4 if and only if for any $\phi \in Incr(\Gamma)$, $\phi(i_1) \leq \phi(i_2) + n$.

Proof. It is easy to check that all these rules are sound. Completeness requires more care as the rules RIGHT/MAX1, RIGHT/MAX2, and RIGHT/VAR are not invertible. The conclusion follows from the following technical lemmas:

- Given a judgment $\Gamma \vdash i \leq_n u$ with $u \geq i' \in \Gamma$, if RIGHT/VAR is the only applicable rule, either the judgment does not hold, or $\Gamma \vdash i \leq_n i'$ holds.
- Given a judgment $\Gamma \vdash i \leq_n \text{MAX}(i_1, i_2)$, if RIGHT/MAX1 and RIGHT/MAX2 are the only applicable rules, either the judgment does not hold, or at least one of the judgments $\Gamma \vdash i \leq_n i_1$ and $\Gamma \vdash i \leq_n i_2$ holds.

Definition 8 (Algorithm for Γ-less-than). *Given a context Γ such that $\Gamma \vdash$ ok holds in EPECC and universe expressions i_1 and i_2, we decide $i_1 \leq_\Gamma i_2$ by deciding whether $\Gamma \vdash i_1 \leq_0 i_2$ can be derived. This is done by applying the rules given figure 4 with the following strategy:*

- *If no rule is applicable, it can not be derived.*
- *If there are applicable rules other than RIGHT/VAR, RIGHT/MAX1, and RIGHT/MAX2, we pick any of them ("don't care" non-determinism). If it has no premise, the test succeeds, otherwise, we recursively test whether its premise holds.*
- *If RIGHT/VAR is the only applicable rule, we recursively decide whether its premise holds.*
- *If RIGHT/MAX1 and RIGHT/MAX2 are the only applicable rules, we pick any of them and check whether its premise holds. If it holds, the test succeeds, otherwise, we decide whether the premise of the other one holds ("don't know" non-determinism).*

Definition 9 (Algorithm for Γ-Equality). *Given a context Γ and terms i_1 and i_2, we decide $i_1 =_\Gamma i_2$ by testing $i_1 \leq_\Gamma i_2$ and $i_2 \leq_\Gamma i_1$.*

Theorem 7. *Algorithms for deciding Γ-inequality and Γ-equality of i_1 and i_2 terminate when $\Gamma \vdash$ ok, $\Gamma \vdash i_1 : \mathsf{univ}$, and $\Gamma \vdash i_2 : \mathsf{univ}$ hold in EPECC. They are sound and complete.*

Proof. Soundness and completeness come from Theorem 6. Termination is proved by defining two measures μ and ν on well-formed universes of Γ:

$$\nu(0) = 1 \qquad \nu(u) = 1 + \nu(i) \text{ where } u \geq i \in \Gamma$$
$$\nu(i + n) = 1 + \nu(i) \qquad \nu(\mathsf{MAX}(i_1, i_2)) = 1 + \nu(i_1) + \nu(i_2)$$

μ is defined similarly, excepted for variables : $\mu(u) = 1$ for any variable u. One can show by induction on the derivation that for any Γ such that $\Gamma \vdash \mathsf{ok}$ and any i such that $\Gamma \vdash i : \mathsf{univ}$, $\mu(i)$ and $\nu(i)$ are well-defined. Then, we define the measure of a judgment $\Gamma \vdash i_1 \leq_n i_2$ as $\mu(i_1) \times \nu(i_2)$. For each rule given in figure 4, the premises are clearly smaller than the conclusion.

The measures defined in this proof also give a complexity measure of the decision algorithms: the strategy of Definition 8 for deciding $\Gamma \vdash i_1 \leq_n i_2$ requires at most $\mu(i_1) \times \nu(i_2)$ application of inference rules.

5 Implementation

We extended œuf, our prototype implementation of a module calculus over the Calculus of Constructions [3,4] with the constructs and rules of EPECC. This prototype is written in Objective Caml with the help of a home-made tool compiling typing rules into Objective Caml code. œuf with polymorphic universes weights about 2800 lines of code. Adding polymorphic universes to œuf was quite simple: we added or changed about 280 lines of code. œuf is available at `http://www.lri.fr/~jcourant/01/oeuf`.

Developments in œuf enjoy the separate development property and are modular in the sense of Section 2. Errors are detected locally, although error reporting in œuf is in an early stage for the moment. We did not benchmarked it as performance does not seems really relevant for universe constraints checking — the major performance issue for type-checking the Calculus of Constructions is the efficiency of reduction. Moreover, our prototype does not implements inductive types yet, which prevents it to compete with Coq on realistic examples for the moment.

6 Comparison with Other Works

We now briefly review other systems and implementations addressing universes with respect to our initial motivations:

– Luo's ECC provides a formal typing system with universes. Although ECC does not deal with the notion of modules nor definitions, it would be quite easy to add a modular layer on top of it. The resulting system would allow separate checking, would be modular and errors could be reported accurately. However, ECC is not flexible enough: each time one makes use of Type, one has to choose which instance of Type one wants to work with. This prevents the development of reusable proof libraries over universes.

- Coq's implicit universes are more flexible as the user does not have to tell in advance on which universe level she is working, but they forbid separate checking; they are non-modular and errors cannot be reported accurately. Also, we claim that implementing implicit universes is much more difficult than EPECC's polymorphic universes: algorithms involved in universe inference are complex while the algorithm we use to check universe constraints is quite straightforward. Moreover, in Coq, the code dealing with implicit universes is scattered all around the type-checker: actually, all the typing judgments Coq implements involve universe constraints; in the implementation, this means that each typing function returns universe constraints in addition to the type it infers and each rule has to deal carefully with these constraints.While the extension of CC with explicit polymorphic universes is quite orthogonal to other extensions such as modules, implicit universes have a large interaction with Coq's modules and inductive types. A minor difference between CC an Coq's underlying theory is the presence of two predicative sorts in Coq, Prop and Set; extending EPECC to this system should be straightforward.
- LEGO proposes an implicit universes mechanism similar to Coq's. However LEGO's is more flexible: when the type-checker introduces some constraints on a defined constant, LEGO duplicates the constant so that subsequent constraints do not interfere with the previous ones. Hence LEGO accepts the development given Section 2.2 while Coq rejects it. Thus LEGO behaves a bit better than Coq with respect to modularity. Unfortunately, this behavior is not perfect since LEGO's trick works only with defined constants, not with variables, and examples containing only variable declarations and similar in spirit to the one of Section 2.2 can be given.
- Nuprl [7] allows to index occurrences of Type by universe level expressions very similar to the ones we use in this paper. A brief explanation of the semantics of these expressions is given in Jackson's thesis [8]. Nuprl's universe polymorphism is less expressive than EPECC's as it does not allow any constraints over universe variables; the effect of some of EPECC constraints can be simulated though. Nuprl allows one to prove theorems which are implicitly quantified over universe level, in Hindley-Milner style. An advantage of this approach over ours is that this style of quantification is more fine-grained than quantification at the module level. An interesting future work would be to define a combination of this style of quantification with EPECC constraints and develop its metatheory.

7 Conclusion

Coq and LEGO mislead the naive user with the illusion that Type has type Type while implementing a universe hierarchy internally. We deliberately designed

EPECC to tell the user the whole truth. Although this is an advantage, one can also argue this steepens the learning curve. However the naive user often does not need to use Type at all: Prop is often enough for her. Moreover, she might ignore universe variables and deal with Type_0, Type_1, *etc.* until she realizes that polymorphism would help her.

Years after universes have been introduced, implementing and using them is still difficult. We think EPECC brings new insights: while much simpler to describe and implement than implicit universes, it behaves nicely with respect to separate checking, enjoys modularity and errors can be detected locally.

References

1. Thierry Coquand and Gérard Huet. The Calculus of Constructions. *Inf. Comp.*, 76:95–120, 1988.
2. Thierry Coquand. An analysis of Girard's paradox. In *Proceedings of the First Symposium on Logic in Computer Science*, Cambridge, MA, June 1986. IEEE Comp. Soc. Press.
3. Judicaël Courant. A Module Calculus for Pure Type Systems. In *Typed Lambda Calculi and Applications'97*, Lecture Notes in Computer Science, pages 112 – 128. Springer-Verlag, 1997.
4. Judicaël Courant. $\mathcal{MC}_2$: A Module Calculus for Pure Type Systems. Research Report 1292, LRI, September 2001.
5. James G. Hook and Douglas J. Howe. Impredicative Strong Existential Equivalent to Type:Type. Technical Report TR86-760, Cornell University, 1986.
6. R. Harper and R. Pollack. Type checking, universe polymorphism, and typical ambiguity in the calculus of constructions. *Theoretical Computer Science*, 89(1), 1991.
7. Paul B. Jackson. *The Nuprl Proof Development System, Version 4.1 Reference Manual and User's Guide.* Cornell University, Ithaca, NY, 1994.
8. Paul B. Jackson. *Enhancing the Nuprl Proof Development System and Applying it to Computational Abstract Algebra.* PhD thesis, Cornell University, 1995.
9. Zhaohui Luo. ECC: an Extended Calculus of Constructions. In *Proceedings of the Fourth Annual Symposium on Logic in Computer Science*, Pacific Grove, California, 1989. IEEE Comp. Soc. Press.
10. Zhaohui Luo. *An Extended Calculus of Constructions.* PhD thesis, University of Edinburgh, 1990.
11. Robin Milner, Mads Tofte, Robert Harper, and David MacQueen. *The Definition of Standard ML (Revised).* MIT Press, 1997.
12. M. Takahashi. Parallel reductions in λ-calculus. Technical report, Department of Information Science, Tokyo Institute of Technology, 1993. Internal report.
13. L.S. van Benthem Jutting, J. McKinna, and R. Pollack. Checking algorithms for pure type systems. In *Types for Proofs and Programs: International Workshop TYPES'93*, volume 806 of *Lecture Notes in Computer Science*, May 1993.

Formalised Cut Admissibility for Display Logic

Jeremy E. Dawson* and Rajeev Goré**

Department of Computer Science and Automated Reasoning Group
Australian National University, Canberra, ACT 0200, Australia
{jeremy,rpg}@discus.anu.edu.au

Abstract. We use a deep embedding of the display calculus for relation algebras $\boldsymbol{\delta}$**RA** in the logical framework Isabelle/HOL to formalise a machine-checked proof of cut-admissibility for $\boldsymbol{\delta}$**RA**. Unlike other "implementations", we explicitly formalise the structural induction in Isabelle/HOL and believe this to be the first full formalisation of cut-admissibility in the presence of explicit structural rules.

1 Introduction

Display Logic [1] is a generalised sequent framework for non-classical logics. Since it is not really a logic, we prefer the term display calculi and use it from now on. Display calculi extend Gentzen's language of sequents with extra, complex, n-ary structural connectives, in addition to Gentzen's sole structural connective, the "comma". Whereas Gentzen's comma is usually assumed to be associative, commutative and inherently poly-valent, no such implicit assumptions are made about the n-ary structural connectives in display calculi. Properties such as associativity are explicitly stated as structural rules.

Such explicit structural rules make display calculi as modular as Hilbert-style calculi: the logical rules remain constant and different logics are obtained by the addition or deletion of structural rules only. Display calculi therefore provide an extremely elegant sequent framework for "logic engineering", applicable to many (classical and non-classical) logics in a uniform way [11,5]. The display calculus $\boldsymbol{\delta}$**RA** [4], for example, captures the logic for relation algebras. The most remarkable property of display calculi is a generic cut-elimination theorem, which applies whenever the rules for the display calculus satisfy certain, easily checked, conditions. Belnap [1] proves that the cut rule is *admissible* in **all** such display calculi: he transforms a derivation whose only instance of cut is at the bottom, into a cut-free derivation of the same end-sequent. His proof does not use the standard double induction over the cut rank and degree a là Gentzen.

In [2] we implemented a "shallow" embedding of $\boldsymbol{\delta}$**RA** which enabled us to mimic derivations in $\boldsymbol{\delta}$**RA** using Isabelle/Pure. But it was impossible to reason *about* derivations since they existed only as the trace of the particular Isabelle session. In [9], Pfenning has given a formalisation of cut-admissibility for traditional sequent calculi for various nonclassical logics using the logical framework

* Supported by an Australian Research Council Large Grant
** Supported by an Australian Research Council QEII Fellowship

V.A. Carreño, C. Muñoz, S. Tahar (Eds.): TPHOLs 2002, LNCS 2410, pp. 131–147, 2002.

Elf, which is based upon dependent type theory. Although dependent types allow derivations to be captured as terms, they do not enable us to formalise all aspects of a meta-theoretic proof. As Pfenning admits, the Elf formalisation cannot be used for checking the correct use of the induction principles used in the cut-admissibility proof, since this requires a "deep" embedding [9]. The use of such "deep" embeddings to formalise meta-logical results is rare [8,7]. To our knowledge, the only full formalisation of a proof of cut-admissibility is that of Schürmann [10], but the calculi used by both Pfenning and Schürmann contain no explicit structural rules, and structural rules like contraction are usually the bane of cut-elimination. Here, we use a deep embedding of the display calculus **δRA** into Isabelle/HOL to fully formalise the admissibility of the cut rule in the presence of explicit (and arbitrary) structural rules.

The paper is set out as follows. In Section 2 we briefly describe the display calculus **δRA** for the logic of relation algebras. In Section 3 we describe an encoding of **δRA** logical constants, logical connectives, formulae, structures, sequents, rules, derivations and derived rules into Isabelle/HOL. In Section 4 we describe the two main transformations required to eliminate cut. In Section 5 we describe how we mechanised these to prove the cut-elimination theorem in Isabelle/HOL. In Section 6 we present conclusions and discuss further work.

2 The Display Calculus δRA

The following grammar defines the syntax of relation algebras:

$$A ::= p_i \mid \top \mid \bot \mid \neg A \mid A \wedge A \mid A \vee A \mid \mathbf{1} \mid \mathbf{0} \mid \sim A \mid \smile A \mid A \circ A \mid A + A$$

A display calculus for relation algebras called **δRA** can be found in [4]. Sequents of **δRA** are expressions of the form $X \vdash Y$ where X and Y are built from the nullary structural constants E and I and formulae, using a binary comma, a binary semicolon, a unary $*$ or a unary $\bullet$ as structural connectives, according to the grammar below:

$$X ::= A \mid I \mid E \mid *X \mid \bullet X \mid X \; ; \; X \mid X \; , \; X$$

Thus, whereas Gentzen's sequents $\Gamma \vdash \Delta$ assume that Γ and Δ are comma-separated lists of formulae, **δRA**-sequents $X \vdash Y$ assume that X and Y are complex tree-like structures built from formulae and the constants I and E using comma, semicolon, $*$ and $\bullet$.

The defining feature of display calculi is that in all logical rules, the principal formula is always "displayed" as the whole of the right-hand or left-hand side. For example, the rule (**LK-** $\vdash \vee$) below is typical of Gentzen's sequent calculi like **LK**, while the rule (**δRA-** $\vdash \vee$) below is typical of display calculi:

$$\frac{\Gamma \vdash \Delta, P \, , \, Q}{\Gamma \vdash \Delta, P \vee Q}(\mathbf{LK}\text{-} \vdash \vee) \qquad \frac{X \vdash P \, , \, Q}{X \vdash P \vee Q}(\boldsymbol{\delta}\mathbf{RA}\text{-} \vdash \vee)$$

3 A Deep Embedding of δRA in Isabelle/HOL

In [2], we describe our initial attempts to formalise display calculi in various logical frameworks, and describe why we chose Isabelle/HOL for this work. To make the current paper self-contained, we now describe the Isabelle/HOL data structures used to represent formulae, structures, sequents and derivations. We assume that the reader is familiar with ML and logical frameworks in general.

3.1 Representing Formulae, Structures, Sequents, and Rules

An actual derivation in a Display Calculus involves structures containing formulae which are composed of primitive propositions (which we typically represent by p, q, r). It uses rules which are *expressed* using structure and formula variables, typically X, Y, Z and A, B, C respectively, to represent structures and formulae made up from primitive propositions. Nonetheless, in deriving theorems or derived rules we will often use a rule instance where the original variables in the rule are replaced by other variables, rather than actual formulae. We may, for example, have to take the cut rule as shown below left and substitute $B \wedge C$ for A, substitute (Z, D) for X and substitute $C \vee D$ for Y to get the cut rule instance shown below right, and reason about this instance.

$$\text{(cut)}\ \frac{X \vdash A \quad A \vdash Y}{X \vdash Y} \qquad\qquad \frac{Z, D \vdash B \wedge C \quad B \wedge C \vdash C \vee D}{Z, D \vdash C \vee D}$$

Our Isabelle formulation must allow this since variables such as X, Y, Z and A, B, C are not part of the language of a Display Calculus, but are part of the meta-language used when reasoning about Display Calculi.

Formulae of **δRA** are therefore represented by the datatype below:

```
datatype formula = Btimes formula formula ("_ && _" [68,68] 67)
  | Rtimes formula formula ("_ oo _" [68,68] 67)
  | Bplus formula formula ("_ v _" [64,64] 63)
  | Rplus formula formula ("_ ++ _" [64,64] 63)
  | Bneg formula ("--_" [70] 70) | Rneg formula ("_^" [75] 75)
  | Btrue ("T") | Bfalse("F") | Rtrue ("r1") | Rfalse("r0")
  | FV string | PP string
```

The constructors FV represents formula variables which appear in the statement of a rule or theorem, and which are instantiated to actual formulae of **δRA** when constructing derivations. The constructor PP represents a primitive proposition variable p: once again this lives at the meta-level.

Structures of **δRA** are represented by the datatype below:

```
datatype structr = Comma structr structr | SemiC structr structr
  | Star structr | Blob structr |I|E| Structform formula | SV string
```

The operator `Structform` "casts " a formula into a structure, since a formula is a special case of a structure (as in the premises of the cut rule given above).

The constructor SV represents structure variables which appear in the statement of a rule or theorem, and which are instantiated to actual structures of **δRA** when constructing derivations. Since we must reason about arbitrary derivations, we have to allow derivations to contain structure variables and we must reason about the instantiations explicitly. We therefore cannot use Isabelle's built-in unification facility for instantiating its "scheme variables" as explained in more detail in [2]. Likewise, formulae of **δRA** are represented by a datatype which include a constructor FV for formula variables which can be instantiated to actual formulae of **δRA**, and a constructor PP for a primitive proposition variable p.

The notation in parentheses in the definition of datatype `formula` describe an alternative infix syntax, closer to the actual syntax of **δRA**. Some complex manipulation of the syntax, available through Isabelle's "parse translations" and "print translations", allows structure variables and constants to be prefixed by the symbol $, and the notations FV, SV and Structform to be omitted. For technical reasons related to this, a different method is used to specify the alternative infix syntax for structures and sequents: details omitted.

Sequents and rules of **δRA** are represented by the Isabelle/HOL datatypes:

```
datatype sequent = Sequent structr structr
datatype rule = Rule (sequent list) sequent
              | Bidi sequent sequent | InvBidi sequent sequent
```

The premises of a rule are represented using a list of sequents while the conclusion is a single sequent. Thus `Rule prems concl` means a rule with premises `prems` and conclusion `concl`. Many single-premise rules of display calculi are defined to be usable from top to bottom as well as from bottom to top: the two constants `Bidi` and `InvBidi` allow us to cater for these. Thus `Bidi prem concl` means an invertible, or "bi-directional" rule (such as the display postulates) and `InvBidi prem concl` means the rule `Bidi prem concl` used in the inverted sense to derive the conclusion `prem` from the premise `concl`.

A sequent `(Sequent X Y)` can also be represented as `$X |- $Y`. Thus the term `Sequent (SV ''X'') (Structform (FV ''A''))` is printed, and may be entered, as `($''X'' |- ''A'')`. Functions `premsRule` and `conclRule` return the premise list and the conclusion of a rule respectively. A structure expression is *formula-free* if it does not contain any formula as a sub-structure: that is, if it does not contain any occurrence of the operator `Structform`. A formula-free sequent is defined similarly. The constant `rls` represents the set of rules of **δRA**, encoded using the datatypes just described: we omit the details of this code.

3.2 Handling Substitutions Explicitly

Since a "deep" embedding requires handling substitution explicitly, we now give definitions relating to substitution for structure and formula variables. We first give some type abbreviations, and then the types of a sample of the functions.

```
fSubst = "(string * formula) list"
sSubst = "(string * structr) list"
```

```
fsSubst = "fSubst * sSubst"
sFind       :: "sSubst => string => structr"
ruleSubst :: "fsSubst => rule => rule"
seqSubst  :: "fsSubst => sequent => sequent"
```

To substitute for a variable, for example `SV ''X''`, in some object, using the substitution `(fsubs, ssubs)`, we use `sFind` to obtain the first pair (if any) in `ssubs` whose first component is `''X''`. If that pair is `(''X'',` X`)`, then `sFind` returns X, and each occurrence of `SV ''X''` in the given object is replaced by X. There are functions which substitute for every formula or structure variable in a derivation tree (defined below), rule, sequent, structure or formula.

3.3 Representing Derivations as Trees

We use the term "derivation" for a proof *within* the sequent calculus, reserving the term "proof" for a meta-theoretic proof of a theorem *about* the sequent calculus. We model a derivation tree (type `dertree`) using the following datatype:

```
datatype dertree = Der sequent rule (dertree list) | Unf sequent
```

In `Der seq rule dts` the subterm `seq` is the sequent at the root (bottom) of the tree, and `rule` is the rule used in the last (bottom) inference. If the tree represents a real derivation, sequent `seq` will be an instance of the conclusion of `rule`, and the corresponding instances of the premises of `rule` will be the roots of the trees in the list `dts`. We say that the root "node" of such a tree is *well-formed.* The trees in `dts` are the *immediate* subtrees of `Der seq rule dts`.

The leaves of a derivation tree are either axioms with no premises, or "`Unf`inished" sequents whose derivations are currently unfinished. The derivation tree for a derivable sequent will therefore have no `Unf` leaves and we call such a derivation tree *finished.* The derivation tree for a derived rule will have the premises of the rule as its `Unf` leaf sequents.

Display calculi typically use the initial sequent $p \vdash p$, using primitive propositions only. It is then proved that the sequent $A \vdash A$ is derivable for all formulae A by induction on the size of A, where A *stands for* a formula composed of primitive propositions and logical connectives. We proved this as the theorem `idfpp`. However we also need to reason about the derivation trees of derived rules; such trees may contain formula and structure variables as well as primitive propositions, and may use the (derived) rule $A \vdash A$, for arbitrary formula A. We therefore sometimes must treat $A \vdash A$ (where A is a formula variable) as an axiom. Thus the derivation tree `Der (''A'' |- ''A'') idf []` stands for a finished derivation, which uses the `idfpp` lemma that $A \vdash A$ is derivable for all A, whereas the derivation tree `Unf (''A'' |- ''A'')` stands for an unfinished derivation with unfinished premise $A \vdash A$.

For example, the unfinished derivation tree shown below at left is represented as the Isabelle/HOL term shown below at right where `''A'' |- PP p && ''A''`

```
allDT         :: "(dertree => bool) => dertree => bool"
allNextDTs    :: "(dertree => bool) => dertree => bool"
wfb           :: "dertree => bool"
frb           :: "rule set => dertree => bool"
premsDT       :: "dertree => sequent list"
conclDT       :: "dertree => sequent"
IsDerivable   :: "rule set => rule => bool"
IsDerivableR  :: "rule set => sequent set => sequent => bool"
```

Fig. 1. Functions for reasoning about derivations

stands for $A \vdash p \wedge A$ and `cA` and `ands` are the contraction and $(\vdash \wedge)$ rules, and `idf` is the derived rule $A \vdash A$:

$$\dfrac{\dfrac{A \vdash p \quad A \vdash A}{A, A \vdash p \wedge A}(\vdash \wedge)}{A \vdash p \wedge A}(ctr)$$

```
Der (''A'' |- PP p && ''A'') cA
  [Der (''A'', ''A'' |- PP p && ''A'') ands
  [Unf (''A'' |- PP p),
   Der (''A'' |- ''A'') idf []]]
```

3.4 Reasoning about Derivations and Derivability

In this section we describe various functions which allow us to reason about derivations in $\boldsymbol{\delta}$**RA**. The types for these functions are shown in Figure 1.

`allDT f dt` holds if property `f` holds for every sub-tree in the tree `dt`.

`allNextDTs f dt` holds if property `f` holds for every proper sub-tree of `dt`.

`wfb (Der concl rule dts)` holds if sequent rule `rule` has an instantiation with conclusion instance `concl` and premise instances which are the conclusions of the derivation trees in the list `dts`. ("`wfb`" stands for *well-formed*).

`allDT wfb dt` holds if every sub-tree of the derivation tree `dt` satisfies `wfb` (ie, if every node in `dt` is well-formed). Such a derivation is said to be *well-formed.*

`frb rules (Der concl rule dts)` holds when the lowest rule `rule` used in a derivation tree `Der concl rule dts` belongs to the set `rules`.

`allDT (frb rules) dt` holds when every rule used in a derivation tree `dt` belongs to the set `rules`.

`premsDT dt` returns a list of all "premises" (unfinished leaves) of the derivation tree `dt`. That is, the sequents found in nodes of `dt` of the form `Unf seq`.

`conclDT dt` returns the end-sequent of the derivation tree `dt`. That is, the conclusion of the bottom-most rule instance.

So `wfb (Der seq rule dts)` means that the bottom node of the derivation tree `Der seq rule dts` is *well-formed.* We say a derivation tree `dt` is *well-formed* if every node in it is well-formed, and express this as `allDT wfb dt`, since `allDT f dt` means that property `f` holds for every sub-tree in the derivation tree `dt`. Also, `allNextDTs f dt` means that every proper sub-tree of `dt` satisfies `f`.

The property `allDT (frb rules)` holds when every rule used in a derivation tree belongs to the set `rules`. The function `premsDT` returns a list of all "premises" (unproved assumptions) of the derivation tree, that is, the sequents found in nodes of the form `Unf seq`.

A tree representing a real derivation in a display calculus naturally is well-formed and uses the rules of the calculus. Further, a tree which derives a sequent (rather than a derived rule) is finished, that is, it has no unfinished leaves.

The cut-elimination procedure involves transformations of derivation trees; in discussing these we will only be interested in derivation trees which actually derive a sequent, so we make the following definition.

Definition 1. *A derivation tree* `dt` *is* valid *if it is well-formed, it uses rules in the set of rules* `rules` *of the calculus, and it has no unfinished leaves.*

```
valid_def = "valid ?rules ?dt ==
   allDT wfb ?dt & allDT (frb ?rules) ?dt & premsDT ?dt = []"
```

We have explicitly added question marks in front of `rules` and `dt` to flag that they are free variables, even though the question mark would be absent in the Isabelle/HOL theory file itself: we follow this practice throughout this paper.

Definition 2 (`IsDerivableR`). *`IsDerivableR rules prems' concl` holds iff there exists a derivation tree* `dt` *which uses only rules contained in the set* `rules`*, is well-formed, has conclusion* `concl`*, and has premises from set* `prems'`*.*

```
"IsDerivableR ?rules ?prems' ?concl == (EX dt.
      allDT (frb ?rules) dt & allDT wfb dt &
      conclDT dt = ?concl & set (premsDT dt) <= ?prems')"
```

Here, `set` is a function that allows us to treat its argument as a set rather than a list, and `<=` is the subset relation $\subseteq$.

Finally, `IsDerivable rules rule` holds iff `rule` may be obtained as a derived rule, from the (unordered) set `rules`. That is, if `rule` has premise list `prems` and conclusion `concl`, then `IsDerivable rules rule` is equivalent to `IsDerivableR rules (set prems) concl`.

3.5 Reasoning about Derivability

Among the results we have proved about the derivability relation are the following theorems. The first is a transitivity result, relating to a derivation of a conclusion from premises which are themselves derived.

Theorem 1. *If* `concl` *is derivable from* `prems'` *and each sequent* `p` *in* `prems'` *is derivable from* `prems` *then* `concl` *is derivable from* `prems`*.*

```
IsDerivableR_trans = "[| IsDerivableR ?rules ?prems' ?concl ;
     ALL p:?prems'. IsDerivableR ?rules ?prems p |] ==>
     IsDerivableR ?rules ?prems ?concl" : thm
```

$$\dfrac{\dfrac{\dfrac{\Pi_{ZAB}}{Z \vdash A, B}}{Z \vdash A \vee B}(\vdash \vee) \qquad \dfrac{\dfrac{\Pi_{AX}}{A \vdash X} \quad \dfrac{\Pi_{BY}}{B \vdash Y}}{A \vee B \vdash X, Y}(\vee \vdash)}{Z \vdash X, Y}(cut)$$

Fig. 2. Principal cut on formula $A \vee B$

The appellation ": `thm`" indicates a statement that has been proved in Isabelle/HOL as a theorem, from previous Isabelle/HOL definitions: we follow this practice for theorems and lemmata throughout this paper.

The second is a a different sort of transitivity result, relating to a derivation using rules which are themselves derived.

Theorem 2 (`IsDerivableR_deriv`). *If each* `rule` *in* `rules'` *is derivable using* `rules`*, and* `concl` *is derivable from* `prems` *using the set* `rules'`*, then* `concl` *is derivable from* `prems` *using* `rules`*.*

```
IsDerivableR_deriv = "[| ALL rule:?rules'.
 IsDerivable ?rules rule ; IsDerivableR ?rules' ?prems ?concl |]
            ==>  IsDerivableR ?rules ?prems ?concl" : thm
```

In another reported formalisation of the notion of derivations in a logical calculus [8], these two properties were, in effect, stated rather than proved. The disadvantage of proceeding that way is the possibility of stating them incorrectly. For example, [8] defines `IsDerivable` inductively as a relation which is transitive in both the senses of the results above; see the second and third clauses of the definition on [8, page 302]. However in the third clause, which deals with the case of a result being provable using derived rules, inappropriate use of an existential quantifier leads to the incorrect result that $P \to Q$ could be used as a derived rule on the grounds that one instance of it, say *True* $\to$ *True*, is provable.

4 An Operational View of Cut-Elimination

We now give an operational view of cut-elimination, to explain the steps involved in the overall cut-elimination procedure a là Belnap [1]. We assume familiarity with notions like "parametric ancestors" of a cut formula [1].

4.1 Principal Cuts and Belnap's Condition C8

Definition 3. *An application of (cut) is* left-principal *[*right-principal*] if the cut-formula is the principal formula of the left [right] premise of the cut rule.*

Given a derivation (tree) with one principal cut, such as in Figure 2, Belnap's condition (C8) on the rules of a Display Calculus ensures that the given

$$\dfrac{\dfrac{\dfrac{\dfrac{\dfrac{\dfrac{\dfrac{\Pi_{ZAB}}{Z \vdash A, B}}{*A, Z \vdash B}\,(cs1) \qquad \dfrac{\Pi_{BY}}{B \vdash Y}}{*A, Z \vdash Y}\,(cut)}{Z \vdash A, Y}\,(\overline{cs1})}{Z, *Y \vdash A}\,(cs2) \qquad \dfrac{\Pi_{AX}}{A \vdash X}}{Z, *Y \vdash X}\,(cut)}{Z \vdash X, Y}\,(\overline{cs2})$$

Fig. 3. Transformed principal cut on formula $A \vee B$

$$\dfrac{A \vdash A \qquad \dfrac{\Pi_{AY}}{A \vdash Y}\,(\text{intro-}A)}{A \vdash Y}\,(cut) \qquad \text{becomes} \qquad \dfrac{\Pi_{AY}}{A \vdash Y}$$

Fig. 4. Principal cut where cut-formula is introduced by identity axiom

derivation can be transformed into one whose cuts are on smaller formulae. For example, the principal cut on $A \vee B$ shown in Figure 2 can be replaced by the derivation shown in Figure 3, where $(cs1)$, $(cs2)$, $(\overline{cs1})$ and $(\overline{cs1})$ are two of the display postulates and their inverses respectively. The replacement derivation contains cuts only on A and B, which are smaller formulae than $A \vee B$.

There is one such transformation for every connective and this is the basis for a step of the cut-elimination proof which depends on induction on the structure or size of the cut-formula. The base case of this induction is where the cut-formula is introduced by the identity axiom. Such a cut, and its removal, are shown in Figure 4. We return to the actual mechanisation in Section 5.1.

The transformation of a principal cut on A into one or more cuts on strict subformulae of A is known as a "principal move". We now need a way to turn arbitrary cuts into principal ones.

4.2 Transforming Arbitrary Cuts into Principal Ones

In the case of a cut that is not left-principal, say we have a tree like the one on the left in Figure 5. Then we transform the subtree rooted at $X \vdash A$ by simply changing its root sequent to $X \vdash Y$, and proceeding upwards, changing all ancestor occurrences of A to Y. In doing this we run into difficulty at each point where A is introduced: at such points we insert an instance of the cut rule. The diagram on the right hand side of Figure 5 shows this in the case where A is introduced at just one point.

In Figure 5, the notation $\Pi_L[A]$ and $Z[A]$ means that the sub-derivation Π_L and structure Z may contain occurrences of A which are parametric ancestors

$$\dfrac{\dfrac{\dfrac{\dfrac{\Pi[A]}{Z[A] \vdash A}\,(\text{intro-}A)}{\Pi_L[A]}\,(\pi)}{X \vdash A}\,(\rho) \qquad \dfrac{\Pi_R}{A \vdash Y}}{X \vdash Y}\,(cut) \qquad\qquad \dfrac{\dfrac{\dfrac{\dfrac{\Pi'[Y]}{Z[Y] \vdash A}\,(\text{intro-}A) \qquad \dfrac{\Pi_R}{A \vdash Y}}{Z[Y] \vdash Y}\,(cut)}{\Pi_L[Y]}\,(\pi)}{X \vdash Y}\,(\rho)$$

Fig. 5. Making a cut left-principal

of the cut-formula A: thus (intro-A) is the lowest rule where A is the principal formula on the right of $\vdash$. The notation $\Pi_L[Y]$ and $Z[Y]$ means that all such "appropriate" instances of A are changed to Y: that is, instances of A which can be traced to the instance displayed on the right in $X \vdash A$. The rules contained in the new sub-derivation $\Pi_L[Y]$ are the same as the rules used in Π_L; thus it remains to be proved that $\Pi_L[Y]$ is well-formed. The resulting cut in the diagram on the right of Figure 5 is left-principal. Notice that the original sub-derivation Π may be transformed into a *different* sub-derivation Π' during this process since the parametric ancestors of A occurring in $\Pi[A]$ will in turn need to be "cut away" below where they are introduced, and replaced by Y.

Belnap's conditions guarantee that where A is introduced by an introduction rule, it is necessarily displayed in the succedent position, as above the top of Π_L in the left branch of the left hand derivation in Figure 5. Other conditions of Belnap (*e.g.* a formula is displayed where it is introduced, and each structure variable appears only once in the conclusion of a rule) ensure that a procedure can be formally defined to accord with the informal description above: the procedure removes a cut on A which is not left-principal and creates (none, one or more) cut(s) on A which are left-principal.

This construction generalises easily to the case where A is introduced (in one of the above ways) at more than one point (*e.g.* arising from use of one of the rules where a structure variable, whose instantiation contains occurrences of A, appears twice in the premises) or where A is "introduced" by use of the weakening rule. Our description of the procedure is very loose and informal – the formality and completeness of detail is reserved for the machine proof!

Subsequently, the "mirror-image" procedure is followed, to convert a left-principal cut into one or more (left- and right-)principal cuts.

The process of making a cut left-principal, or of making a left-principal cut (left and right) principal is called a "parametric move".

5 Functions for Reasoning about Cuts

We therefore need functions, with the following types, for reasoning about derivations which end with a cut:

```
cutOnFmls     :: "formula set => dertree => bool"
cutIsLP       :: "formula => dertree => bool"
cutIsLRP      :: "formula => dertree => bool"
```

Each require the bottom node of the derivation tree to be of the form `Der seq rule dts`, and that if `rule` is (*cut*), then: for `cutOnFmls s` the cut is on a formula in the set `s`; for `cutIsLP A` the cut is on formula `A` and is left-principal; and for `cutIsLRP A` the cut is on formula `A` and is (left- and right-)principal.

Note that it also follows from the actual definitions that a derivation tree satisfying any of `allDT (cutOnFmls s)`, `allDT (cutIsLP A)` and `allDT (cutIsLRP A)` has no unfinished leaves: we omit details.

5.1 Dealing with Principal Cuts

For each logical connective and constant in the calculus, we prove that a derivation ending in a (left and right) principal cut, where the main connective of the cut formula is that connective, can be transformed into another derivation of the same end-sequent, using only cuts (if any) on formulae which are strict subformulae of the original cut-formula. Some SML code is used to do part of the work of finding these replacement derivation trees. But the proof that such a replacement derivation tree is well-formed, for example, has to be done using the theorem prover. Here is the resulting theorem for $\vee$: there is an analogous theorem for every logical connective and logical constant.

Theorem 3 (`orC8`)**.** *Assume we are given a valid derivation tree* `dt` *whose only instance of cut (if any) is at the bottom, and that this cut is principal with cut-formula $A \vee B$. Then there is a valid derivation tree* `dtn` *with the same conclusion as* `dt`*, such that* each *cut (if any) in* `dtn` *has A or B as cut-formula.*

```
orC8 = "[| allDT wfb ?dt; allDT (frb rls) ?dt;
    cutIsLRP (?A v ?B) ?dt; allNextDTs (cutOnFmls {}) ?dt |]
  ==> EX dtn. conclDT dtn = conclDT ?dt & allDT wfb dtn &
      allDT (frb rls) dtn & allDT (cutOnFmls {?B, ?A}) dtn" : thm
```

5.2 Making a Cut (Left) Principal

For boolean `b`, structures `X`, `Y` and sequents `seq1` and `seq2`, the expression `seqRep b X Y seq1 seq2` is true iff `seq1` and `seq2` are the same, except that (possibly) one or more occurrences of `X` in `seq1` are replaced by corresponding occurrences of `Y` in `seq2`, where, when `b` is `True` [`False`], such differences occur only in succedent [antecedent] positions. For two lists `seql1` and `seql2` of sequents, `seqReps b X Y seql1 seql2` holds if each nth member of `seql1` is related to the nth member of `seql2` by `seqRep b X Y`.

Next come the main theorems used in the mechanised proof based on making cuts (left and right) principal. Several use the relation `seqRep pn (Structform A) Y`, since `seqRep pn (Structform A) Y seqa seqy` holds when `seqa` and `seqy` are corresponding sequents in the trees $\Pi_L[A]$ and $\Pi_L[Y]$ from Figure 5.

Theorem 4 (seqExSub1). *If sequent* pat *is formula-free and does not contain any structure variable more than once, and can be instantiated to obtain sequent* seqa, *and* seqRep pn (Structform A) Y seqa seqy *holds, then* pat *can be instantiated to obtain sequent* seqy.

```
seqExSub1 = "[| ~ seqCtnsFml ?pat; noDups (seqSVs ?pat);
    seqSubst (?fs, ?suba) ?pat = ?seqa;
    seqRep ?pn (Structform ?A) ?Y ?seqa ?seqy |]
  ==> EX suby. seqSubst (?fs, suby) ?pat = ?seqy" : thm
```

To see why pat must be formula-free, suppose that pat contains Structform (FV ''B''), which means that pat is not formula-free. Then, this part of pat can be instantiated to Structform A, but not to an arbitrary structure Y as desired. The condition that a structure variable may not appear more than once in the conclusion of a rule is one of Belnap's conditions [1].

The stronger result seqExSub2 is similar to seqExSub1, except that the antecedent [succedent] of the sequent pat may contain a formula, provided that the whole of the antecedent [succedent] is that formula.

The result seqExSub2 is used in proceeding up the derivation tree $\Pi_L[A]$, changing A to Y: if pat is the conclusion of a rule, which, instantiated with (fs, suba), is used in $\Pi_L[A]$, then that rule, instantiated with (fs, suby), is used in $\Pi_L[Y]$. This is expressed in the theorem extSub2, which is one step in the transformation of $\Pi_L[A]$ to $\Pi_L[Y]$.

To explain theorem extSub2 we define bprops rule to hold if the rule rule satisfies the following three properties, which are related (but do not exactly correspond) to Belnap's conditions (C3), (C4) and (C5):

- the conclusion of rule has no repeated structure variables
- if a structure variable in the conclusion of rule is also in a premise, then it has the same "cedency" (ie antecedent or succedent) there
- if the conclusion of rule has formulae, they are displayed (as the whole of one side)

Theorem 5 (extSub2). *Suppose we are given a rule* rule *and an instantiation* ruleA *of it, and given a sequent* conclY, *such that (i)* seqRep pn (Structform A) Y (conclRule ruleA) conclY *holds; (ii)* bprops rule *holds; (iii) if the conclusion of* rule *has a displayed formula on one side then* conclrule ruleA *and* conclY *are the same on that side. Then there exists* ruleY, *an instantiation of* rule, *whose conclusion is* conclY *and whose premises* premsY *are, respectively, related to* premsRule ruleA *by* seqRep pn (Structform A) Y.

```
extSub2 = "[| conclRule rule = Sequent pant psuc ;
 conclRule ruleA = Sequent aant asuc ; conclY = Sequent yant ysuc ;
 (strIsFml pant & aant = Structform A --> aant = yant) ;
 (strIsFml psuc & asuc = Structform A --> asuc = ysuc) ;
 ruleMatches ruleA rule ; bprops rule ;
 seqRep pn (Structform A) Y (conclRule ruleA) conclY |]
```

```
    ==> (EX subY. conclRule (ruleSubst subY rule) = conclY &
        seqReps pn (Structform A) Y (premsRule ruleA)
        (premsRule (ruleSubst subY rule)))" : thm
```

This theorem is used to show that, when a node of the tree $\Pi_L[A]$ is transformed to the corresponding node of $\Pi_L[Y]$, then the next node(s) above can be so transformed. But this does not hold at the node whose conclusion is $X' \vdash A$ (see Fig.5); condition (iii) above reflects this limitation.

5.3 Turning One Cut into Several Left-Principal Cuts

To turn one cut into several left-principal cuts we use the procedure described above. This uses `extSub2` to transform $\Pi_L[A]$ to $\Pi_L[Y]$ up to each point where A is introduced, and then inserting an instance of the (cut) rule. It is to be understood that the derivation trees have no (unfinished) premises.

Theorem 6 (`makeCutLP`). *Given cut-free derivation tree* `dtAY` *deriving* `(A |- Y)` *and* `dtA` *deriving* `seqA`*, and given* `seqY`*, where* `seqRep True (Structform A) Y seqA seqY` *holds (ie,* `seqY` *and* `seqA` *are the same except (possibly) that A in a succedent position in* `seqA` *is replaced by Y in* `seqY`*), there is a derivation tree deriving* `seqY` *whose cuts are all left-principal on A.*

```
makeCutLP = "[| allDT (cutOnFmls {}) ?dtAY; allDT (frb rls) ?dtAY;
 allDT wfb ?dtAY; conclDT ?dtAY = (?A |- $?Y);
 allDT (frb rls) ?dtA; allDT wfb ?dtA; allDT (cutOnFmls {}) ?dtA;
 seqRep True (Structform ?A) ?Y (conclDT ?dtA) ?seqY |]
      ==> EX dtY. conclDT dtY = ?seqY & allDT (cutIsLP ?A) dtY &
           allDT (frb rls) dtY & allDT wfb dtY" : thm
```

Function `makeCutRP` is basically the symmetric variant of `makeCutLP` ; so `dtAY` is a cut-free derivation tree deriving $(Y \vdash A)$. But with the extra hypothesis that A is introduced at the bottom of `dtAY`, the result is that there is a derivation tree deriving `seqY` whose cuts are all (left- and right-) principal on A.

These were the most difficult to prove in this cut-elimination proof. The proofs proceed by structural induction on the initial derivation tree, where the inductive step involves an application of `extSub2`, except where the formula A is introduced. If A is introduced by an introduction rule, then the inductive step involves inserting an instance of (cut) into the tree, and then applying the inductive hypothesis. If A is introduced by the axiom (*id*), then (and this is the base case of the induction) the tree `dtAY` is substituted for $A \vdash A$.

Next we have the theorem expressing the transformation of the whole derivation tree, as shown in the diagrams.

Theorem 7 (`allLP`). *Given a valid derivation tree* `dt` *containing just one cut, which is on formula A and is at the root of* `dt`*, there is a valid tree with the same conclusion (root) sequent, all of whose cuts are left-principal and are on A.*

```
allLP = "[| cutOnFmls {?A} ?dt; allDT wfb ?dt; allDT (frb rls) ?dt;
            allNextDTs (cutOnFmls {}) ?dt |]
    ==> EX dtn. conclDT dtn = conclDT ?dt & allDT (cutIsLP ?A) dtn &
          allDT (frb rls) dtn & allDT wfb dtn" : thm
```

allLRP is a similar theorem where we start with a single left-principal cut, and produce a tree whose cuts are all (left- and right-) principal.

5.4 Putting It All Together

A monolithic proof of the cut-admissibility theorem would be very complex, involving either several nested inductions or a complex measure function. For the transformations above replace one arbitrary cut by many left-principal cuts, one left-principal cut by many principal cuts and one principal cut by one or two cuts on subformulae, whereas we need, ultimately, to replace many arbitrary cuts in a given derivation tree. We can manage this complexity by considering how we would write a program to perform the elimination of cuts from a derivation tree. One way would be to use a number of mutually recursive routines, as follows:

`elim` eliminates a single arbitrary cut, by turning it into several left-principal cuts and using `elimAllLP` to eliminate them ...
`elimAllLP` eliminates several left-principal cuts, by repeatedly using `elimLP` to eliminate the top-most remaining one ...
`elimLP` eliminates a left-principal cut, by turning it into several principal cuts and using `elimAllLRP` to eliminate them ...
`elimAllLRP` eliminates several principal cuts, by repeatedly using `elimLRP` to eliminate the top-most remaining one ...
`elimLRP` eliminates a principal cut, by turning it into several cuts on smaller cut-formulae, and using `elimAll` to eliminate them ...
`elimAll` eliminates several arbitrary cuts, by repeatedly using `elim` to eliminate the top-most remaining one ...

Such a program would terminate because any call to `elim` would (indirectly) call `elim` only on smaller cut-formulae.

We turn this program outline into a proof. Each routine listed above, of the form "routine P does ... and uses routine Q" will correspond to a theorem which will say essentially "if routine Q completes successfully then routine P completes successfully" (assuming they are called with appropriately related arguments).

We define two predicates, `canElim` and `canElimAll`, whose types and meanings (assuming valid trees) are given below. When we use them, the argument `f` will be one of the functions `cutOnFmls`, `cutIsLP` and `cutIsLRP`.

Definition 4. *`canElim` f holds for property f if, for any valid derivation tree `dt` satisfying f and containing at most one cut at the bottom, there is a valid cut-free derivation tree which is* equivalent *to (has the same conclusion as) `dt`.*

`canElimAll` f means that if every subtree of a given valid tree `dt` satisfies f, then there is a valid cut-free derivation tree `dt'` equivalent to `dt` such that if the bottom rule of `dt` is not (cut), then the same rule is at the bottom of `dt'`.

```
canElim      :: "(dertree => bool) => bool"
canElimAll   :: "(dertree => bool) => bool"

"canElim ?f == (ALL dt. ?f dt & allNextDTs (cutOnFmls {}) dt &
    allDT wfb dt & allDT (frb rls) dt -->
  (EX dtn. allDT (cutOnFmls {}) dtn & allDT wfb dtn &
    allDT (frb rls) dtn & conclDT dtn = conclDT dt))"

"canElimAll ?f ==
  (ALL dt. allDT ?f dt & allDT wfb dt & allDT (frb rls) dt -->
  (EX dt'. (botRule dt ~= cutr --> botRule dt' = botRule dt) &
    conclDT dt' = conclDT dt & allDT (cutOnFmls {}) dt' &
    allDT wfb dt' & allDT (frb rls) dt'))"
```

We restate allLP and allLRP using canElim and canElimAll.

Theorem 8 (allLP', allLRP')**.**

(a) If we can eliminate any number of left-principal cuts on A from a valid tree dt*, then we can eliminate a single arbitrary cut on A from the bottom of* dt*.*
(b) If we can eliminate any number of principal cuts on A from a valid tree dt*, then we can eliminate a single left-principal cut on A from the bottom of* dt*.*

```
allLP' = "canElimAll (cutIsLP ?A) ==> canElim (cutOnFmls {?A})":thm
allLRP' = "canElimAll (cutIsLRP ?A) ==> canElim (cutIsLP ?A)":thm
```

Now if we can eliminate one arbitrary cut (or left-principal cut, or principal cut) then we can eliminate any number, by eliminating them one at a time starting from the top-most cut. This is easy because eliminating a cut affects only the proof tree above the cut. (There is just a slight complication: we need to show that eliminating a cut does not change a lower cut from being principal to not principal, but this is not difficult). This gives the following three results:

Theorem 9. *If we can eliminate one arbitrary cut (or left-principal cut, or principal cut) from any given derivation tree, then we can eliminate any number of such cuts from any given derivation tree.*

```
elimLRP= "canElim (cutIsLRP ?A) ==> canElimAll (cutIsLRP ?A)":thm
elimLP=  "canElim (cutIsLP ?A) ==>  canElimAll (cutIsLP ?A)":thm
elimFmls="canElim (cutOnFmls ?s) ==> canElimAll (cutOnFmls ?s)":thm
```

We also have the theorems such as orC8 (see §4.1) dealing with a tree with a single (left- and right-) principal cut on a given formula.

Theorem 10. *A tree with a single (left- and right-) principal cut on a given formula can be replaced by a tree with arbitrary cuts on the immediate subformulae (if any) of that formula.*

These theorems (one for each constructor for the type formula) are converted to a list of theorems thC8Es', of which an example is

```
"canElimAll (cutOnFmls {?B,?A}) ==> canElim (cutIsLRP (?Av?B))":thm
```

We have one such theorem for each logical connective or constant, and one for the special formulae `FV str` and `PP str`. These latter two arise in the trivial instance of cut-elimination when Y is A and Π_{AY} is empty in Figure 4.

Together with the theorems `elimLRP`, `allLRP'`, `elimLP`, `allLP'` and `elimFmls`, we now have theorems corresponding to the six routines described above. As noted already, a call to `elim` would indirectly call `elim` with a smaller cut-formula as argument, and so the program would terminate, the base case of the recursion being the trivial instance of Figure 4 mentioned above. Correspondingly, the theorems we now have can be combined to give the following.

Theorem 11. *We can eliminate a cut on a formula if we can eliminate a cut on each of the formula's immediate subformulae.*

```
"canElim (cutOnFmls {?B,?A}) ==> canElim (cutOnFmls {?A v ?B})":thm
```

The proof of Theorem 11 is by cases on the formation of the cut-formula, and here we have shown the case for $\vee$ only. There is one such case for each constructor for the type `formula`. We can therefore use structural induction on the structure of a formula to prove that we can eliminate a cut on any given formula `fml`: that is, we can eliminate any cut.

Theorem 12. *A sequent derivable using (cut) is derivable without using (cut).*

Proof. Using `elimFmls` from Theorem 9, it follows that we can eliminate any number of cuts, as reflected by the following sequence of theorems.

```
canElimFml = "canElim (cutOnFmls {?fml})" : thm
canElimAny = "canElim (cutOnFmls UNIV)" : thm
canElimAll = "canElimAll (cutOnFmls UNIV)" : thm
cutElim    = "IsDerivableR rls {} ?concl ==>
              IsDerivableR (rls - {cut}) {} ?concl" : thm
```

Corollary 1. *The rule (cut) is admissible in* $\boldsymbol{\delta RA}$.

6 Conclusion and Further Work

We have formulated the Display Calculus for Relation Algebra, $\boldsymbol{\delta}$**RA**, in Isabelle/HOL as a "deep embedding", allowing us to model and reason about derivations rather than just performing derivations (as in a shallow embedding). We have proved, from the definitions, "transitivity" results about the composition of proofs. These are results which were omitted – "due to their difficulty" – from another reported mechanised formalisation of provability [8, p. 302].

We have proved Belnap's cut-admissibility theorem for $\boldsymbol{\delta}$**RA**. This was a considerable effort, and could not have been achieved without the complementary features (found in Isabelle) of the extensive provision of powerful tactics, and the powerful programming language interface available to the user.

The most important disadvantage of our approach is the inability to easily produce a program for cut-elimination from our Isabelle/HOL proofs, even when our proofs mimic a programming style (see §5.4).

Results like `IsDerivableR_trans` and `IsDerivableR_deriv` are general results about the structure of derivations, closely resembling facilities available in Isabelle: namely, successive refinement of subgoals, and use of a previously proved lemma. A higher order framework allows us to reason about higher and higher meta-levels, like the `IsDerivable` relation itself, without invoking explicit "reflection principles" [6]. This is the topic of future work.

References

1. N D Belnap. Display logic. *Journal of Philosophical Logic*, 11:375–417, 1982.
2. Jeremy E Dawson and R Goré. Embedding display calculi into logical frameworks: Comparing Twelf and Isabelle. In C Fidge (Ed), *Proc. CATS 2001: The Australian Theory Symposium*, ENTCS, 42: 89–103, 2001, Elsevier.
3. J E. Dawson and R Gore. A new machine-checked proof of strong normalisation for display logic. Submitted 2001.
4. R Goré. Cut-free display calculi for relation algebras. In D van Dalen and M Bezem (Eds), *CSL'96: Selected Papers of the Annual Conference of the European Association for Computer Science Logic*, LNCS 1258:198–210. Springer, 1997.
5. R Goré. Substructural logics on display. *Logic Journal of the Interest Group in Pure and Applied Logic*, 6(3):451–504, 1998.
6. J Harrison. Metatheory and reflection in theorem proving: A survey and critique. Technical Report CRC-053, SRI International Cambridge Computer Science Research Centre, 1995.
7. S Matthews. Implementing FS_0 in Isabelle: adding structure at the metalevel. In J Calmet and C Limongelli, editors, *Proc. Disco'96*. Springer, 1996.
8. A Mikhajlova and J von Wright. Proving isomorphism of first-order proof systems in HOL. In J Grundy and M Newey, editors, *Theorem Proving in Higher-Order Logics*, LNCS 1479:295–314. Springer, 1998.
9. F Pfenning. Structural cut elimination. In *Proc. LICS 94*, 1994.
10. C Schürmann. *Automating the Meta Theory of Deductive Systems.* PhD thesis, Dept. of Comp. Sci. , Carnegie Mellon University, USA, CMU-CS-00-146, 2000.
11. H Wansing. *Displaying Modal Logic*, volume 3 of *Trends in Logic*. Kluwer Academic Publishers, Dordrecht, August 1998.

Formalizing the Trading Theorem for the Classification of Surfaces

Christophe Dehlinger and Jean-François Dufourd

Laboratoire des Sciences de l'Image, de l'Informatique
et de la Télédétection (UMR CNRS 7005)
Université Louis-Pasteur de Strasbourg
Pôle API, boulevard S. Brant, 67400 Illkirch, France
{dehlinger,dufourd}@lsiit.u-strasbg.fr

Abstract. We study the formalization and then the proof of a well-known theorem of surface topology called the trading theorem. This is the first major work undertaken with our Coq specification of the generalized maps, which may be used as a model for surfaces subdivisions. We explain how we expressed in terms of subdivisions the notion of topological equivalence, and how we used this notion to prove the trading theorem, while giving a quick look at the specification we have built.

1 Introduction

Geometry modelling involves a large variety of objects and algorithms that are often complex from a mathematical point of view, and the crafting and understanding of which heavily rely on intuition. Thus, geometric modellers are often based on representations and processes that are likely, but not surely, right. This is typically a domain that could really benefit from an effort in formalization, as Knuth pointed out in "Axioms and Hulls" [12].

This work is part of a rigorous study of mathematical models that have been implemented and used in the geometry modelling community for over ten years. For this purpose, Puitg and Dufourd [17] have specified combinatory maps, that can represent subdivisions of closed orientable surfaces. That work eventually featured the definition of a planarity criterion and to the proof of a combinatorial form of the Jordan theorem. This success led us to carry on by focussing on generalized maps, an extension of combinatory maps that also allow representation of subdivisions of open or non-orientable surfaces. This model is the topological foundation of the modeller TOPOFIL [3].

In this paper, we will show how to express and prove a fundamental theorem on surfaces that Griffiths called the trading theorem [9]. This theorem actually is one half of the surface classification theorem. Formally proving this kind of properties for a model strengthens three convictions: that the model is well adapted to the represented objects, that the formal specification of the model fits it well, and that the represented objects do verify the considered property.

V.A. Carreño, C. Muñoz, S. Tahar (Eds.): TPHOLs 2002, LNCS 2410, pp. 148–163, 2002.

After this introduction, we briefly present some related works in Sect. 2. In Sect. 3, we state precisely what we mean by "surfaces", emphasizing the link between surface subdivisions and generalized maps. We also introduce the trading theorem and its place in the classification theorem. In Sect. 4, we describe intuitively and then formally the model of generalized maps, as well as important related basic notions. In Sect. 5, we describe a set of operations on generalized maps that we call conservative, and use it to define topological equivalence. In Sect 6, we describe what we call the normal maps. In Sect. 7, we outline our proof of the trading theorem. In Sect. 8, we give a quick description of the structure of our formal specification, and we conclude in Sect. 9.

Our theoretical framework has entirely been formalized in the Calculus of Inductive Constructions (CIC) using the language Gallina, and all proofs have been carried with the help of Coq, a proof assistant based on CIC. However, due to space constraints, we will use a standard mathematical language instead of the Gallina language, which would require too long an introduction.

2 Related Work

The geometry modelling aspects of this work are mostly related to generalized maps. Theses maps are a variation of Jacques's combinatory maps [10], which have been extended by Cori into hypermaps [5] and then by Lienhardt into generalized maps [13], and later specified by Bertrand and Dufourd [3].

There have been several different approaches to the trading theorem (see for instance [7] for a word-based combinatorial approach). We took our inspiration in the works of Griffiths's [9] and Fomenko's [8], as they can be adapted to generalized maps relatively easily.

Generalized maps offer one constructive model for geometry. Another approche of constructive geometry has been explored by Von Plato [18] and formalized in Coq by Kahn [11]. A similar experiment has been undertaken by Dehlinger et al [6]. Also using Coq, Pichardie and Bertot [16] have formalized Knuth's plane orientation and location axioms, and then proved correct algorithms of computation of convex hulls.

Proof assistant Coq [1] is based on the Calculus of Inductive Constructions [4], an intuitionnistic type theory [14]. In addition to being a powerful tool for proofs, it features the extraction of certified programs from proof terms [15].

3 The Theorem of Classification

The theorem that we are trying to prove here is a subpart of the classification theorem, which states that all surfaces can be classified with respect to their topology using only three characteristic numbers. We shall however introduce the rest of the classification theorem, as it will influence our specification. For now, we restrict ourselves to open surfaces (i.e. surfaces with at least one boundary).

Our definition of surfaces is rather classic and intuitive. For us, a surface is built from simple basic elements called *panels*. A panel is a surface that is homeomorphic to a disc, that is the image of disc by a bijective and bicontinuous transformation. In other words, a panel is the interior of a closed compact curve without self-intersection, also known as a *Jordan curve.* By definition, a panel has a single boundary. A disc, a polygon and a punctured sphere all are panels. Neither a sphere nor a ring is a panel.

We call *surface* any "patchwork" of panels glued along each other's boundaries with an imaginary sticky tape, each edge glued at most to only one other boundary (Fig. 1). Surfaces may be *open* or *closed*, i.e. with or without boundary, and orientable or not. A disc and a ring are open and orientable. A torus and a sphere are closed and orientable. A Moebius strip is open and non orientable. A Klein bottle is closed and non orientable. We impose as a definition that two surfaces are in the same class if they are homeomorphic to each other.

Obviously, surfaces are classified on the basis of a criterion that is geometric, i.e. that depends on the actual objects, while we want to only manipulate their structures. The challenge is to express it in topological and combinatory terms at the level of the surface subdivisions in a coherent and satisfying manner. Thus, the classification theorem states: "Any open surface belongs to at least one class, the members of which are homeomorphic to each other. Each class is characterized by a triplet of natural numbers (p, q, r) with $r \leq 2$."

Definition 1. *The normal surfaces $P_{p,q,r}$ are defined exclusively by:*

- $P_{0,0,0}$ *is a disc;*
- $P_{p+1,0,0}$ *is made of* $P_{p,0,0}$ *glued to an* ear, *i.e. a ring;*
- $P_{p,q+1,0}$ *is made of* $P_{p,q,0}$ *glued to a* bridge, *i.e. a punctured torus, at the only possible location: the puncture;*
- $P_{p,q,r+1}$ *is made of* $P_{p,q,r}$ *glued to a* twisted ear, *i.e. a Möbius strip.*

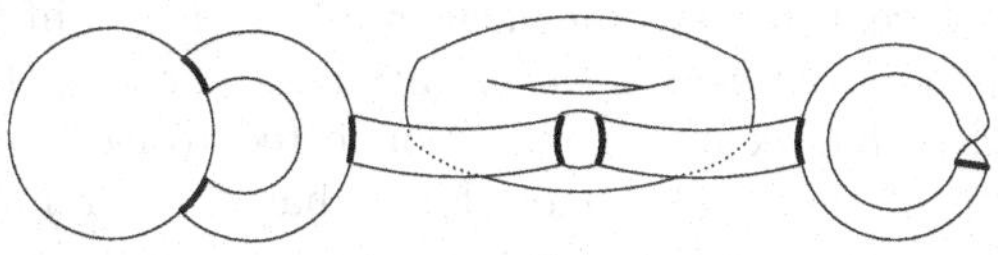

Fig. 1. Surface $P_{1,1,1}$: a disc, a ring, a punctured torus and a Möbius strip glued together along their boundaries

The surfaces $P_{p,q,r}$ are *geometric.* The latter term is used to emphasize the "metric" aspect of the surface. In contrast, the term "combinatory" denotes a surface that is regarded only from a combinatory point of view, meaning that only the relations between the elements of this surface are considered. With the help of these surfaces, we reformulate the classification theorem by splitting it into two independent halves. This article only deals with the second one:

- *Normalization theorem*: "for any open surface S there exists a triplet of naturals (p, q, r) such that S is homeomorphic to $P_{p,q,r}$"

- *Trading theorem*: "for any triplet (p, q, r) such that $r \geq 1$, surface $P_{p,q+1,r}$ is homeomorphic to surface $P_{p,q,r+2}$"

4 Generalized Maps

The notion of *generalized maps*, or *g-maps*, encompasses a series of combinatorial models used to represent the topology of different classes of objects. A generalized map's, foremost characteristic is its dimension, an integer greater or equal to -1. The type of represented objects varies with the dimension: g-maps of dimension -1, or -1-g-maps, represent isolated vertices, 0-g-maps isolated edges, 1-g-maps simple curves, 2-g-maps surfaces, 3-g-maps volumes, etc. The formal specification that we have developed includes all types of generalized maps and has allowed us to find new results on this mathematical models. In order to simplify the presentation, we shall focus on 2-g-maps, which are the only ones needed to prove the classification theorem.

A 2-g-map is built from basic abstract elements called *darts*. We note by *dart* the type of darts. The mathematical definition of generalized maps of dimension 2 is the following:

Definition 2. *A* generalized map of dimension 2*, or* 2-g-map*, is a quadruplet* $(D, \alpha_0, \alpha_1, \alpha_2)$ *where* D *is a finite subset of dart, and where the* α_i *are involutions on* D*, such that* α_0 *and* α_1 *have no fixpoint and that* $\alpha_0 \circ \alpha_2$ *is also an involution:* $\forall D \subset dart,\ D\ finite,\ \forall \alpha_0, \alpha_1, \alpha_2 : D \to D,\ G = (D, \alpha_0, \alpha_1, \alpha_2)$ *is a 2-g-map if*

- $\forall x \in D, \forall i \leq 2, \alpha_i^2(x) = x;$
- $\forall x \in D, \forall i \leq 1, \alpha_i(x) \neq x;$
- $\forall x \in D, (\alpha_0 \circ \alpha_2)^2(x) = x;$

A dart x is said to be *sewn at dimension* k, or k*-sewn*, to dart y if $\alpha_k(x) = y$. As the α_i are involutive, hence symmetrical, x is k-sewn to y iff y is k-sewn to x. Dart y is also said to be the k*-neighbor* of dart x if $\alpha_k(x) = y$. Intuitively, darts can be understood as half-edges that are connected to each other in different ways with the help of the α_i. Each of the α_i has a different meaning: α_0 is used to make up edges, α_1 simple curves, and α_2 surfaces. In general, for any given k, α_k is used to make up cells of dimension k. The conditions imposed on the α_i enforce the consistency and completeness of cells: involutivity of the α_i guarantees that each dart has exactly one k-neighbor for each k. The lack of fixpoints for α_0 and α_1 prevents the presence of dangling darts. As we will see later, fixpoints of α_2 are the darts that belong to a boundary. Finally, forcing $\alpha_0 \circ \alpha_2$ to be involutive ensures that whenever a dart is 2-sewn to another, their respective 0-neighbors are also 2-sewn. Thus, only whole edges are 2-sewn. Fig. 2 shows the standard representation for darts and sewings. In this figure, x and y are darts. A dart pictured without k-neighbor is implicitely k-sewn to itself.

A priori, there is no condition on *dart*. However, we will have to make a couple of assumptions on it in order to construct proofs. First, we need to assume that we can compare any two darts, which is not trivial in CIC's intuitionnistic logic:

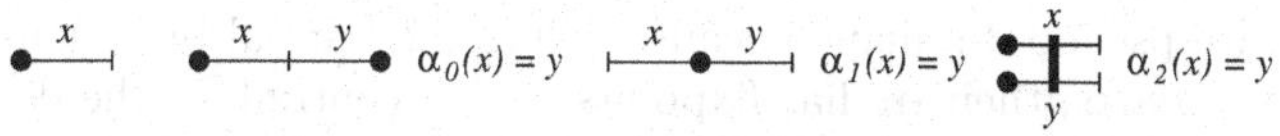

Fig. 2. Standard graphic representation of darts

Axiom 1. *Dart equality is decidable:*

$\forall x, y \in dart,\ x = y \vee \neg x = y$

Besides, we will need darts that do not belong to the manipulated maps in order to build intermediary 2-g-maps. In order to make sure that there are always new darts available, we impose that *dart* is infinite. To do so, we assume that we know an injection from natural numbers into the darts set. We call this function *idg* (injective dart generator), its type is $I\!N \to dart$. Note that there may very well be darts that are not images of *idg*.

Axiom 2. *Function idg is injective:*

$\forall n, n' \in I\!N,\ idg(n) = idg(n') \Rightarrow n = n'.$

These two axioms are the only ones in our specification, D and *idg* are its only parameters. Before defining the usual notions of topology in the formalism of 2-g-maps, we need to introduce the mathematical notion of *orbit*:

Definition 3. *Let D be a set and $f_1, f_2, \ldots, f_n$ functions on D. For any $x \in D$, the* orbit of $f_1, f_2, \ldots, f_n$ at x *is defined to be the smallest subset of D containing x and stable by all functions f_i. It is noted $< f_1, f_2, \ldots, f_n > (x)$.*

From now on, $G = (D, \alpha_0, \alpha_1, \alpha_2)$ will be a 2-g-map we use to give our definitions. With the notion of orbit, connected components and cells are easily defined as 2-g-maps derived from an initial 2-g-map:

Definition 4. *The* connected component *of G* incident to dart $x \in D$ *is defined to be the 2-g-map $G' = (D', \alpha'_0, \alpha'_1, \alpha'_2)$ satisfying*

- *$D' =< \alpha_0, \alpha_1, \ldots, \alpha_n > (x)$;*
- *$\forall i \mid 0 \le i \le 2$, α'_i is the restriction of α_i to D'.*

Definition 5. *For any $x \in D$, we call orbit $< \alpha_1, \alpha_2 > (x)$ the* vertex of G incident to x, *or* 0-cell of G incident to x. *Similarly, we call orbit $< \alpha_0, \alpha_2 > (x)$ the* edge of G incident to x, *or* 1-cell of G incident to x; *and we call orbit $< \alpha_0, \alpha_1 > (x)$ the* face of G incident to x, *or* 2-cell of G incident to x. *We define the* map of k-cells *of G to be the pseudo-2-g-map obtained from G by ripping all k-sewings. It is noted G_k. Thus, an edge may have either 2 darts (*open *edge) or 4 darts (*closed *edge).*

The G_k are pseudo-2-g-maps because, while they are also built using some α_i and a set of darts, their α_k has fixpoints, which contradicts the definition of 2-g-maps for $k = 0$ and $k = 1$. Obviously, two darts belong to the same k-cell of G iff these two darts belong to the same connected component of G_k. Fig. 3 shows an example 2-g-map as well as its maps of cells; in this example, G has two connected components. The standard embedding (i.e. the projection of topological objects into a representation space, here the euclidean space) of g-maps consists in embedding all darts from the same topological vertex into the same geometrical vertex. Similarly, all darts from the same topological edge are embedded into the same geometric edge and all darts from the same topological face are embedded into the same geometric face. Incident topological objects are embedded into incident geometrical objects. The 2-g-map $\delta(G)$ is the *2-g-map of boundaries* of G: its darts are the darts of G that are incident to a boundary, each of them being 1-sewn in $\delta(G)$ to its boundary-neighbor in G, the notions of boundary and boundary neighborhood being defined as:

Definition 6. *A dart $x \in D$ is said to be incident to a* boundary *of G if $\alpha_2(x) = x$. A dart incident to a boundary is called* external, *it is called* internal *else.*

Definition 7. *For any dart x, dart $y \in D$ is called* boundary-neighbor *of x if:*

- *y is incident to a boundary (i.e. $\alpha_2(y) = y$) and $y \neq x$;*
- *there is a natural number k such that y can be written as $y = (\alpha_2 \circ \alpha_1)^k(x)$, k being the smallest such number.*

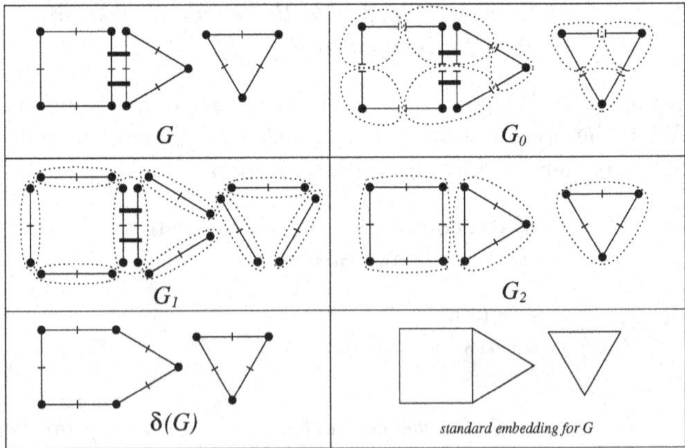

Fig. 3. An example of a 2-g-map and its maps of cells, 2-g-map of boundaries and standard embedding

5 Conservative Operations on 2-g-Maps

The main purpose of our specification is to prove the classification theorem for surfaces represented by 2-g-maps, i.e. to prove that any surface is topologically equivalent to a "normal" surface identified by only three numbers. Thus, we have to find a way to express that surfaces are topologically equivalent, or homeomorphic. As we said before, this is a geometric condition, so we have to find a way to express it in combinatory terms. Thus, we pose that two 2-g-maps are topologically equivalent if one is obtained from the other by applying a series of operations believed to preserve the topology of their arguments, called *conservative operations*. They are willingly kept as simple as possible, so that we can be convinced that they actually are topological equivalents of homeomorphisms. We shall now list these operations as well as the conditions under which we allow their use. From our point of view, this axiomatic definition of topological equivalence is the most obvious objection to the validity of this work.

5.1 Removal of an Edge

Given a 2-g-map and a dart from it, operation *rmedge* (**remove edge**) removes from this 2-g-map the edge incident to this dart and joins the two edges to which this edge was incident (Fig. 4). This operation may also be interpreted as the fusion of neighboring vertices that are connected with the edge incident to the dart. It can be used to remove open as well as closed edges.

Definition 8. *Removing the edge incident to dart $x \in D$ yields 2-g-map $rmedge(x, G) = (D', \alpha'_0, \alpha'_1, \alpha'_2)$ such that:*

- $D' = D - < \alpha_0, \alpha_2 > (x)$;
- $\alpha'_1(y) = \alpha_1(\alpha_0(\alpha_1(y)))$ *for* $y \in \alpha_1(< \alpha_0, \alpha_2 > (x))$
- $\alpha'_i(y) = \alpha_i(y)$ *for all other combinations of* y *and* i.

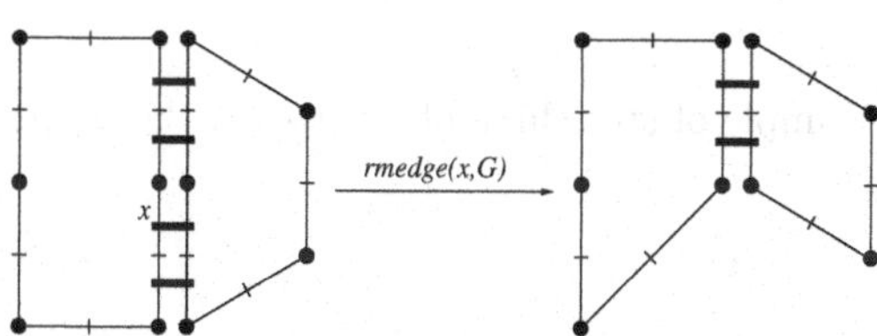

Fig. 4. Example of edge removal

The application of *rmedge* is considered to conserve topology in both the easily recognized following cases:

1. the vertex incident to x is an *internal* vertex, which means that none of its darts are incident to a boundary. The removed edge is then always closed. The operation is a fusion of two neighboring vertices, one of which is internal;
2. the edge incident to x is open, and isn't a boundary by itself. In other words, it is part of a boundary but isn't an edge both ends of which are 1-sewn to darts incident to the same vertex.

Besides, the use of *rmedge* is forbidden on some degenerate edges, namely dangling edges, crossed or not, and elementary loops, 2-sewn or not.

5.2 Vertex Stretching with an Open Edge

This operation, called *stro* (**str**etch with **o**pen edge), "stretches" the vertex incident to a dart by inserting an open edge between this dart and its 1-neighbor (Fig. 5). It can be understood as the inverse of *rmedge* restricted to open edges. It takes four arguments: a 2-g-map, the dart used to locate the vertex and the two extra darts that will make up the new edge. Stretching of a vertex with an open edge is allowed only if the vertex is incident to a boundary, thus ensuring that it will only enlarge the boundary and not create a new one.

Definition 9. *Stretching the vertex incident to dart $x \in D$ with an edge made up of distinct darts $x_1, x_2 \notin D$ yields 2-g-map $stro(x, x_1, x_2, G) = (D', \alpha'_0, \alpha'_1, \alpha'_2)$ such that*

- $D' = D \cup \{x_1, x_2\}$;
- $\alpha'_0(x_1) = x_2$, $\alpha'_1(x_1) = x$, $\alpha'_1(x_2) = \alpha_1(x)$;
- $\alpha'_2(y) = y$ *for* $y \in \{x_1, x_2\}$;
- $\alpha'_i(y) = \alpha_i(y)$ *for all other combinations* y *of* i.

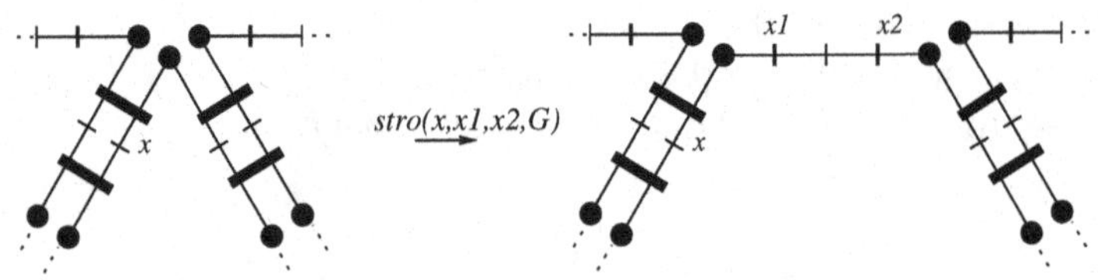

Fig. 5. Example of stretching of a vertex with an open edge

5.3 Vertex Stretching with a Closed Edge

This operation consists in stretching a vertex in two places with open edges, and then 2-sew the two new edges together (Fig. 6). In other words, the operation splits the vertex in half, and then joins both new vertices with a closed edge, hence its name: *strc* (**str**etch with a **c**losed edge). In reality, this operation is the combination of two applications of *stro* followed by a 2-sewing of edges:

Definition 10. *Stretching the vertex incident to $x \in D$ in positions located by $x, x' \in < alpha_1, \alpha_2 > (x)$ with the closed edge made up of distinct darts $x_1, x_2, x_3, x_4 \notin D$ yields 2-g-map $strc(x, x', x_1, x_2, x_3, x_4, G) = (D', \alpha'_0, \alpha'_1, \alpha'_2)$ such that:*

- $D' = D \cup \{x_1, x_2, x_3, x_4\}$;
- $\alpha'_2(x_1) = x_3$, $\alpha'_2(x_2) = x_4$;
- *the other α'_i have the same values as the α_i of 2-g-map $stro(x', x_3, x_4, stro(x, x_1, x_2, G))$*

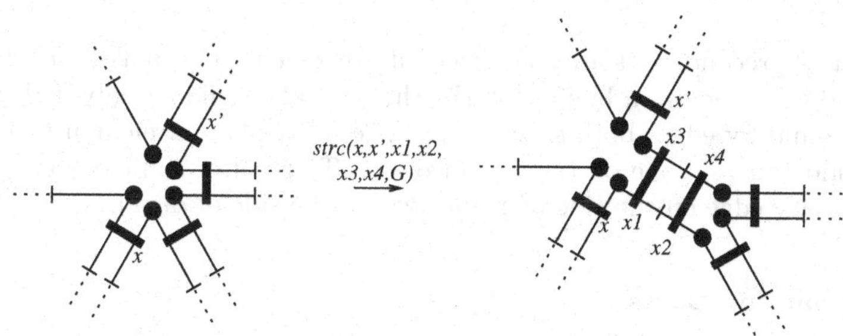

Fig. 6. Example of stretching with a closed edge

The precondition stems from the fact that if x and x' are inappropriately chosen, the new edges might end up sewn in the wrong way, resulting in an unwanted twist in the surface. The following criterion allows us to identify the cases where such twists would appear:

Definition 11. *Operation strc may be applied to x and x' only if there is a $k \in \mathbb{N}$ such that $x' = \alpha_2 \circ (\alpha_2 \circ \alpha_1)^k(x)$ with, for any $k' \leq k$, $x' \neq (\alpha_2 \circ \alpha_1)^{k'}(x)$.*

5.4 Sliding along a Boundary

Consider two distinct edges (x_1, x_2) and (x_3, x_4) that are 2-sewn together in the manipulated 2-g-map (thus forming a single closed edge), such that x_1 has a boundary neighbor. This operation "slides" the sewing along this boundary, which consists in 2-unsewing the two edges, and then 2-sewing the second edge to the edge incident to x_1's boundary neighbor (Fig. 7).

Definition 12. *Sliding the edge incident to x towards its boundary neighbor x' yields map $slide(x, G) = (D, \alpha_0, \alpha_1, \alpha'_2)$ such that:*

- $\alpha'_2(y) = y$ *for* $y \in \{x, \alpha_0(x)\}$
- $\alpha'_2(x') = \alpha_0(\alpha_2(x))$, $\alpha'_2(\alpha_0(x')) = \alpha_2(x)$
- $\alpha'_2(y) = \alpha_2(y)$ *for* $y \in D - \{x, x', \alpha_0(x), \alpha_0(x')\}$

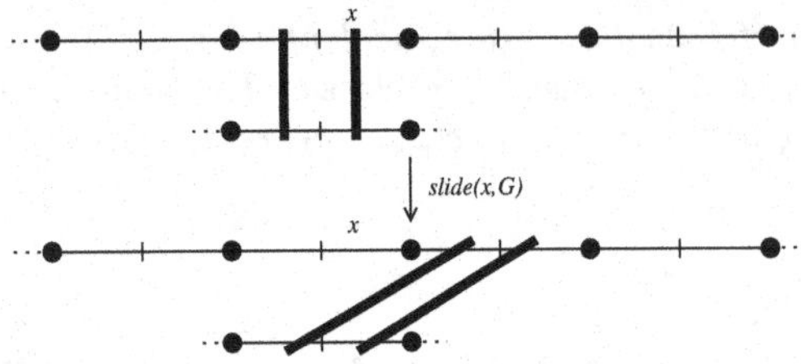

Fig. 7. Example of sliding

As far as preconditions are concerned, all we have to do is make sure that we neither add nor remove a boundary. For that, we allow sliding only if there are enough boundary edges both sides of the edge incident to x, enough being two on the side of x and one on the side of $\alpha_0(x)$. Thus, after sliding, there is still one boundary edge on the side of x and two on the side of $\alpha_0(x)$.

5.5 Renaming Darts

This operation allows to apply an injective function f of type *dart* $\rightarrow$ *dart* to all darts in a 2-g-map. The only precondition, injectivity of f, is required to make sure that f is inversible and thus avoid any confusion of darts.

Definition 13. *Renaming the darts of G with injective function f yields 2-g-map* $rename(f, G) = (f(D), \alpha'_0, \alpha'_1, \alpha'_2)$ *such that, for any* $i \leq 2$ *and any* $x \in f(D)$, *the equality* $\alpha'_i(x) = f(\alpha_i(f^{-1}(x)))$ *stands.*

5.6 Topological Equivalence

Definition 14. *Topological equivalence on 2-g-maps is the smallest reflexive symmetrical transitive relation on 2-g-maps that is stable by application of any conservative operations, provided all preconditions are verified at every step.*

6 Normal Maps

6.1 Definitions and Choices

The next step of our specification consists in formally describing the topology of surfaces $P_{p,q,r}$ with 2-g-maps. We call *normal 2-g-maps* a particular set of 2-g-maps that represent the topology of surfaces $P_{p,q,r}$. Indeed, a 2-g-map represents a subdivision of surface, and there is an infinity of subdivisions for each surface. Thus, there is an infinity of 2-g-maps that represent each surface $P_{p,q,r}$. Therefore, we must pick one such 2-g-map for each surface $P_{p,q,r}$. The selection is made in two steps.

First, we try to find an "easy" subdivision of the surface, meaning a subdivision which we are convinced that it fits the surface. For example, for $P_{1,1,1}$ (the disc, an ear, a bridge and a twisted ear), we choose the 2-g-map depicted in Fig. 8, which features a single panel. On this figure, the two shades of grey

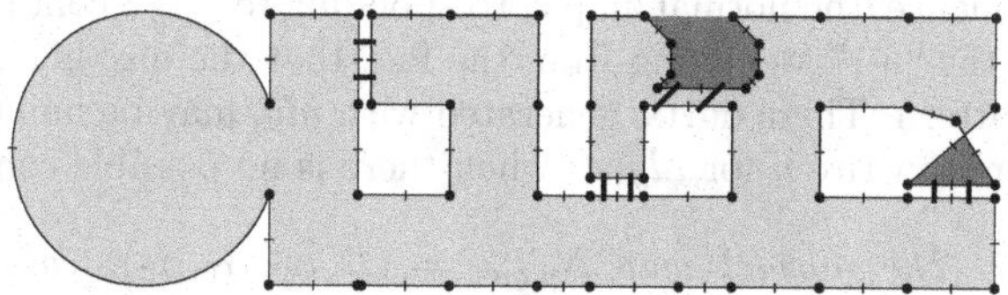

Fig. 8. Rough 2-g-map representing $P_{1,1,1}$

are used to separate the recto from the verso of the panel. It's easy to see how a similar 2-g-map could be built for each $P_{p,q,r}$, with patterns for the disc, ears, bridges and twisted ears. The problem with these 2-g-maps is that they contain way too many darts, which makes them too bulky. As we work modulo topological equivalence, we can simplify them by repeatedly applying conservative operations Applying as many edge removals as possible and then a single vertex stretching with a closed edge to the 2-g-map in Fig. 8 yields the 2-g-map in Fig. 9. The 2-g-maps in Figs. 8 and 9 are topologically equivalent. The vertex

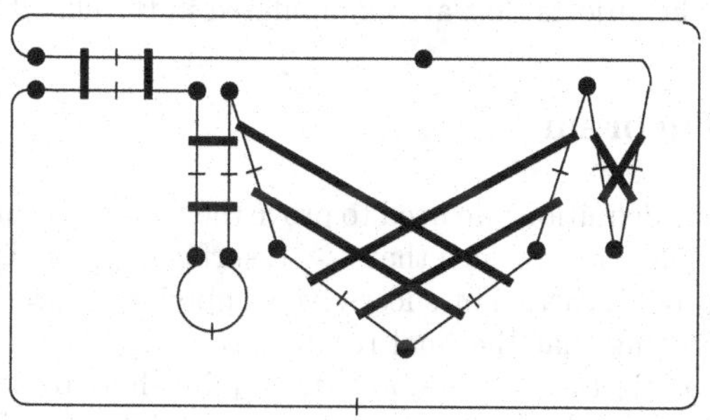

Fig. 9. Normal map for $P_{1,1,1}$

stretching occurred at the level of the disc pattern, and allows us to replace the disc pattern by simply another ear pattern. Thus, instead of four, there are only three different pattern types, at the inexpensive cost of four additionnal darts.

Looking at the normal map in Fig. 9, it's easy to see that it's made up of a single cycle of darts that are alternately 0- and 1-sewn, in which some darts have later been 2-sewn in order to make up the basic patterns. An ear pattern is made up of 6 consecutive darts from the cycle, such that the first has been 2-sewn to the sixth and the third to itself. Similarly, a bridge pattern is made up of 8 consecutive darts such that the first and third dart are respectively 2-sewn to the sixth and eighth dart. A twisted ear pattern is made up of 4 consecutive darts such that the first and third dart are 2-sewn. The sewings of all unmentioned darts can be deduced from the fact that only entire edges are sewn.

As a consequence, the normal map corresponding to $P_{p,q,r}$ contains $6(p+1)+8q+4r$ darts, the "+1" stemming from the fact that the disc is represented by an extra ear pattern. These darts, generated with idg, may be numbered from 0 on. We will simply write n for $idg(n)$ when there is no possible confusion:

Definition 15. *The normal map* $N_{p,q,r} = (D, \alpha_0, \alpha_1, \alpha_2)$*, corresponding to* $P_{p,q,r}$*, is defined by:*

1. $D = \{idg(n) | n < (6(p+1) + 8q + 4r)\}$
2. $\forall i \leq ((6(p+1) + 8q + 4r)/2)\ \alpha_0(2i) = (2i+1)$
3. $\forall 0 < i \leq ((6(p+1) + 8q + 4r)/2)\ \alpha_1(2i) = (2i-1)$
4. $\alpha_1(0) = \alpha_1(6(p+1) + 8q + 4r - 1)$
5. $\forall i \leq p,\ \alpha_2(6i) = (6i+5)$
6. $\forall i \leq p,\ \alpha_2(6i+2) = (6i+2)$
7. $\forall i < q,\ \alpha_2(6p+8i) = (6p+8i+6)$
8. $\forall i < q,\ \alpha_2(6p+8i+3) = (6p+8i+8)$
9. $\forall i < r,\ \alpha_2(6p+8q+4i) = (6p+8q+4i+3)$

Relations 2, 3 and 4 build the cycle of darts. Relations 5 and 6 build ear patterns, relations 7 and 8 bridge patterns and relation 9 twisted ear patterns. Normal maps are the objects that are manipulated in the classification theorem.

7 Trading Theorem

Now, we have all the definitions we need to prove the trading theorem. Recall that it reads "for any triplet (p, q, r) such that $r \geq 1$, surface $P_{p,q+1,r}$ is homeomorphic to $P_{p,q,r+2}$". The proof is carried by successively applying conservative operations to $N_{p,q+1,r}$ and proving that the final result is $N_{p,q,r+2}$. This proof could have been further automatized, but was not as it is quite short to do manually. The formal version of the trading theorem, the text of which precisely tells which operations to perform, states:

Theorem 1. *Let* (p, q, r) *be a triplet of naturals such that* $r \geq 1$*. Then the following equality stands:*
$N_{p,q,r+2} = rename(f,$
$strc(idg(2), idg(3), idg(1), idg(0), idg(3), idg(2),$
$rmedge(idg(6p+8q+6),$
$rmedge(idg(6p+8q+2),$
$slide(idg(6p+8q+3),$
$slide(idg(6p+8q+11),$
$slide(idg(6p+8q+7),$

$slide(idg(3),$
$slide(idg(6p+8q),$
$stro(idg(6p+8q+7), x_2, x_3,$
$stro(idg(6p+8q+11), x_0, x_1,$
$rmedge(idg(0), N_{p,q+1,r})))))))))))$ *with* f *and* x_i *well chosen.*

The first step is the removal of the edge incident to $idg(0)$. This edge was the edge that had been added to the disc pattern so that it could be replaced by an ear pattern. Removing it makes all darts that are incident to a twisted ear or a bridge incident to an external vertex. The 11th step reverses this removal. Steps 2 and 3 are vertex stretchings with open edges, that are later reversed in steps 9 and 10. Steps 4 through 8 are the core of the proof. They all are slides that occur within the first three twisted ears patterns; they are used to rearrange them into a bridge and a twisted ear patterns. Step 12 is a simple dart renaming. All steps are depicted on Fig. 9. On each subfigure, the darts that are used as arguments of the next transformation are dashed.

8 Formal Specification Outline

Our Gallina specification assumes the form of a three-level hierarchy of increasingly constrained types. A similar approach in a simpler context has already been used in [17] for combinatory maps. Their specification and ours even share the lowest (least constrained) level, the level of free maps.

The type of free maps is noted *fmap*. A *fmap* is simply a set of darts, represented by a list, with a set of sewings, represented by a list of triplets made up of the dimension of the sewing and the two sewn darts. The *fmap* are very simple, without any constraints, and it is easy to define operations on them and to perform structural induction to prove their properties, but in a way they are too general as we are really interested in generalized maps, while free maps can also be used to represent lots of other models.

In order to retain the simplicity of use of *fmap* while at the same time restraining ourselves to generalized maps, we use the following technique to define the second level of our specification, the level of generalized maps: the type of generalized maps, noted *ngmap*, is defined as the type of pairs made up of a *fmap*, called the *support* of the g-map, and a proof that the support satisfies a predicate of *well-formedness*. This predicate is a straightforward adaptation of the mathematical definition of g-maps given in Sect. 4.

Coq's type coercions allow to apply *fmap* selectors and transformations to a *ngmap* by implicitly applying them to its support. Note that such a transformation will still yield a *fmap* when applied to a *ngmap*. In order to make it a real *ngmap* transformation, it must first be proved to preserve well-formedness. This way, it is possible to define operations on a complex type like *ngmap* by expressing it intuitively in *fmap* and then proving that it preserves well-formedness.

The third level is the level of *sewn cells maps* of type *smap*. They are defined at the free map level as roughly the set of maps obtained by any number of

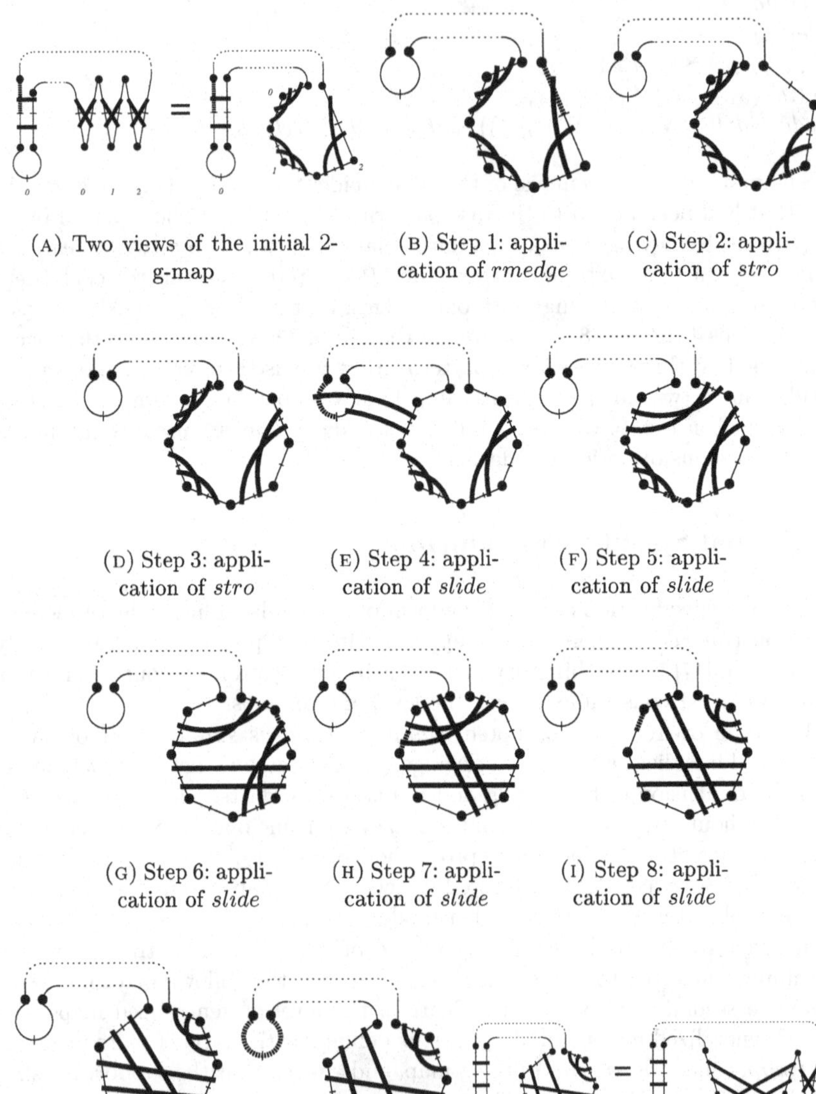

(A) Two views of the initial 2-g-map

(B) Step 1: application of *rmedge*

(C) Step 2: application of *stro*

(D) Step 3: application of *stro*

(E) Step 4: application of *slide*

(F) Step 5: application of *slide*

(G) Step 6: application of *slide*

(H) Step 7: application of *slide*

(I) Step 8: application of *slide*

(J) Step 9: application of *rmedge*

(K) Step 10: application of *rmedge*

(L) Step 11: application of *strc* + Step 12: renaming

applications of a complex cell sewing operation, also defined at the free map level, under some preconditions. Much like well-formedness, this property is expressed by a predicate of *well-constructedness*. Thus, a *smap* is a pair made of a *fmap*, its support, and a proof of the support's well-constructedness. A useful theorem that we have proved is that well-constructedness entails well-formedness, hence any *smap* support is also a *ngmap* support. We use coercions again to use *smaps* as *ngmaps*.

We have also proved a major theorem, stating that well-formedness implies well-constructedness, modulo reordering of the elements of the support. This theorem used to be a conjecture but was widely used nonetheless. Its computer-assisted proof turned out to be very useful, as it allowed to reveal a formerly missing precondition to the sewing of cells. The order of the elements of a free map being semantically irrelevant, we deduce from these two theorems that any meaningful result on *ngmap* can be extended to *smap* and vice versa.

The biggest gain of this equivalence is of being able to work in *smap*, a cleaner, more constrained subset of *ngmap*, in order to prove properties of *ngmap* as a whole. In particular, the structure of the *smap* is very well adapted for structural induction, whereas the structure of *ngmap* isn't at all. This ability to reason by induction on generalized maps will be crucial to adapt the fundamentally inductive proof of the normalization theorem.

9 Conclusion

In this article, we have shown a way to express and prove a well-known surface classification theorem, called the trading theorem, in the formalism of generalized maps of dimension 2. For that, we had to find a satisfying way to express the metric-related notion of topological equivalence in the framework of surface subdivisions. Our solution was to choose a set of operations that are likely to preserve the topology of their arguments, and declare that two 2-g-maps are topologically equivalent if one is the image of the other by any number of application of these operations. With the help of this definition, we have formally proved the trading theorem following a proof scheme proposed by Fomenko.

This result further confirms that generalized maps of dimension 2 are a good model for surfaces subdivisions, in which reasoning and object building is rather natural, as we did it with normal surfaces. On the opposite point of view, if one is convinced that 2-g-maps are a good model for surface subdivisions, one can be convinced that we have formally proved the trading theorem for surfaces. This success also shows the quality of our formal specification of generalized maps, that, while being very close to the mathematical specification of the maps, turns out to be powerful and simple enough to prove this hard-to-formalize theorem.

Of course, this work isn't over, as the normalization theorem hasn't been proved yet. Besides, months of experience in the use of our specification has made clear that some of our early decisions were not optimal. A "cleaning up" and fine-tuning step would be a good preliminary step to undertaking any major theorem studies. This will take some time, as the specification is already very large: it features 222 definitions, 1458 theorems for a total of 37000 script lines. Afterwards, it will also be interesting to make sure that the operations that we use really preserve the topology of their arguments by formalizing topology [13].

In the long term, we would like to use these results in the field of discrete geometry, for instance by representing each pixel by a square face and studying the obtained corresponding 2-g-maps. A good definition of discrete surfaces and irrefutable properties would be very welcome in the field of 3D imaging in order to define, build and manipulate "hulls" of voxels [2].

References

1. Barras, B. et al.: The Coq proof assistant reference manual. `http://coq.inria.fr/doc/main.html`
2. Bertrand, G., Malgouyres, R.: Some topological properties of surfaces in Z^3. Journal of Mathematical Imaging and Vision no 11 (1999) 207–221
3. Bertrand, Y., Dufourd, J.-F.: Algebraic specification of a 3D-modeler based on hypermaps. Graphical Models and Image Processing (1994) 56(1) 29–60
4. Coquand, T., Huet, G.: Constructions: a higher order proof system for mechanizing mathematics. EUROCAL (1985) Springer-Verlag LNCS **203**
5. Cori, R.: Un Code pour les Graphes Planaires et ses Applications. Société Math. de France, Astérisque 27 (1970)
6. Dehlinger, C., Dufourd, J.-F., Schreck, P.: Higher-Order Intuitionistic Formalization and Proofs in Hilbert's Elementary Geometry. Automated Deduction in Geometry (2000) Springer-Verlag LNAI **2061**
7. Firby, P. A., Gardiner, C. F.: Surface Topology. Ellis Horwood Ltd. (1982)
8. Fomenko, A. T.: Differential Geometry and Topology. Consultant Associates (1987)
9. Griffiths, H.: Surfaces. Cambridge University Press (1981)
10. Jacques, A.: Constellations et graphes topologiques. Combinatorial Theory And Applications (1970) 657–673
11. Kahn, G.: Elements of Constructive Geometry, Group Theory, and Domain Theory. Coq contribution (`http://coq.inria.fr/contribs-eng.html`)
12. Knuth, D.E.: Axioms and Hulls. Springer-Verlag (1992) LNCS **606**
13. Lienhardt, P.: Subdivisions of N-Dimensional Spaces and N-Dimensional Generalized Maps. Computational Geometry ACM Symp. (1989) 228–236
14. Martin-Löf, P.: Intuitionistic Type Theory. Bibliopolis (1984)
15. Parent, C.: Synthesizing proofs from programs in the calculus of inductive constructions. Mathematics of Program Construction (1995) Springer-Verlag LNCS **947**
16. Pichardie, D., Bertot, Y.: Formalizing Convex Hulls Algorithms. TPHOL (2001) Springer-Verlag LNCS **2152** 346–361
17. Puitg, F., Dufourd, J.-F.: Formal specifications and theorem proving breakthroughs in geometric modelling. TPHOL (1998) Springer-Verlag LNCS **1479** 401–427
18. Von Plato, J.: The axioms of constructive geometry. Annals of pure and applied logic **76** (1995) 169–200

Free-Style Theorem Proving*

David Delahaye

Programming Logic Group
Chalmers University of Technology
Department of Computing Science, S-412 96 Gothenburg, Sweden
delahaye@cs.chalmers.se
http://www.cs.chalmers.se/~delahaye

Abstract. We propose a new proof language based on well-known existing styles such as procedural and declarative styles but also using terms as proofs, a specific feature of theorem provers based on the Curry-Howard isomorphism. We show that these three styles are really appropriate for specific domains and how it can be worth combining them to benefit from their advantages in every kind of proof. Thus, we present, in the context of the Coq proof system, a language, called $\mathcal{L}_{pdt}$, which is intended to make a fusion between these three styles and which allows the user to be much more free in the way of building his/her proofs. We provide also a formal semantics of $\mathcal{L}_{pdt}$ for the Calculus of Inductive Constructions, as well as an implementation with a prototype for Coq, which can already run some relevant examples.

1 Introduction

In theorem provers, we can generally distinguish between two kinds of languages: a proof[1] language, which corresponds to basic or more elaborated primitives and a tactic language, which allows the user to write his/her own proof schemes. In this paper, we focus only on the first kind of language and we do not deal with the power of automation of proof systems. Here, we are interested in expressing proofs by considering different kinds of styles, finding the "good" criteria and thinking about combining these criteria according to the proof.

Proof languages for theorem provers have been extensively studied and, currently, two styles have gained prominence: the procedural style and the declarative style. John Harrison makes a complete comparison between these two styles in [5], where the declarative style is rather enhanced but also moderated. Indeed, the author shows how it is difficult to make a clear division between these two styles and this can be seen in the Mizar mode [6] he developed for HOL. In the same way, Markus Wenzel has implemented a generic declarative layer, called Isar [11], on top of Isabelle. From a more declarative-dedicated viewpoint, there

* This work has been realized within the LogiCal project (INRIA-Rocquencourt, France).

[1] The word "proof" may be open to several interpretations. Here, we mostly use "proof" in the sense of a script to be presented to a machine for checking.

V.A. Carreño, C. Muñoz, S. Tahar (Eds.): TPHOLs 2002, LNCS 2410, pp. 164–181, 2002.

are the work of Don Syme with his theorem prover Declare [8] and Vincent Zammit's thesis [12]. Closer to natural language, there is also the Natural package of Yann Coscoy [2] in Coq, where it is possible to produce *natural* proofs from proof terms and to write *natural* proof scripts directly. Along the same lines, in Alfa [1] (successor of ALF), we can produce *natural* proofs in several languages.

A third kind of proof language, quite specific to the logic used by the proof system, could be called "term language". This is a language which uses the coding of proofs into terms and propositions into types. This semantic (Brouwer-Heyting-Kolmogorov) deals only with intuitionistic logics (without excluded middle) and significantly modifies the way proofs are checked. In such a context, the logic is seen as a type system and checking if a proof (term) corresponds to a proposition (type) is only typechecking (Curry-Howard's isomorphism). The first versions of Coq used that kind of language before having a procedural style even if it is always possible to give terms as proofs. In Lego or Nuprl, we have exactly the same situation. Currently, only Alfa uses a direct manipulation of terms to build proofs.

The main idea in this paper is to establish what the contributions of these three kinds of languages are, especially in which frameworks, and to make a fusion which could be interesting in practice. Before presenting our proposition, we consider a short example of a proof to allow the reader to have some idea of these three proving styles. It also illustrates the various features of these languages and shows the domains (in terms of the kinds of proofs) in which they are suitable. Next, we give the syntax of the language we suggest in the context of Coq [9], and which has been called $\mathcal{L}_{pdt}$ ("pdt" are initials denoting the fusion between the three worlds: "p" is for procedural, "d" is for declarative and "t" is for term), as well as some ideas regarding the formal semantics that has been designed. This last point is also a major originality of this work and as far as the author is aware, ALF is the only system, which has a formally described proof language [7]. However, this semantics deals only with proof terms and, here, we go further trying to give also, in the same context, a formal semantics to procedural parts, as well as to declarative features. From this semantics, a prototype has been carried out and, finally, we consider some examples of use, which show how $\mathcal{L}_{pdt}$ can be appropriate, but also what kind of improvements can be expected.

2 Proof Examples

The example we chose to test the three kinds of languages is to show the decidability[2] of equality on the natural numbers. It can be expressed as follows:

$$\forall n, m \in \mathbb{N}. n = m \vee \neg(n = m)$$

[2] This proposition is called "decidability of equality" because we will give an intuitionistic proof in every case. Thus, we can always *realize* the proof of this lemma toward a program which, given two natural numbers, answers "yes" if the two numbers are equal and "no" otherwise.

Informally, the previous proposition can be shown in the following way:

1. We make an induction on n.
   The basis case gives us $0 = m \vee \neg(0 = m)$ to be shown:
   (a) We reason by case on m.
   First, we must show that $0 = 0 \vee \neg(0 = 0)$, which is trivial because we know that $0 = 0$.
   (b) Next, we must show that $0 = m + 1 \vee \neg(0 = m + 1)$, which is trivial because we know that $\neg(0 = m + 1)$.
2. For the inductive case, we suppose that we have $n = m \vee \neg(n = m)$ (H) and we must show that $n + 1 = m \vee \neg(n + 1 = m)$:
   (a) We reason by case on m.
   First, we must show that $n+1 = 0 \vee \neg(n+1 = 0)$, which is trivial because we know that $\neg(n + 1 = 0)$.
   (b) Next, we must show that $n + 1 = m + 1 \vee \neg(n + 1 = m + 1)$. Thus, we reason by case with respect to the inductive hypothesis H:
      i. Either $n = m$: we can conclude because we can deduce that $n+1 = m+1$.
      ii. Or $\neg(n = m)$: we can conclude in the same way because we can deduce that $\neg(n + 1 = m + 1)$.

2.1 Procedural Proof

As a procedural style, we decided to choose the Coq proof system [9]. A possible proof script for the previous proposition is the following:

```
Lemma eq_nat:(n,m:nat)n=m\/~n=m.
Proof.
  Induction n.
  Intro;Case m;[Left;Auto|Right;Auto].
  Intros;Case m;[Right;Auto|Intros].
  Case (H n1);[Intro;Left;Auto|Intro;Right;Auto].
Save.
```

Briefly, we can describe the instructions involved: `Intro[s]` introduces products in the local context (hypotheses), `Induction` applies induction schemes, `Case` makes reasoning by case, `Left`/`Right` applies the first/second constructor of an inductive type with two constructors and `Auto` is a resolution tactic which mainly tries to apply lemmas of a data base.

2.2 Declarative Style

For the declarative proof, we used Mizar [10], which seems to be very representative of this proving style[3]. The proof is partially described in the following script (the complete proof is a little too long, so we have only detailed the basis case of the first induction):

[3] Indeed, the Mizar project developed a very impressive library essentially in mathematics. So, we can easily consider the stability of this language with respect to its very widespread use.

```
theorem
  eq_nat: n = m or not (n = m)
  proof
    A1: for m holds 0 = m or not (0 = m)
    proof
      A2: 0 = 0 or not (0 = 0)
      proof
        A3: 0 = 0;
        hence thesis by A3;
      end;
      A4: (0 = m0 or not (0 = m0)) implies
          (0 = m0 + 1 or not (0 = m0 + 1))
      proof
        assume 0 = m0 or not (0 = m0);
        A5: not (0 = m0 + 1) by NAT_1:21;
        hence thesis by A5;
      end;
      for k holds 0 = k or not (0 = k) from Ind(A2,A4);
      hence thesis;
    end;
    A6: (for m holds (n0 = m or not (n0 = m))) implies
        (for m holds ((n0 + 1) = m or not ((n0 + 1) = m)))
    ...
```

where `by`/`from` is used to perform a more or less powerful resolution, **`hence thesis`**, to conclude the current proof (between `proof` and `end`) and `assume`, to suppose propositions.

2.3 Term Refinement

The third solution to solve the previous proposition is to give the proof term more or less directly. This can be done in a very easy and practical way by the prover Alfa [1], which is a huge graphic interface to Agda (successor of ALF). Figure 1 shows a part of the proof term which has been built gradually (filling placeholders). As for the Mizar proof, the complete proof is a bit too large, so, in the same way, we have only printed the basis case of the first induction. The rest is behind the "..." managed by Alfa itself, which allows parts of proofs to be hidden. Even if the explicit pretty-print allows well-known constructions (λ-abstractions, universal quantifiers, ...) to be recognized, some comments can be made to understand this proof. The universal quantifier $\forall$ has a special coding and to be introduced/eliminated, specific constants $\forall$I/$\forall$E are to be used. $\vee$I1/$\vee$I2 allow the side of a $\vee$-proposition to be chosen. Finally, *natrec* is the recursor (proof term) of the induction scheme on natural numbers, *eqZero* is the proof that $0 == 0$ and *nat_ discr1* is the lemma $\forall a \in Nat.\neg(0 == a+1)$.

2.4 Comparison

The procedural script in Coq shows that it is quite difficult to read procedural proofs. The user quickly becomes lost and does not know what the state of the

```
File | Edit | View | Options | Utils

import Examples/Satslogik
import Examples/Predikatlogik
import Examples/Nat
import Examples/EqNat
[ natrec (C ∈ Nat → Prop, bc ∈ C 0,
          ic ∈ (n ∈ Nat) → C n → C (succ n), m ∈ Nat) ∈ C m
    natrec C bc ic 0 ≡ bc
[   natrec C bc ic (n + 1) ≡ ic n (natrec C bc ic n)
[ postulate nat_discr1   ∈ ∀a ∈ Nat.¬ (0 == (a + 1))
[ postulate nat_discr2   ∈ ∀a ∈ Nat.¬ ((a + 1) == 0)
[ postulate not_eqSucc (a, b ∈ Nat, aeqb ∈ ¬ (a == b)) ∈ ¬ ((a + 1) == (b + 1))
[ eq_nat_dec   ∈ ∀a ∈ Nat.∀a' ∈ Nat.a == a' ∨ ¬ (a == a')
    eq_nat_dec ≡ ∀I (λa → [∀a' ∈ Nat.a == a' ∨ ¬ (a == a')]
                         natrec
                           (λh → ∀a' ∈ Nat.h == a' ∨ ¬ (h == a'))
                           (∀I (λa' → natrec
                                        (λh → 0 == h ∨ ¬ (0 == h))
                                        (∨I1 eqZero)
                                        (λn h → ∨I2 (∀E nat_discr1))
                                        a'))
                           (λn h → ∀I ...)
[                          a)

Import Examples/Satslogik
```

Fig. 1. Proof of the decidability of equality on natural numbers in Alfa

proof is. Without the processing of Coq, the user cannot understand the proof and this is enforced by, for example, the apparition of free variables (as `H` or `n1` here) which are only bound during the checking of the proof. So, that kind of script cannot be written in a batch way but only and very easily in interaction with Coq. Thus, procedural proofs are rather backward-oriented because there is not much information provided by user's instructions and everything depends on the current goal to be proved. As concerns maintainability, obviously, we can imagine that these scripts are quite sensitive to system changes (naming conventions for example).

The Mizar script shows a very readable and so understandable proof. The declarations of auxiliary states allows the user to know what is proved and how to conclude. The proof can be written in batch mode and an interactive loop is not required at all to build this kind of proof. Indeed, the state of the proof may be *far* from the initial goal to be proved and we know how to prove it only at the end when we know how to make the link between this initial goal and the auxiliary lemmas we have proved to succeed, so that, declarative scripts are rather forward-oriented. A direct inconvenience of this good readability is that it is rather verbose and the user may well find it tedious to repeat many propositions (even with copy-paste). These many (forced) repetitions make declarative scripts fragile with respect to specification changes (permutations of arguments in a definition, for example) and we cannot imagine specific edition tools which could deal with every kind of change.

The proof in Alfa is rather readable. Of course, it requires a little practice to read terms as proofs but it is quite full of information especially when you reveal type annotations. In the same way, it is easy to write such scripts[4], both in an interactive proof loop as well as in batch mode, because, when building terms, we can ignore type annotations, which are filled by the proof system with respect to the goal. Thus, that kind of style is completely backward-oriented. Regarding robustness, system changes does not interfere because we do not give instructions to build a proof term but a proof term directly. But, as declarative scripts, specification changes may involve many changes in the term even without type signatures.

Following on the previous observations, we can easily tell where these styles are useful and also when it is interesting to use them. We will use procedural proofs for small proofs, known to be trivial and realized interactively in a backward mode, for which, we are not interested in the formal details. These proofs must be seen as black boxes. Declarative style will be used for more complex proofs we want to build more in a forward style (as in a natural proof in mathematics), rather in a batch way and very precisely[5], i.e. with much information for the reader. Finally, proof terms will be also for complex proofs, but backward-oriented, built either interactively or in batch mode (this is adapted for both methods), and for which we can choose the level of detail (put all type signatures, some of them or none). Thus, we can notice that the three styles correspond to specific needs and it could be a good idea to amalgamate them to benefit from their advantages in every kind of proof.

3 Presentation of $\mathcal{L}_{pdt}$

As said previously, $\mathcal{L}_{pdt}$ was developed in the context of the Coq proof system and there are several reasons for that choice. First, in this theorem prover, the procedural part comes for free because the proof language is purely procedural. But, the main reason is certainly that Coq uses the Curry-Howard isomorphism to code proofs into terms and the adaptation to make Coq accept terms as proofs will be rather easy[6]. Only declarative features are not quite natural for insertion in Coq but we will see how they can be simulated on a procedural proof

---

[4] We do not consider here the useful graphical interface which allows placeholders to be filled very easily (especially for beginners).

[5] However, this depends on the level of automation (that we do not want to consider here). Too great, an automation can strictly break readability even if we ask decision procedures to give details of the proofs they built because automated proofs are not really minimal and thus as clear as "human"-built proofs.

[6] We can already give complete terms (`Exact` tactic) or partial terms (`[E]Apply`, `Refine` tactics) to prove lemmas, but no means is provided to build these terms gradually or to have a total control on the placeholders we want to fill, so most of these tactics are only used internally.

machine. Although the choice for suitable systems might seem rather restricted (procedural systems based on Curry-Howard's isomorphism), such an extension could be also possible and quite appropriate for systems like Lego or Nuprl.

3.1 Definition

The syntax of $\mathcal{L}_{pdt}$ is shown in figure 2, presented in a BNF-like style and where *<proof>* is the start entry. *<ident>* and *<int>* are respectively the entries for identifiers and integers. *<tac>* includes all the current tactic language of Coq. Here, we have deliberately simplified the *<term>* entry. In particular, the `Cases` construction has been reduced compared to the real version and the `Fix`/`CoFix` forms has been removed (we will not use them directly in our examples and we will consider recursors just as constants in the way of Martin-Löf's type theory).

3.2 Semantics

As said previously, one of the major novelties of this work is to try to give a formal semantics to a proof language, here $\mathcal{L}_{pdt}$, not only for proof terms as for ALF in [7], but also for procedural and declarative parts. We chose to build a natural semantics (big steps), which is very concrete and so very close to a possible implementation. This semantics is correct in the context of the Calculus of Inductive Constructions (CIC) but without universes and to deal with the complete theory with universes, we only have to verify the universes constraints at the end with a specific procedure we did not formalize (this allows us to skip universe constraints from the evaluation rules and to keep *light* rules which are easier to read). Thus, in this context, to give a value to a proof script will consist in giving a term whose type is convertible to the lemma to be proved.

Preliminaries. In figure 2, we call *proof script* or simply *script*, every expression of the *<proof>* entry. A *sentence* is an expression of the *<proof-sen>* entry. We call *term*, every expression of the *<term>* entry and *pure term*, every expression of the *<term>* entry, the *<proc>* and *<decl>* entries excluded. We call *procedural part*, every expression of the *<proc>* entry and declarative part, every expression of the *<decl>* entry. Finally, a tactic is an expression of the *<tac>* entry.

A *local environment* or *context* is an ordered list of hypotheses. An hypothesis is of the form $(x : T)$, where x is an identifier and T, a pure term called type of x. A *global environment* is an ordered list of hypotheses and definitions. A definition may be inductive or not. A non-inductive definition is of the form $c := t : T$, where c is an identifier called constant, t, a pure term called body of c and T, a pure term called type of c or t. An inductive definition is of the form $\mathsf{Ind}[\Gamma_p](\Gamma_d := \Gamma_c)$, where Γ_p, Γ_d and Γ_c are contexts corresponding to the parameters, the defined inductive types and the constructors. A *goal* is made of a global environment Δ, a local environment Γ and a pure term T (type).

```
<proof>         ::= (<proof-sen>)+

<proof-sen>     ::= <proof-top>.

<proof-top>     ::= Let [[?<int>]] <ident> : <term>
                  | Let <let-clauses>
                  | ?<int> [: <term>]:= <term>
                  | Proof
                  | Qed | Save

<let-clauses>   ::= [[?<int>]] <ident> [: <term>] := <term>
                    (And [[?<int>]] <ident> [: <term>] := <term>)*

<term>          ::= [<binders>]<term>
                  | <term> -> <term>
                  | (<ident>(, <ident>)* : <term>)<term>
                  | ((<term>)*)
                  | [<<term>>] Cases <term> of (<rules>)* end
                  | Set | Prop | Type
                  | <ident>
                  | ? | ?<int>
                  | <proc>
                  | <decl>

<binders>       ::= <ident> (,<ident>)* [: <term>](; <binders>)*

<rules>         ::= [|] <pattern> => <term> (| <pattern> => <term>)*

<pattern>       ::= <ident> | ((<ident>)+)

<proc>          ::= <by <tac>>

<tac>           ::= Intro | Induction <ident> | ...

<decl>          ::= Let <let-clauses> In <term>
```

Fig. 2. Syntax of $\mathcal{L}_{pdt}$

The *evaluated terms* are defined by the *<term-eval>* entry of figure 3. The set of evaluated terms is noted $\mathcal{T}_{\mathcal{E}}$. We call *implicit argument*, every term of the form ?, or every evaluated term of the form $?_{in}$, where n is an integer. The set of evaluated terms which are implicit arguments is noted $\mathcal{I}_{\mathcal{E}}$. We call *metavariable*, every term or every evaluated term of the form $?n$, where n is an integer. Finally, every term or every evaluated term of the form `Set`, `Prop` or `Type` is a *sort* and the sort set is noted S.

```
<term-eval> ::= [<binders>]<term-eval>
              | <term-eval> -> <term-eval>
              | (<ident>(, <ident>)* : <term-eval>)<term-eval>
              | ((<term-eval>)*)
              | [<<term-eval>>] Cases <term-eval> of (<rules>)* end
              | Set | Prop | Type
              | <ident>
              | ?i<int> | ?<int>
```

Fig. 3. Evaluated terms

Term Semantics. This consists in typechecking incomplete[7] terms of CIC (pure terms) and interpreting some other parts (procedural or declarative parts) which give indications regarding the way the corresponding pure term has to be built. Procedural and declarative parts which are in the context of a term are replaced by metavariables and instantiations (of those metavariables) where they occur at the root.

The values are:

- either $(t, \Delta[\Gamma \Vdash T], m, \sigma)$, where $t \in \mathcal{T}_{\mathcal{E}}$, $\Delta[\Gamma \Vdash T]$ is a goal, m is a set of couples metavariable numbers-goals $(i, \Delta[\Gamma_i \Vdash T_i])$, σ is a substitution from $\mathcal{I}_{\mathcal{E}}$ to $\mathcal{T}_{\mathcal{E}}$, s.t. $\nexists j$, $?j \in \Delta[\Gamma \Vdash T]$ and $?j \in \Delta[\Gamma_i \Vdash T_i]$,
- or Error.

The set of those values is noted $\mathcal{V}_{\mathcal{T}}$.

Thus, a value is given by a proof term t, the type of this proof term $\Delta[\Gamma \Vdash T]$, a set of metavariables (bound to their types) m occurring in t and an instantiation of some implicit arguments (coming from the typechecking of t) σ. Moreover, there are two restrictions: there is no metavariable in the type of t ($\Delta[\Gamma \Vdash T]$), as well as in the types of the metavariables (in m).

Two modes of evaluation are possible if we have the type of the term or not: the verification mode and the inference mode. Initially, we have the type of the term (coming from the goal) and we use the verification mode, but some terms (like applications) can only be evaluated in the inference mode.

Due to lack of space here, we will not be able to give all the evaluation rules (with the corresponding error rules) of the several semantics. The reader can refer to [4][8] for a complete description (figures 5.3–5.16, pages 57–73) and we will just give some examples of rules. In particular, we will not consider the error rules, which can be also found in [4] (appendix A).

[7] Indeed, pure terms may contain some metavariables.

[8] A version of this document can be found at: `http://logical.inria.fr/~delahaye/these-delahaye.ps`.

Pure Terms. A pure term t is evaluated, in the inference mode (resp. in the verification mode), into v, with $v \in \mathcal{V}_T$, in the undefined goal $\Delta[\Gamma]$ (resp. in the goal $\Delta[\Gamma \Vdash T]$), iff $(t, \Delta[\Gamma]) \rhd v$ (resp. $(t, \Delta[\Gamma \Vdash T]) \rhd v$) holds.

As an example of rule (to derive the relation $\rhd$), the evaluation of a λ-abstraction *à la Curry* in an undefined goal is the following:

$$\frac{\begin{array}{c}(t, \Delta[\Gamma, (x : ?_{in})]) \rhd (t_1, \Delta[\Gamma_1, (x : T_1) \Vdash T_2], m, \sigma) \\ x \notin \Delta[\Gamma] \qquad \mathsf{new\_impl}(n)\end{array}}{([x : ?]t, \Delta[\Gamma]) \rhd ([x : T_1]t_1, \Delta[\Gamma_1 \Vdash (x : T_1)T_2], m, \sigma)} \; (\lambda_{\mathsf{UCurry}})$$

where $\mathsf{new\_impl}(n)$ ensures that the integer n numbering an implicit argument is unique.

This rule consists in giving a number (n) to the implicit argument corresponding to the type of x, verifying that x is not already in the (global and local) environments and typechecking t.

Declarative Parts. We extend the relation $\rhd$ to deal with the declarative parts: a pure term or a declarative part t is evaluated, in the inference mode (resp. in the verification mode), into v, with $v \in \mathcal{V}_T$, in the undefined goal $\Delta[\Gamma]$ (resp. in the goal $\Delta[\Gamma \Vdash T]$), iff $(t, \Delta[\Gamma]) \rhd v$ (resp. $(t, \Delta[\Gamma \Vdash T]) \rhd v$) holds.

For instance, a `Let` ... `In` in an undefined goal is evaluated as follows:

$$\frac{\begin{array}{c}(T_1, \Delta[\Gamma]) \rhd (T_2, \Delta[\Gamma_1 \Vdash s], m_1, \sigma_1) \qquad s \in S \\ (t_1, \Delta[\Gamma_1 \Vdash T_2]) \rhd (t_3, \Vdash [\Gamma_2 \Vdash T_3], m_2, \sigma_2) \\ (t_2, \Delta[\Gamma_2, (x : T_3)]) \rhd (t_4, \Delta[\Gamma_3, (x : T_4) \Vdash T_5], m_3, \sigma_3)\end{array}}{\begin{array}{c}(Let\ x : T_1 := t_1\ In\ t_2, \Delta[\Gamma]) \rhd \\ (([x : T_4]t_4\ t_3\sigma_3), \Delta[\Gamma_3 \Vdash T_5[x \backslash t_3\sigma_3]], (m_1\sigma_2 \cup m_2)\sigma_3 \cup m_3, \sigma_1\sigma_2\sigma_3)\end{array}} \; (\mathsf{ULetIn})$$

In this rule, first, the type of the binding (T_1) is typechecked, next, the body of the binding (t_1) is typechecked with the previous type where some instantiations (of implicit arguments) may have been performed (T_2) and, finally, the body of the `Let` ... `In` is typechecked adding x as well as the new type of x (T_3) in the local context. We can notice that this rule is sequential (the rule must be read from the top to the bottom) and instead of allowing *parallel* evaluations, which would require a specific mechanism to merge the several constraints, we chose to impose an evaluation order giving the constraints from an evaluation to the next one.

Procedural Parts. Finally, to deal with all the terms, we have to include the procedural parts in the relation $\rhd$. To do so, we have to define the evaluation of tactics. This evaluation can be done only in verification mode because most of tactics do not provide a pure term as proof but a method to build a pure term from goals which they are applied to.

A tactic t is evaluated into v, with $v \in \mathcal{V}_T$, in the goal $\Delta[\Gamma \Vdash T]$, which is noted $(t, \Delta[\Gamma \Vdash T]) \blacktriangleright v$, iff $(t, \Delta[\Gamma \Vdash T]) \blacktriangleright v$ holds.

For example, the Cut tactic, which is used to make a cut of a term, is evaluated by means of the following rule:

$$\frac{(T_1, \Delta[\Gamma]) \rhd (T_3, \Delta[\Gamma_1 \Vdash s], \emptyset, \sigma_1) \qquad s \in S \qquad \mathsf{new\_meta}(n_1, n_2)}{\begin{array}{c}(\texttt{Cut}\ T_1, \Delta[\Gamma \Vdash T_2]) \unrhd ((?n_1\ ?n_2), \Delta[\Gamma_1 \Vdash T_2\sigma_1],\\ \{(n_1, \Delta[\Gamma_1 \Vdash (x : T_3)T_2\sigma_1]), (n_2, \Delta[\Gamma_1 \Vdash T_3])\}, \sigma_1)\end{array}}\ (\mathsf{Cut})$$

where $\mathsf{new\_meta}(n_1, n_2, \ldots, n_p)$ ensures that the integers $n_1, n_2, \ldots, n_p$, numbering some metavariables, are unique.

As this rule generates a cut, that is to say an application in the proof term, this produces two subgoals, bound to metavariables $?n_1$ and $?n_2$, and the corresponding proof term is $(?n_1\ ?n_2)$. Once the term to be cut (T_1) evaluated (into T_3), the first goal (bound to $?n_1$), which corresponds to the head function, is a product type $((x : T_3)T_2\sigma_1)$ and the second subgoal (bound ot $?n_2$), which corresponds to the argument, is the result of the previous evaluation (T_3).

Now, using the relation $\unrhd$[9], we can give the complete evaluation of the terms, including the procedural parts: a term t is evaluated, in the inference mode (resp. in the verification mode), into v, with $v \in \mathcal{V}_T$, in the undefined goal $\Delta[\Gamma]$ (resp. in the goal $\Delta[\Gamma \Vdash T]$), iff $(t, \Delta[\Gamma]) \rhd v$ (resp. $(t, \Delta[\Gamma \Vdash T]) \rhd v$) holds.

Sentence Semantics. A sentence allows us to instantiate a metavariable or to add a lemma in the context of a metavariable. The values are the following:

- $(t, \Delta[\Gamma \Vdash T], m, p, d)$, where t is a pure term, $\Delta[\Gamma \Vdash T]$ is a goal, m is a set of couples metavariable numbers-goals $(i, \Delta[\Gamma_i \Vdash T_i])$, p is a stack of values ($[\,]$ is the empty stack and $\lhd$ is the concatenation operator), $d \in \mathcal{D}$ with $\mathcal{D} = \{\mathsf{Lemma}; \mathsf{Let}_i; \mathsf{Lemma}_b; \mathsf{Let}_{i,b}\}$, s.t. $\nexists j$, $?j \in \Delta[\Gamma \Vdash T]$ and $?j \in \Delta[\Gamma_i \Vdash T_i]$,
- Error

The set of those values is noted $\mathcal{V}_\mathcal{P}$ and a value will be also called a *state*.

Here, as for the term semantics, a value is given by a proof term t, the type of this proof term $(\Delta[\Gamma \Vdash T])$ and the set of the metavariables (bound to their types) m occurring in t. A difference is that we do not keep an instantiation of the implicit arguments and, in particular, this means that all the implicit arguments involved by a sentence must be solved during the evaluation of this sentence (we cannot use the next sentences to do so). Moreover, in a value, we have also a stack of values, which will be used to store the current proof session when opening new proof sessions during the evaluation of Let's which provide no proof term. Finally, in the same way, we add a declaration flag, which deals with the several possible declarations (Lemma or Let). Initially, this flag is set to Lemma and becomes Lemma_b when Proof is evaluated. This is similar for Let

[9] The relation $\unrhd$ is used by the relation $\rhd$ during the evaluation of a <by ... > expression (rule By in [4]).

with the flags Let_i and $\mathsf{Let}_{i,b}$, where we give also a tag (here i) which is the metavariable corresponding to the upcoming proof of the Let.

A sentence "$p.$" is evaluated into v, with $v \in \mathcal{V}_{\mathcal{P}}$, in the state (or value) $(t, \Delta[\Gamma \Vdash T], m, p, d)$, iff $(p., t, \Delta[\Gamma \Vdash T], m, p, d) \twoheadrightarrow v$ holds.

As an example, let us consider the evaluation of a Let in the current goal:

$$\frac{\begin{array}{c} d = \mathsf{Lemma}_b \text{ ou } d = \mathsf{Let}_{i,b} \\ n = min\{i \mid (?i, b) \in m\} \qquad (?n, \Delta[\Gamma_n \Vdash T_n]) \in m \qquad x \notin \Gamma_n \\ (T, \Delta[\Gamma_n]) \rhd (T_2, \Delta[\Gamma_1 \Vdash s], \emptyset, \sigma) \qquad \mathsf{new\_meta}(k) \end{array}}{\begin{array}{c} (\mathtt{Let}\ x : T, t, \Delta[\Gamma \Vdash T_1], m, p, d) \Rrightarrow \\ (?1, \Delta[\Gamma_1 \Vdash T_2], \{(?1, \Delta[\Gamma_1 \Vdash T_2])\}, \\ \lfloor((\,[x : T_2]?n\ ?k), \Delta[\Gamma \Vdash T_1]\sigma, (m - \{(?n, \Delta[\Gamma_n \Vdash T_n])\})\sigma \cup \\ \{(?n, \Delta[\Gamma_n\sigma, (x : T_2) \Vdash T_n\sigma]; (?k, \Delta[\Gamma_1 \Vdash T_2])\}, p\sigma, d)\rfloor \lhd p\sigma, \mathsf{Let}_k) \end{array}} \ (\mathsf{LetCur})$$

In this rule[10], the first line consists only in verifying that we are in the context of a opened proof (introduced by Lemma or another toplevel Let, and where a Proof command has been evaluated). Next, we look for the current goal, which this Let is applied to. Our convention is to choose this goal such that it is bound to the lowest metavariable number (n). The type of the Let (T) is then evaluated (into T_2) and we generate a new metavariable ($?k$), which will correspond to the proof given to the Let (once Qed reached, this metavariable will be replaced by the actual proof). The result of the evaluation shows that we obtain a new goal bound to metavariable $?1$ (this will be also when opening a new proof session with Lemma). This goal has the context of the current goal (after the evaluation of T, i.e. Γ_1) and the evaluated type introduced by the Let (T_2). The current proof state is put in the stack. In this state, as expected, the global proof term is an application $(([x : T_2]?n\ ?k))$. Here, contrary to the previous tactic Cut, the type of the cut is introduced automatically (this is an expectable behavior) and this is why the proof term has a λ-abstraction in head position. Finally, the Let_k flag is set to declare that we are in the context of Let bound to metavariable $?k$. To accept other Let instructions or instantiations, a Proof command has to be processed to transform this flag into the flag $\mathsf{Let}_{k,b}$.

Script Semantics. Finally, the values of the script semantics are:

- $(t, \Delta[\Vdash T])$, where t is a pure term, $\Delta[\Vdash T]$ is a goal, s.t. $\nexists j,\ ?j, ?_{ij} \in t, T$.
- Error

The set of those values is noted $\mathcal{V}_{\mathcal{S}}$.

As a script must provide a complete proof, a value is a proof term t and the type of this proof term $((t, \Delta[\Vdash T]))$, where there is no metavariable or implicit argument. The context of the type is empty because we suppose that a script is a proof of a lemma introduced in the toplevel of the global environment (Δ) by the command Lemma. Hence, we call *lemma*, every goal $\Delta[\Vdash T]$, where $\Delta[\Vdash T]$ is a goal introduced by Lemma.

---

[10] Here, we use the relation $\Rrightarrow$ which is related to the relation $\twoheadrightarrow$ just adding a dot (".") at the end of the sentence (rule Sen in [4]).

A script "$p_1.\ p_2.\ \ldots\ p_n.$" of the lemma $\Delta[\Vdash T]$ is evaluated into v, with $v \in \mathcal{V}_S$, iff $(p_1.\ p_2.\ \ldots\ p_n., \Delta[\Vdash T]) \downarrow v$ holds.

This evaluation consists mainly in iterating the rules for sentences.

4 Examples

We have implemented $\mathcal{L}_{pdt}$ for the latest versions V7 of Coq. Currently, this is a prototype which is separate from the official release and all the following examples have been carried out in this prototype.

4.1 Decidability of the Equality on Natural Numbers

Now, we can go back to the example we introduced to compare the three proving styles. Using $\mathcal{L}_{pdt}$, the proof could look like the following script:

```
Lemma eq_nat:(n,m:nat)n=m\/~n=m.
Proof.
  Let b_n:(m:nat)(0)=m\/~(0)=m.
  Proof.
    ?1 := [m:nat]
          <[m:nat](0)=m\/~(0)=m>
          Cases m of
          | O => <by Left;Auto>
          | (S n) => <by Right;Auto>
          end.
  Qed.
  Let i_n:(n:nat)((m:nat)n=m\/~n=m)->(m:nat)(S n)=m\/~(S n)=m.
  Proof.
    ?1 := [n:nat;Hrec;m:nat]
          <[m:nat](S n)=m\/~(S n)=m>
          Cases m of
          | O => <by Right;Auto>
          | (S n0) => ?2
          end.
    ?2 := <[_:n=n0\/~n=n0](S n)=(S n0)\/~(S n)=(S n0)>
          Cases (Hrec n0) of
          | (or_introl _) => <by Left;Auto>
          | (or_intror _) => <by Right;Auto>
          end.
  Qed.
  ?1 := (nat_ind ([n:nat](m:nat)n=m\/~n=m) b_n i_n).
Save.
```

In this proof, we have mixed the three *proof worlds*. The first induction on `n` is done declaratively with two toplevel `Let`'s for the basis (`b_n`) and the inductive (`i_n`) cases. The two cases are used to perform the induction on `n`, at the end of the script, by an instantiation (of `?1`) with a term using `nat_ind`, the induction scheme on natural numbers (the first argument of `nat_ind` is the predicate to be proved by induction). The proofs of `b_n` and `i_n` are realized by instantiations with terms and trivial parts are settled by small procedural scripts, which are clearly identified (using `<by ...>`). In particular, reasonings by case are dealt with `Cases` constructions and the term between `<...>` is the corresponding predicate (the first λ-abstraction is the type of the term to be destructured and the rest is the type of the branches, i.e. the type of the `Cases`). Currently, these predicates have to be provided (by the user) because all the `Cases` used in this script are incomplete (presence of metavariables, even with procedural scripts, see section 3.2) and the typechecker (due to possible dependences) needs to be helped to give a type to metavariables.

More precisely, to solve the basis case (`b_n`), we use directly an instantiation (of `?1`). We *assume* the variable `m` (with a λ-abstraction `[...]`) and we perform a reasoning by case (with `Cases`) on `m`. If `m` is equal to `O`, then this is trivial and we can solve it in a procedural way. We choose the left-hand side case with the tactic `Left` and we have to prove `O=O`, which is solved by the tactic `Auto`. Otherwise, if `m` is of the form `(S n)`, it is quite similar and we choose the right-hand side case with the tactic `Right` to prove `~(S n)=O` with `Auto`.

For the inductive case (`i_n`), we use also directly an instantiation (of `?1`). We assume the induction hypothesis `Hrec` (coming from the induction on `n`) and we perform another reasoning by case on `m`. If `m` is equal to `O`, this is trivial and we solve it procedurally in the same way as in the basis case. Otherwise, if `m` is equal to `(S n0)`, this is a bit more complicated and we decide to *delay* the proof using a metavariable (`?2`). Next, this metavariable is instantiated performing a reasoning by case on `Hrec` applied to `n0`. The two cases correspond to either `n=n0` or `~n=n0` (`or_introl/r` is the first/second constructor of `\/`), which are trivially solved in a procedural way as in the basis case.

As said above, the last part of the script consists in solving the global lemma using an instantiation (of `?1`) and performing the induction on `n` with the induction scheme `nat_ind`. This scheme takes three arguments: the predicate to be proved by induction, the proof of the basis case and the proof of the inductive case. The two last proofs have been already introduced by two toplevel `Let`'s and they are directly used.

4.2 Stamp Problem

As another example, let us consider the stamp problem, coming from the PVS tutorial [3] and asserting that every postage requirement of 8 cents or more can be met solely with stamps of 3 and 5 cents. Formally, this means that every natural number greater or equal to 8 is the sum of some positive multiple of 3 and some positive multiple of 5. This lemma can be expressed and proved (using $\mathcal{L}_{pdt}$) in the following way:

```
Lemma L3_plus_5:(n:nat)(EX t:Z|(EX f:Z|`(inject_nat n)+8=3*t+5*f`)).
Proof.
  Let cb:(EX t:Z|(EX f:Z|`(inject_nat O)+8=3*t+5*f`)).
  Proof.
    ?1 := (choose `1` (choose `1` <by Ring>)).
```

```
  Qed.
  Let ci:(n:nat)(EX t:Z|(EX f:Z|'(inject_nat n)+8=3*t+5*f'))->
                (EX t:Z|(EX f:Z|'(inject_nat (S n))+8=3*t+5*f')).
  Proof.
    ?1 := [n;Hrec](skolem ?2 Hrec).
    ?2 := [x;H:(EX f:Z|'(inject_nat n)+8=3*x+5*f')](skolem ?3 H).
    ?3 := [x0;H0:'(inject_nat n)+8 = 3*x+5*x0']?4.
    Let cir:(EX t:Z|(EX f:Z|'3*x+5*x0+1=3*t+5*f')).
    Proof.
      ?1 := <[_:?](EX t:Z|(EX f:Z|'3*x+5*x0+1=3*t+5*f'))>
            Cases (dec_eq 'x0' '0') of
            | (or_introl H) => ?2
            | (or_intror _) => ?3
            end.
      ?2 := (choose 'x-3' (choose '2' <by Rewrite H;Ring>)).
      ?3 := (choose 'x+2' (choose 'x0-1' <by Ring>)).
    Qed.
    ?4 := <by Rewrite inj_S;Rewrite Zplus_S_n;Unfold Zs;Rewrite H0;
              Exact cir>.
  Qed.
  ?1 := (nat_ind ([n:nat]?) cb ci).
Save.
```

where Z is the type of integers, EX is the existential quantifier, inject_nat is the injection from nat to Z (this injection is needed because we have set this equality over Z in order to be able to use the decision procedure for Abelian rings, i.e. the tactic Ring), choose is syntactical sugar[11] for (ex_intro ? ?) (ex_intro is the single constructor of EX), skolem is syntactical sugar for (ex_ind ? ? ?) (ex_ind is the induction scheme of EX) and (dec_eq x_1 x_2) is equivalent to x_1=x_2\/~x_1=x_2.

This proof is done by induction (on n). The basis case is declared with the name cb, as well as the inductive case with the name ci. For cb, this is trivial, the instantiations are 1 and 1 (given by the term choose), since we have 0+8=3*1+5*1 (which is solved by the tactic Ring). For the inductive case, in the first instantiation (of ?1), we introduce the induction hypothesis (Hrec), which is skolemized (with skolem). Then, in the second instantiation (of ?2), we give a name for the skolem symbol (x) and we introduce the corresponding skolemized hypothesis (H), which is skolemized again. In the third instantiation (of ?3), in the same way, we give another skolem name (x0) and we introduce the skolemized hypothesis (H0). At that point, we can notice that the conclusion can be transformed (in particular, moving the successor S outside inject_nat) in order to obtain the left-hand side member of the skolemized hypothesis as a subterm. So, we carry out a cut (cir) of the conclusion after this *trivial step* (which require mainly rewritings). This auxiliary lemma is proved by case analysis (Cases term) according to the fact that x0=0 or not. If x0=0 (this is the first case, bound to metavariable ?2), the instantiations are x-3 and 2, then Ring

[11] This is done by means of toplevel syntactic definitions (see [9] for more details).

solves the goal. Otherwise (this is the second case, bound to metavariable `?3`), the instantiations are `x+2` and `x0-1`, then the equality is directly solved by `Ring` again. Once `cir` proved, we can perform, in a procedural way, the rewriting steps in the conclusion (we do not detail all the tactics and lemmas involved, which may require an advanced knowledge of Coq, and the reader only has to know what transformation is carried out by this procuderal part) to obtain the type of `cir` and to be able to conclude with the direct use of `cir` (with the tactic `Exact`). Finally, the induction over `n` is done using the term `nat_ind` directly with the corresponding proofs `cb` and `ci` as arguments.

Regarding proof style, we can notice that rewriting is, *a priori*, compatible with $\mathcal{L}_{pdt}$. The user only has to make a cut of the expression where the rewritings have been done and this cut is then directly used in the procedural part, which carries out the rewritings. This method allows us to use rewriting in a procedural way and in the middle of a proof. This seems to be a good method if rewriting is not heavily used. Otherwise, for example, for proofs which are built exclusively using rewritings (lemmas proved over axiomatizations), this could be a bit rigid and an interesting extension of $\mathcal{L}_{pdt}$ could consist in providing an appropriate syntax for equational reasoning.

5 Conclusion

5.1 Summary

In this paper, several points have been achieved:

- We have compared three proof styles, namely, the declarative, the procedural and the term style. This has been done by means of three well-known theorem provers, which are respectively, Coq, Mizar and Alfa. It has appeared that these three styles do not deserve to be opposed but were dedicated to specific kinds of proofs. A merging could allow the user greater freedom in the presentation of his/her proofs.
- In this view, a uniting language, called $\mathcal{L}_{pdt}$, has been designed in the context of the Coq proof system. In particular, the declarative features are available by means of `Let` ... `In`'s and toplevel `Let`'s. The procedural parts are directly inherited from the tactic language of Coq. Finally, the term language uses the intuitionistic Curry-Howard isomorphism in the context of Coq and is based on a language of incomplete proof terms (i.e. with metavariables), which can be refined by some instantiations.
- As a major originality of $\mathcal{L}_{pdt}$, a formal semantics has been given. This semantics consists essentially in building a proof term (always using Curry-Howard's isomorphism) corresponding to a given script. Here, the novelty has been to give also such a semantics for declarative and procedural parts.
- An implementation for versions V7 of Coq has been also realized. Some examples have been described and can be already evaluated in this prototype. In particular, those examples have shown that these three kinds of languages could naturally coexist in the same proof.

5.2 Extensions and Future Work

Some evolutions or improvements for $\mathcal{L}_{pdt}$ can be expected:

- As seen in the example of the stamp problem (subsection 4.2), $\mathcal{L}_{pdt}$ could be more appropriate to build easily proofs by rewritings. Such proofs can be made in $\mathcal{L}_{pdt}$, but this must be rather sporadic and especially trivial enough in such a way that the corresponding cut can be easily related to the initial goal. Thus, another method remains to be found to deal with proofs which use equational reasoning heavily. An idea could be to add a feature, which allows us to iterate some equalities, as e.g., in the Mizar-mode for HOL.
- $\mathcal{L}_{pdt}$ provides various `Let`'s, which add declarative features to the language (especially the toplevel `Let`'s). However, the way the lemmas introduced by some `Let`'s are combined to prove some other more complex lemmas can be improved in order to get a more declarative behavior. Indeed, as can be seen in the examples we have described (section 4), the auxiliary lemmas, which are declared by some `Let`'s, are simply and directly used. No specific automation is used to combine them and to solve. It would be interesting to add some automations here, because this is also a significative feature of declarative systems[12]. A first step, almost for free, could consist in using naturally the tactic `Auto` and adding systematically the lemmas declared by some `Let`'s in the database of this tactic. A more ambitious step would be to make `Auto` perform automatically some inductions.

5.3 Generalization and Discussion

To conclude, we can wonder how such a proof language could be generalized to other logical frameworks and, in particular, how it could be applied to other proof systems. As can be seen, in $\mathcal{L}_{pdt}$, the term part has been clearly emphasized and, in a way, Curry-Howard's isomorphism is *brought to light*. Indeed, here, λ-terms are not only seen as a way of coding (or programming) proofs and giving a computational behavior to proofs, but also as a way of expressing proofs. So, $\mathcal{L}_{pdt}$ can be applied, almost for free, to other systems based on the (intuitionistic) Curry-Howard isomorphism, such as Lego or Nuprl. We can also include Alfa, although Alfa does not provide any tactic language yet (to add such a language to Alfa would certainly be a significative work, but more from a practical point of view than for theoretical reasons). Next, if we want to deal with other systems, the problem becomes a bit more difficult due to the term part. First, a good idea would be to consider only procedural systems (as said previously, to provide a full tactic language is a quite significative and tedious task), such as PVS or HOL. As those systems are generally based on classical logic, this means we have to consider a classical Curry-Howard isomorphism and we have to design another term language using some λC-calculi or $\lambda\mu$-calculi. Such an extension could be

[12] For instance, ACL2 (successor of Nqthm) provides a huge automation and even Mizar, which is less automated, has a non trivial deduction system, which is hidden behind the keyword `by`.

very interesting because as far as the author knows, no classical proof system based on such λ-calculi has been ever designed. Finally, if we deal with pure declarative systems, like Mizar or ACL2, the task is quite harder because we have to build the term part, as well as the procedural part (but again, this must be considered as a problem in practice and not from a theoretical point of view).

References

1. Thierry Coquand, Catarina Coquand, Thomas Hallgren, and Aarne Ranta. The Alfa Home Page, 2001.
`http://www.md.chalmers.se/~hallgren/Alfa/`.
2. Yann Coscoy. A Natural Language Explanation for Formal Proofs. In C. Retoré, editor, *Proceedings of Int. Conf. on Logical Aspects of Computational Linguistics (LACL), Nancy*, volume 1328. Springer-Verlag LNCS/LNAI, September 1996.
3. Judy Crow, Sam Owre, John Rushby, Natarajan Shankar, and Mandayam Srivas. A Tutorial Introduction to PVS. In *Workshop on Industrial-Strength Formal Specification Techniques*, Boca Raton, Florida, April 1995.
4. David Delahaye. *Conception de langages pour décrire les preuves et les automatisations dans les outils d'aide à la preuve: une étude dans le cadre du système* Coq. PhD thesis, Université Pierre et Marie Curie (Paris 6), Décembre 2001.
5. John Harrison. Proof Style. In Eduardo Giménez and Christine Paulin-Mohring, editors, *Types for Proofs and Programs: International Workshop TYPES'96*, volume 1512 of *LNCS*, pages 154–172, Aussois, France, 1996. Springer-Verlag.
6. John Harrison. A Mizar Mode for HOL. In *J. von Wright, J. Grundy, and J. Harrison, editors, Theorem Proving in Higher Order Logics: TPHOLs'96*, volume 1125 of *LNCS*, pages 203–220, 1996.
7. Lena Magnusson. *The Implementation of* ALF*—a Proof Editor Based on Martin-Löf's Monomorphic Type Theory with Explicit Substitution.* PhD thesis, Chalmers University of Technology, 1994.
8. Don Syme. *Declarative Theorem Proving for Operational Semantics.* PhD thesis, University of Cambridge, 1998.
9. The Coq Development Team. *The Coq Proof Assistant Reference Manual Version 7.3.* INRIA-Rocquencourt, May 2002. `http://coq.inria.fr/doc-eng.html`.
10. Andrzej Trybulec. The Mizar-QC/6000 logic information language. In *ALLC Bulletin (Association for Literary and Linguistic Computing)*, volume 6, pages 136–140, 1978.
11. Markus Wenzel. Isar - A Generic Interpretative Approach to Readable Formal Proof Documents. In Yves Bertot, Gilles Dowek, André Hirschowitz, Christine Paulin-Mohring, and Laurent Théry, editors, *Theorem Proving in Higher Order Logics: TPHOLs'99*, volume 1690 of *LNCS*, pages 167–184. Springer-Verlag, 1999.
12. Vincent Zammit. *On the Readability of Machine Checkable Formal Proofs.* PhD thesis, University of Kent, Canterbury, October 1998.

The predicate D_f cannot be defined simply by saying that it is the domain of definition of f, since, in type theory, we need to define it before we can define f. Therefore, D_f must be given before and independently from f. One way to do it is to characterize D_f as the predicate satisfied by those elements of A for which the iteration technique converges to the same value for every initial function. This is a good definition when we try to model lazy functional programming languages, but, when interpreting strict programming languages or imperative languages, we find that this predicate would be too weak, being satisfied by elements for which the associated program diverges, as we have seen at the end of the previous section.

Sometimes the domain of definition of a function can be characterized independently of the function by an inductive predicate called *accessibility* [11, 7, 5, 3]. This simply states that an element of a can be proved to be in the domain if the application of f on a calls f recursively on elements that have already been proved to be in the domain. For example, if in the recursive definition of f there is a clause of the form

$$f(e) := \cdots f(e_1) \cdots f(e_2) \cdots$$

and a matches e, that is, there is a substitution of variables ρ such that $a = \rho(e)$; then we add a clause to the inductive definition of Acc of type

$$\mathsf{Acc}(e_1) \to \mathsf{Acc}(e_2) \to \mathsf{Acc}(e).$$

This means that to prove that a is in the domain of f, we must first prove that $\rho(e_1)$ and $\rho(e_2)$ are in the domain.

This definition does not always work. In the case of nested recursive calls of the function, we cannot eliminate the reference to f in the clauses of the inductive definition Acc. If, for example, the recursive definition of f contains a clause of the form

$$f(e) := \cdots f(f(e')) \cdots$$

then the corresponding clause in the definition of Acc should be

$$\mathsf{Acc}(e') \to \mathsf{Acc}(f(e')) \to \mathsf{Acc}(e)$$

because we must require that all arguments of the recursive calls of f satisfy Acc to deduce that also e does. But this definition is incorrect because we haven't defined the function f yet and so we cannot use it in the definition of Acc. Besides, we need Acc to define f, therefore we are locked in a vicious circle.

In our case, we have two instances of nested recursive clauses, for the sequential composition and *while* loops. When trying to give a semantics of the commands, we come to the definition

$$[\![c_1; c_2]\!]_\sigma := [\![c_2]\!]_{[\![c_1]\!]_\sigma}$$

for sequential composition and

$$[\![\mathsf{while}\ b\ \mathsf{do}\ c]\!]_\sigma := [\![\mathsf{while}\ b\ \mathsf{do}\ c]\!]_{[\![c]\!]_\sigma}$$

for a *while* loop, if the interpretation of b in state σ is true.

Both cases contain a nested occurrence of the interpretation function $[\![-]\!]$.

An alternative solution, presented in [4], exploits the extension of type theory with simultaneous induction-recursion [6]. In this extension, an inductive type or inductive

Two-Level Meta-reasoning in Coq

Amy P. Felty

School of Information Technology and Engineering
University of Ottawa, Ottawa, Ontario K1N 6N5, Canada
afelty@site.uottawa.ca

Abstract. The use of *higher-order abstract syntax* is central to the direct, concise, and modular specification of languages and deductive systems in a logical framework. Developing a framework in which it is also possible to reason about such deductive systems is particularly challenging. One difficulty is that the use of higher-order abstract syntax complicates reasoning by induction because it leads to definitions for which there are no monotone inductive operators. In this paper, we present a methodology which allows Coq to be used as a framework for such meta-reasoning. This methodology is directly inspired by the two-level approach to reasoning used in the $FO\lambda^{\Delta\mathbb{N}}$ (pronounced *fold-n*) logic. In our setting, the Calculus of Inductive Constructions (CIC) implemented by Coq represents the highest level, or *meta-logic*, and a separate *specification logic* is encoded as an inductive definition in Coq. Then, in our method as in $FO\lambda^{\Delta\mathbb{N}}$, the deductive systems that we want to reason about are the *object logics* which are encoded in the specification logic. We first give an approach to reasoning in Coq which very closely mimics reasoning in $FO\lambda^{\Delta\mathbb{N}}$ illustrating a close correspondence between the two frameworks. We then generalize the approach to take advantage of other constructs in Coq such as the use of direct structural induction provided by inductive types.

1 Introduction

Higher-order abstract syntax encodings of object logics are usually expressed within a typed meta-language. The terms of the untyped λ-calculus can be encoded using higher-order syntax, for instance, by introducing a type tm and two constructors: abs of type $(\boldsymbol{tm} \rightarrow tm) \rightarrow tm$ and app of type $tm \rightarrow tm \rightarrow tm$. As this example shows, it is often useful to use negative occurrences of the type introduced for representing the terms of the object logic. (Here the single negative occurrence is in boldface.) Predicates of the meta-logic are used to express judgments in the object logic such as "term M has type t". Embedded implication is often used to represent *hypothetical judgments*, which can result in negative occurrences of such predicates. For example the following rule which defines typing for λ-abstraction in the object logic

$$\frac{\begin{array}{c}(x : \tau_1)\\ M : \tau_2\end{array}}{\lambda x.M : \tau_1 \rightarrow \tau_2}$$

V.A. Carreño, C. Muñoz, S. Tahar (Eds.): TPHOLs 2002, LNCS 2410, pp. 198–213, 2002.

can be expressed using the *typeof* predicate in the following formula.

$$\forall M : tm \rightarrow tm. \forall \tau_1, \tau_2 : tm.$$
$$(\forall x : tm.(typeof\ x\ \tau_1) \supset (typeof\ (M\ x)\ \tau_2))$$
$$\supset (typeof\ (abs\ M)\ (\tau_1 \rightarrow \tau_2))$$

The Coq system [21] implements the Calculus of Inductive Constructions (CIC) and is one of many systems in which such negative occurrences cause difficulty. In particular, the inductive types of the language cannot be used directly for this kind of encoding of syntax or inference rules.

$FO\lambda^{\Delta\mathbb{N}}$ is a logical framework capable of specifying a wide variety of deductive systems [13]. It is one of the first to overcome various challenges and allow both specification of deductive systems and reasoning about them within a single framework. It is a higher-order intuitionistic logic with support for natural number induction and definitions. A rule of definitional reflection is included and is central to reasoning in the logic [8]. This rule in particular represents a significant departure from the kinds of primitive inference rules found in Coq and a variety of other systems that implement similar logics. Our methodology illustrates that, for a large class of theorems, reasoning via this rule can be replaced by reasoning with inductive types together with a small number of assumptions about the constants that are introduced to encode a particular deductive system.

We define both the specification logic and the object logic as inductive definitions in Coq. Although there are no inductive definitions in $FO\lambda^{\Delta\mathbb{N}}$, our Coq definitions of specification and object logics closely resemble the corresponding $FO\lambda^{\Delta\mathbb{N}}$ definitions of the same logics. The use of a two-level logic in both $FO\lambda^{\Delta\mathbb{N}}$ and Coq solves the problem of inductive reasoning in the presence of negative occurrences in hypothetical judgments. Hypothetical judgments are expressed at the level of the object logic, while inductive reasoning about these object logics takes place at the level of the specification logic and meta-logic. More specifically, in $FO\lambda^{\Delta\mathbb{N}}$, a combination of natural number induction and definitional reflection provides induction on the height of proofs in the specification logic. For the class of theorems we consider, we can mimic the natural number induction of $FO\lambda^{\Delta\mathbb{N}}$ fairly directly in Coq. In addition, the Coq environment provides the extra flexibility of allowing reasoning via direct induction using the theorems generated by the inductive definitions. For example, we can use direct structural induction on proof trees at both the specification level and the object-level.

One of our main goals in this work is to provide a system that allows programming and reasoning about programs and programming languages within a single framework. The Centaur System [3] is an early example of such a system. We are interested in a proof and program development environment that supports higher-order syntax. In particular, we are interested in the application of such a system to building proof-carrying code (PCC) systems. PCC [17] is an approach to software safety where a producer of code delivers both a program and a formal proof that verifies that the code meets desired safety policies. We have built prototype PCC systems [1,2] in both λProlog [16] and Twelf [19] and have found higher-order syntax to be useful in both programming and expressing safety properties. Definitional reflection as in $FO\lambda^{\Delta\mathbb{N}}$ is difficult to program directly in λProlog and Twelf. On the other hand, support for inductive types similar to that of Coq is straightforward to implement. We hope to carry over the methodology we describe here to provide more flexibility in constructing proofs in the PCC setting.

In this paper, after presenting the Calculus of Inductive Constructions in Sect. 2, we begin with the example proof of subject reduction for the untyped λ-calculus from McDowell and Miller [13] (also used in Despeyroux et al. [5]). For this example, we use a sequent calculus for a second-order minimal logic as our specification logic. We present a version of the proof that uses natural number induction in Sect. 3. By using natural number induction, we are able to mimic the corresponding $FO\lambda^{\Delta\mathbb{N}}$ proof, and in Sect. 4 we discuss how the $FO\lambda^{\Delta\mathbb{N}}$ proof illustrates the correspondence in reasoning in the two systems. In Sect. 5, we present an alternate proof which illustrates reasoning by direct structural induction in Coq. In Sect. 6, we conclude as well as discuss related and future work.

2 The Calculus of Inductive Constructions

We assume some familiarity with the Calculus of Inductive Constructions. We note here the notation used in this paper, much of which is taken from the Coq system. Let x represent variables and M, N represent terms of CIC. The syntax of terms is as follows.

$$\begin{gathered} Prop \mid Set \mid Type \mid x \mid MN \mid \lambda x:M.N \mid \\ \forall x:M.N \mid M \to N \mid M \wedge N \mid M \vee N \mid \\ \exists x:M.N \mid \neg M \mid M = N \mid True \mid Ind\ x:M\ \{N_1|\cdots|N_n\} \mid \\ Rec\ M\ N \mid Case\ x:M\ of\ M_1 \Rightarrow N_1,\ldots,M_n \Rightarrow N_n \end{gathered}$$

Here ∀ is the dependent type constructor and the arrow (→) is the usual abbreviation when the bound variable does not occur in the body. Of the remaining constants, *Prop*, *Set*, *Type*, λ, *Ind*, *Rec*, and *Case* are primitive, while the others are defined. *Prop* is the type of logical propositions, whereas *Set* is the type of data types. *Type* is the type of both *Prop* and *Set*. *Ind* is used to build inductive definitions where M is the type of the class of terms being defined and $N_1,\ldots,N_n$ where $n \geq 0$ are the types of the constructors. *Rec* and *Case* are the operators for defining recursive and inductive functions, respectively, over inductive types. Equality on *Set* (=) is Leibnitz equality.

A constant is introduced using the Parameter keyword and ordinary definitions which introduce a new constant and the term it represents are defined using the Definition keyword. Inductive definitions are introduced with an Inductive declaration where each constructor is given with its type separated by vertical bars. When an inductive definition is made, Coq automatically generates operators for reasoning by structural induction and for defining recursive functions on objects of the new type. We use the section mechanism of the system which provides support for developing theories modularly. The Variable keyword provides a way to introduce constants that will be discharged at the end of a section. Axiom is used to introduce formulas that are assumed to hold and Theorem introduces formulas which are immediately followed by a proof or a series of commands (*tactics*) that indicate how to construct the proof.

3 An Example: Subject Reduction for the Untyped λ-Calculus

A variety of specification logics can be defined. In this paper, we use a simple minimal logic taken from McDowell and Miller [13]. In Coq, we introduce the type *prp* to encode formulas of the specification logic, and the type *atm* (left as a parameter at this stage) to encode the atomic formulas of the object logic.

Variable $atm : Set$.
Variable $tau : Set$.
Inductive $prp : Set :=$
$\langle\rangle : atm \to prp \mid tt : prp \mid \& : prp \to prp \to prp \mid$
$\Rightarrow: atm \to prp \to prp \mid \bigwedge : (tau \to prp) \to prp \mid \bigvee : (tau \to prp) \to prp.$

The operator $\langle\rangle$ is used to coerce objects of type *atm* to *prp*. The other constructors of the inductive definition of *prp* define the logical connectives of the specification logic. We use a higher-order syntax encoding of the quantifiers $\bigwedge$ (forall) and $\bigvee$ (exists), i.e., each quantifier takes one argument which is a λ-term so that binding of quantifiers in the specification logic is encoded as λ-binding at the meta-level. Note that we parameterize the quantification type; this version of the specification logic limits quantification to a single type *tau*. This is not a serious restriction here since we encode all syntactic objects in our examples using the single type *tm*; also, it can be extended to include other types if necessary. Here, we freely use infix and prefix/postfix operators, without discussing the details of using them in Coq.

For illustration purposes, we show the induction principle generated by Coq resulting from the above definition of *prp*.

$$\begin{array}{l}\forall P : prp \to Prop.\\ \quad [(\forall A : atm.P\langle A\rangle) \to\\ \quad P(tt) \to\\ \quad (\forall B : prp.PB \to \forall C : prp.PC \to P(B\&C)) \to\\ \quad (\forall A : atm.\forall B : prp.PB \to P(A \Rightarrow B)) \to\\ \quad (\forall B : tau \to prp.(\forall x : tau.P(Bx)) \to P(\bigwedge B)) \to\\ \quad (\forall B : tau \to prp.(\forall x : tau.P(Bx)) \to P(\bigvee B))] \to \forall B : prp.PB\end{array}$$

After closing the section containing the above definitions, *prp* will have type $Set \to Set \to Set$ because *atm* and *tau* are discharged.

The Coq inductive definition in Fig. 1 is a direct encoding of the specification logic. The predicate *prog* is used to declare the object-level deductive system. It is a parameter at this stage. A formula of the form $(prog\ A\ b)$ as part of the

Variable $prog : atm \rightarrow prp \rightarrow Prop.$
Inductive $seq : nat \rightarrow list\ atm \rightarrow prp \rightarrow Prop :=$
$sbc : \forall i : nat. \forall A : atm. \forall L : list\ atm. \forall b : prp.$
$(prog\ A\ b) \rightarrow (seq\ i\ L\ b) \rightarrow (seq\ (S\ i)\ L\ \langle A \rangle)$
$\mid sinit : \forall i : nat. \forall A, A' : atm. \forall L : list\ atm.$
$(element\ A\ (A' :: L)) \rightarrow (seq\ i\ (A' :: L)\ \langle A \rangle)$
$\mid strue : \forall i : nat. \forall L : list\ atm. (seq\ i\ L\ tt)$
$\mid sand : \forall i : nat. \forall B, C : prp. \forall L : list\ atm.$
$(seq\ i\ L\ B) \rightarrow (seq\ i\ L\ C) \rightarrow (seq\ (S\ i)\ L\ (B \& C))$
$\mid simp : \forall i : nat. \forall A : atm. \forall B : prp. \forall L : list\ atm.$
$(seq\ i\ (A :: L)\ B) \rightarrow (seq\ (S\ i)\ L\ (A \Rightarrow B))$
$\mid sall : \forall i : nat. \forall B : tau \rightarrow prp. \forall L : list\ atm.$
$(\forall x : tau. (seq\ i\ L\ (B\ x))) \rightarrow (seq\ (S\ i)\ L\ (\bigwedge B))$
$\mid ssome : \forall i : nat. \forall B : tau \rightarrow prp. \forall L : list\ atm.$
$\forall x : tau. (seq\ i\ L\ (B\ x)) \rightarrow (seq\ (S\ i)\ L\ (\bigvee B)).$
Definition $\rhd : list\ atm \rightarrow prp \rightarrow Prop := \lambda l : list\ atm. \lambda B : prp. \exists i : nat. (seq\ i\ l\ B).$
Definition $\rhd_0 : prp \rightarrow Prop := \lambda B : prp. \exists i : nat. (seq\ i\ nil\ B).$

Fig. 1. Definition of the Specification Logic in Coq

object logic means roughly that b implies A where A is an atom. We will see shortly how *prog* is used to define an object logic. Most of the clauses of this definition encode rules of a sequent calculus which introduce connectives on the right of a sequent. For example, the *sand* clause specifies the following $\wedge$-R rule.

$$\frac{L \longrightarrow B \qquad L \longrightarrow C}{L \longrightarrow B \wedge C}\ \wedge\text{-R}$$

In the Coq definition, the natural number i indicates that the proofs of the premises have height at most i and the proof of the conclusion has height at most $i+1$. (S is the successor function from the Coq libraries.) The *sinit* clause specifies when a sequent is initial (i.e., the formula on the right appears in the list of hypotheses on the left). We omit the definition of *element*, which is straightforward. The *sbc* clause represents backchaining. A backward reading of this rule states that A is provable from hypotheses L in at most $i+1$ steps if b is provable from hypotheses L in at most i steps, where "A implies B" is a statement in the object logic. The definitions of $\rhd$ and $\rhd_0$ at the end of the figure are made for convenience in expressing properties later. The former is written using infix notation.

Theorems which *invert* this definition can be directly proved using the induction and recursion operators for the type *seq*. For example, it is clear that if a proof ends in a sequent with an atomic formula on the right, then the sequent was either derived using the rule for *prog* (*sbc*) or the atom is an element of the list of formulas on the left (*sinit*). This theorem is expressed as follows.

Theorem $seq\_atom\_inv : \forall i : nat. \forall A : atm. \forall l : list\ atm. (seq\ i\ l\ \langle A \rangle) \rightarrow$
$[\exists j : nat. \exists b : prp. (i = (S\ j)) \wedge (prog\ A\ b) \wedge (seq\ j\ l\ b)) \vee$
$\exists A' : atm. \exists l' : list\ atm. (l = (A' :: l')) \wedge (element\ A\ l))].$

Induction principles generated by Coq are also useful for reasoning by case analysis. For example, case analysis on *seq* can be used to prove the *seq_cut* property below, which is an essential part of our proof development.

Theorem $seq\_cut : \forall a : atm.\forall b : prp.\forall l : list\ atm.(a :: l) \rhd b \rightarrow l \rhd \langle a \rangle \rightarrow l \rhd b.$

This theorem can also be proven by case analysis on *prp* using the induction principle shown earlier. In fact, for this particular theorem, case analysis on *prp* leads to a somewhat simpler proof than case analysis on *seq*.

Our object logic consists of untyped λ-terms, types, and rules for assigning types to terms. Terms and types are encoded using the parameter declarations below.

Parameter $tm : Set.$
Parameter $gnd : tm.$ Parameter $abs : (tm \rightarrow tm) \rightarrow tm.$
Parameter $arr : tm \rightarrow tm \rightarrow tm.$ Parameter $app : tm \rightarrow tm \rightarrow tm.$
Axiom $gnd\_arr : \forall t, u : tm.\neg(gnd = (arr\ t\ u)).$
Axiom $abs\_app : \forall R : tm \rightarrow tm.\forall M, N : tm.\neg((abs\ R) = (app\ M\ N)).$
Axiom $arr\_inj : \forall t, t', u, u' : tm.(arr\ t\ u) = (arr\ t'\ u') \rightarrow t = t' \wedge u = u'.$
Axiom $abs\_inj : \forall R, R' : tm \rightarrow tm.(abs\ R) = (abs\ R') \rightarrow R = R'.$
Axiom $app\_inj : \forall M, M', N, N' : tm.$
$(app\ M\ N) = (app\ M'\ N') \rightarrow M = M' \wedge N = N'.$

The five axioms following them express properties about distinctness and injectivity of constructors. For example, a term beginning with *abs* is always distinct from one beginning with *app*. Also, if two terms $(abs\ R)$ and $(abs\ R')$ are equal then so are R and R'. For objects defined inductively in Coq, such properties are derivable. Here, we cannot define *tm* inductively because of the negative occurrence in the type of the *abs* constant, so we must include them explicitly. They are the only axioms we require for proving properties about this object logic. The type *tm* is the type which instantiates *tau* in the definitions of *prp* and *seq* above.

Note that by introducing constants and axioms, we are restricting the context in which reasoning in Coq is valid and actually corresponds to reasoning about the deduction systems we encode. For example, we cannot discharge these constants and instantiate them with arbitrary objects such as inductively defined elements of Set. We do not want to be able to prove any properties about these constants other than the ones we assume and properties that follow from them.

The definitions for atomic formulas and for the *prog* predicate, which encode typing and evaluation of our object logic are given in Fig. 2. An example of an inversion theorem that follows from this definition is the following. Its proof requires *seq_atom_inv* above.

Theorem $eval\_nil\_inv : \forall j : nat.\forall M, V : tm.(seq\ j\ nil\ \langle M \Downarrow V \rangle) \rightarrow$
$[(\exists R : tm \rightarrow tm.M = (abs\ R) \wedge V = (abs\ R)) \vee$
$(\exists k : nat.\exists R : tm \rightarrow tm.\exists P, N : tm.j = (S\ (S\ k)) \wedge M = (app\ P\ N) \wedge$
$(seq\ k\ nil\ \langle P \Downarrow (abs\ R) \rangle) \wedge (seq\ k\ nil\ \langle (R\ N) \Downarrow V \rangle))].$

$$\begin{array}{l}
\textsf{Inductive}\ atm : Set := typeof : tm \rightarrow tm \rightarrow atm \mid \Downarrow : tm \rightarrow tm \rightarrow atm. \\
\textsf{Inductive}\ prog : atm \rightarrow prp \rightarrow Prop := \\
\quad tabs : \forall t, u : tm. \forall R : tm \rightarrow tm. \\
\qquad\quad (prog\ (typeof\ (abs\ R)\ (arr\ t\ u)) \\
\qquad\qquad\quad (\bigwedge\ \lambda n : tm.((typeof\ n\ t) \Rightarrow \langle typeof\ (R\ n)\ u\rangle))) \\
\quad \mid tapp : \forall M, N, t : tm. \\
\qquad\quad (prog\ (typeof\ (app\ M\ N)\ t) \\
\qquad\qquad\quad (\bigvee\ \lambda u : tm.(\langle typeof\ M\ (arr\ u\ t)\rangle \& \langle typeof\ N\ u\rangle))) \\
\quad \mid eabs : \forall R : tm \rightarrow tm.(prog\ ((abs\ R) \Downarrow (abs\ R))\ tt) \\
\quad \mid eapp : \forall M, N, V : tm. \forall R : tm \rightarrow tm. \\
\qquad\quad (prog\ ((app\ M\ N) \Downarrow V)\ (\langle M \Downarrow (abs\ R)\rangle \& \langle (R\ N) \Downarrow V\rangle))
\end{array}$$

Fig. 2. Definition of the Object Logic in Coq

We are now ready to express and prove the subject reduction property.

$$\textsf{Theorem}\ sr : \forall p, v : tm.\ \rhd_0 \langle p \Downarrow v\rangle \rightarrow \forall t : tm.\ \rhd_0 \langle typeof\ p\ t\rangle \rightarrow \rhd_0 \langle typeof\ v\ t\rangle.$$

Our proof of this theorem corresponds directly to the one given by Miller and McDowell [13]. We show a few steps to illustrate. After one definition expansion of $\rhd_0$, several introduction rules for universal quantification and implication, and an elimination of the existential quantifier on one of the assumptions, we obtain the sequent

$$(seq\ i\ nil\ \langle p \Downarrow v\rangle), \rhd_0 \langle typeof\ p\ t\rangle \longrightarrow \rhd_0 \langle typeof\ v\ t\rangle. \tag{1}$$

(We display meta-level sequents differently than the Coq system. We omit type declarations and names of hypotheses, and we separate the hypotheses from the conclusion with a sequent arrow.) We now apply complete induction, which comes from the basic Coq libraries and is stated:

$$\begin{array}{l}\textsf{Theorem}\ lt\_wf\_ind : \forall k : nat. \forall P : nat \rightarrow Prop. \\ \quad (\forall n : nat. (\forall m : nat. m < n \rightarrow Pm) \rightarrow Pn) \rightarrow Pk.\end{array}$$

After solving the trivial subgoals, we are left to prove $\forall k.(k < j \supset (IP\ k)) \supset (IP\ j)$, where IP denotes the formula

$$\lambda i : nat. \forall p, v : tm.(seq\ i\ nil\ \langle p \Downarrow v\rangle) \rightarrow \forall t : tm.\ \rhd_0 \langle typeof\ p\ t\rangle \rightarrow \rhd_0 \langle typeof\ v\ t\rangle.$$

After clearing the old assumptions and applying a few more intro/elim rules we get the following sequent.

$$\forall k.k < j \rightarrow (IP\ k), (seq\ j\ nil\ \langle p \Downarrow v\rangle), \rhd_0 \langle typeof\ p\ t\rangle \longrightarrow \rhd_0 \langle typeof\ v\ t\rangle. \tag{2}$$

Note that a proof of $(seq\ j\ nil\ \langle p \Downarrow v\rangle)$ must end with the first clause for *seq* (containing *prog*). Here, we apply the *eval_nil_inv* inversion theorem to obtain

$$\begin{array}{l}\forall k.k < j \rightarrow (IP\ k), \\ [(\exists R : tm \rightarrow tm.p = (abs\ R) \wedge v = (abs\ R)) \vee \\ \ (\exists k' : nat. \exists R : tm \rightarrow tm. \exists P, N : tm.j = (S\ (S\ k')) \wedge p = (app\ P\ N) \wedge \\ \quad (seq\ k'\ nil\ \langle P \Downarrow (abs\ R)\rangle) \wedge (seq\ k'\ nil\ \langle (R\ N) \Downarrow v\rangle))], \\ \rhd_0 \langle typeof\ p\ t\rangle \longrightarrow \rhd_0 \langle typeof\ v\ t\rangle.\end{array}$$

Then after eliminating the disjunction, existential quantifiers, and conjunction, as well as performing the substitutions using the equalities we obtain the following two sequents.

$$\forall k.k < j \rightarrow (IP\ k), \rhd_0\langle typeof\ (abs\ R)\ t\rangle \longrightarrow \rhd_0\langle typeof\ (abs\ R)\ t\rangle \quad (3)$$

$$\forall k.k < (S\ (S\ k')) \rightarrow (IP\ k), (seq\ k'\ nil\ \langle P \Downarrow (abs\ R)\rangle),$$
$$(seq\ k'\ nil\ \langle (R\ N) \Downarrow v\rangle), \rhd_0\langle typeof\ (app\ P\ N)\ t\rangle \longrightarrow \rhd_0\langle typeof\ v\ t\rangle \quad (4)$$

Note that the first is directly provable.

We carry out one more step of the Coq proof of (4) to illustrate the use of a distinctness axiom. In particular, we show the two sequents that result from applying an inversion theorem to the formula just before the sequent arrow in (4) and then applying all possible introductions, eliminations, and substitutions. (We abbreviate $(S\ (S\ (S\ k'')))$ as $(S^3\ k'')$ and similarly for other such expressions.)

$$\forall k.k < (S^5\ k'') \rightarrow (IP\ k), (seq\ (S^3\ k'')\ nil\ \langle P \Downarrow (abs\ R)\rangle),$$
$$(seq\ (S^3\ k'')\ nil\ \langle (R\ N) \Downarrow v\rangle), (app\ P\ N) = (abs\ R), t = (arr\ T'\ U),$$
$$(seq\ k''\ ((typeof\ M\ T') :: nil)\ \langle (typeof\ (R\ M)\ U)\rangle)) \longrightarrow \rhd_0\langle typeof\ v\ t\rangle$$
$$\forall k.k < (S^5\ k'') \rightarrow (IP\ k), (seq\ (S^3\ k'')\ nil\ \langle P \Downarrow (abs\ R)\rangle),$$
$$(seq\ (S^3\ k'')\ nil\ \langle (R\ N) \Downarrow v\rangle), (seq\ k''\ nil\ \langle typeof\ P\ (arr\ u\ t)\rangle)$$
$$(seq\ k''\ nil\ \langle typeof\ N\ u\rangle)) \longrightarrow \rhd_0\langle typeof\ v\ t\rangle$$

Note the occurrence of $(app\ P\ N) = (abs\ R)$ in the first sequent. This sequent must be ruled out using the *abs_app* axiom.

The remainder of the Coq proof continues using the same operations of applying inversion theorems, and using introduction and elimination rules. It also includes applications of lemmas such as *seq_cut* mentioned above.

4 A Comparison to $FO\lambda^{\Delta\mathbb{N}}$

The basic logic of $FO\lambda^{\Delta\mathbb{N}}$ is an intuitionistic version of a subset of Church's Simple Theory of Types with logical connectives $\bot$, $\top$, $\wedge$, $\vee$, $\supset$, $\forall_\tau$, and $\exists_\tau$. Quantification is over any type τ not containing o, which is the type of meta-level formulas. The inference rules of the logic include the usual left and right sequent rules for the connectives and rules that support natural number induction. Note that these sequent rules are at the meta-level, and thus we have sequent calculi both at the meta-level and specification level in our example proof. $FO\lambda^{\Delta\mathbb{N}}$ also has the following rules to support definitions.

$$\frac{\Gamma \longrightarrow B\Theta}{\Gamma \longrightarrow A}\ \mathrm{def}\mathcal{R}, \qquad \text{where } A = A'\Theta \text{ for some clause } \forall\bar{x}[A' =_\Delta B]$$

$$\frac{\{B\Theta, \Gamma\Theta \longrightarrow C\Theta | \Theta \in CSU(A, A') \text{ for some clause } \forall\bar{x}[A' =_\Delta B]\}}{A, \Gamma \longrightarrow C}\ \mathrm{def}\mathcal{L}$$

A definition is denoted by $\forall\bar{x}[A' =_\Delta B]$ where the symbol $=_\Delta$ is used to separate the object being defined from the body of the definition. Here, A' has the form $(p\ \bar{t})$ where p is a predicate constant, every free variable in B is also free in $(p\ \bar{t})$, and all variables free in $\bar{t}$ are contained in the list $\bar{x}$. The first rule provides "backchaining" on a clause of a definition. The second rule is the rule of *definitional reflection* and uses complete sets of unifiers (CSU). When this set is infinite, there will be an infinite number of premises. In practice, such as in the proofs in McDowell and Miller's work [12,13], this rule is used only in finite cases.

Fig. 3 illustrates how *seq* and *prog* are specified as $FO\lambda^{\Delta\mathbb{N}}$ definitions. Each

$$
\begin{array}{rl}
seq\ (SI)\ L\ \langle A\rangle & =_\Delta \exists b.[prog\ A\ b \wedge seq\ I\ L\ b] \\
seq\ I\ (A' :: L)\ \langle A\rangle & =_\Delta element\ A\ (A' :: L) \\
seq\ I\ L\ tt & =_\Delta \top \\
seq\ (SI)\ L\ (B\&C) & =_\Delta seq\ I\ L\ B \wedge seq\ I\ L\ C \\
seq\ (SI)\ L\ (A \Rightarrow B) & =_\Delta seq\ I\ (A :: L)\ B \\
seq\ (SI)\ L\ (\bigwedge_\tau\ B) & =_\Delta \forall_\tau x.[seq\ I\ L\ (Bx)] \\
seq\ (SI)\ L\ (\bigvee_\tau\ B) & =_\Delta \exists_\tau x.[seq\ I\ L\ (Bx)]
\end{array}
$$

$$
\begin{array}{ll}
prog\ (typeof\ (abs\ R)\ (arr\ T\ U)) & \bigwedge_{tm} \lambda N.((typeof\ N\ T) \Rightarrow \langle typeof\ (R\ N)\ U\rangle) \\
prog\ (typeof\ (ap\ M\ N)\ T) & \bigvee_{tm} \lambda U.(\langle typeof\ M\ (arr\ U\ T)\rangle \& \langle typeof\ N\ U\rangle) \\
prog\ ((abs\ R) \Downarrow (abs\ R)) & tt \\
prog\ ((app\ M\ N) \Downarrow V) & \langle M \Downarrow (abs\ R)\rangle \& \langle (R\ N) \Downarrow V\rangle
\end{array}
$$

Fig. 3. Definitions of Specification and Object Logics in $FO\lambda^{\Delta\mathbb{N}}$

of the clauses for *prog* ends in $=_\Delta \top$ which is omitted.

To a large extent, the inversion and case analysis theorems we proved in Coq were introduced to provide the possibility to reason in Coq in a manner which corresponds closely to reasoning directly in $FO\lambda^{\Delta\mathbb{N}}$. In particular, they allow us to mimic steps that are directly provided by definitional reflection in $FO\lambda^{\Delta\mathbb{N}}$. For types that cannot be defined inductively in Coq such as *tm*, the axioms expressing distinctness and injectivity of constructors are needed for this kind of reasoning.

To illustrate the correspondence between proofs in $FO\lambda^{\Delta\mathbb{N}}$ and Coq, we discuss how several of the steps in the proof outlined in Sect. 3 correspond to steps in a $FO\lambda^{\Delta\mathbb{N}}$ proof of the same theorem. For instance, application of sequent rules in $FO\lambda^{\Delta\mathbb{N}}$ correspond directly to introduction and elimination rules in Coq. Thus, we can begin the $FO\lambda^{\Delta\mathbb{N}}$ proof of the *sr* theorem similarly to the Coq proof, in this case with applications of $\forall$-R, $\supset$ -R, $\exists$-L at the meta-level, from which we obtain sequent (1) in Sect. 3.

Complete induction is derivable in $FO\lambda^{\Delta\mathbb{N}}$, so using this theorem as well as additional sequent rules allows us to obtain sequent (2) in the $FO\lambda^{\Delta\mathbb{N}}$ proof similarly to how it was obtained in the Coq proof.

It is at this point that the first use of definitional reflection occurs in the $FO\lambda^{\Delta\mathbb{N}}$ proof. Applying def$\mathcal{L}$ to the middle assumption on the left of the sequent arrow in (2), we see that this formula only unifies with the left hand side of the first clause of the definition of sequents in Fig. 3. We obtain

$$\forall k.k < (S\ j') \rightarrow (IP\ k), \exists d.[(prog\ (p \Downarrow v)\ d) \wedge (seq\ j'\ nil\ d)], \rhd_0\langle typeof\ p\ t\rangle \\ \longrightarrow \rhd_0\langle typeof\ v\ t\rangle.$$

Then applying left sequent rules, followed by def$\mathcal{L}$ on $(prog\ (p \Downarrow v)\ d)$, we get two sequents.

$$\forall k.k < (S\ j') \rightarrow (IP\ k), (seq\ j'\ nil\ tt), \\ \rhd_0\ \langle typeof\ (abs\ R)\ t\rangle \longrightarrow \rhd_0\langle typeof\ (abs\ R)\ t\rangle \\ \forall k.k < (S\ j') \rightarrow (IP\ k), (seq\ j'\ nil\ (\langle P \Downarrow (abs\ R)\rangle \& \langle (R\ N) \Downarrow v\rangle)), \\ \rhd_0\ \langle typeof\ (app\ P\ N)\ t\rangle \longrightarrow \rhd_0\langle typeof\ v\ t\rangle$$

Like sequent (3) in Sect. 3, the first is directly provable. The def$\mathcal{L}$ rule is applied again, this time on the middle assumption of the second sequent. Only the fourth clause of the definition of sequents in Fig. 3 can be used in unification. Following this application by conjunction elimination yields a sequent very similar to (4). The *eval_nil_inv* theorem used in the Coq proof and applied to (2) at this stage encompasses all three def$\mathcal{L}$ applications. Its proof, in fact, uses three inversion theorems. Note that because of the unification operation, there is no need for existential quantifiers and equations as in the Coq version.

The use of inductive types in Coq together with distinctness and injectivity axioms are sufficient to handle most of the examples in McDowell and Miller's paper [13]. One specification logic given there relies on extensionality of equality which holds in $FO\lambda^{\Delta\mathbb{N}}$. In Coq, however, equality is not extensional, which causes some difficulty in reasoning using this specification logic. Assuming extensional equality at certain types in Coq will be necessary. Other axioms may be needed as well. In the application of the definitional reflection rule, higher-order unification is a central operation. Examples which we cannot handle also include those in McDowell's thesis [12] which require complex uses of such unification. For instance, we cannot handle applications for which there are multiple solutions to a single unification problem.

We have claimed that Coq provides extra flexibility by allowing reasoning via direct induction using the induction principles generated by the system. A simple example of this extra flexibility is in the proof of the *seq_cut* theorem mentioned in Sect. 3. The $FO\lambda^{\Delta\mathbb{N}}$ proof is similar to the Coq proof that does case analysis using the induction principle for *seq*. In Coq, we were also able to do a simpler proof via case analysis provided by structural induction on *prp*, which is not possible in $FO\lambda^{\Delta\mathbb{N}}$. The next section illustrates other examples of this extra flexibility.

5 Structural Induction on Sequents

In this section, we discuss an alternate proof of theorem *sr* that uses Coq's induction principle for *seq*. Since we do not do induction on the height of the proof for this example, the natural number argument is not needed, so we omit it and use the definition given in Fig. 4 instead.

$$
\begin{array}{l}
\textsf{Inductive } seq : list\ atm \rightarrow prp \rightarrow Prop :=\\
\quad sbc : \forall A : atm.\forall L : list\ atm.\forall b : prp.\\
\qquad\qquad (prog\ A\ b) \rightarrow (seq\ L\ b) \rightarrow (seq\ L\ \langle A\rangle)\\
\ \mid sinit : \forall A, A' : atm.\forall L : list\ atm.\\
\qquad\qquad (element\ A\ (A' :: L)) \rightarrow (seq\ (A' :: L)\ \langle A\rangle)\\
\ \mid strue : \forall L : list\ atm.(seq\ L\ tt)\\
\ \mid sand : \forall B, C : prp.\forall L : list\ atm.\\
\qquad\qquad (seq\ L\ B) \rightarrow (seq\ L\ C) \rightarrow (seq\ L\ (B \& C))\\
\ \mid simp : \forall A : atm.\forall B : prp.\forall L : list\ atm.\\
\qquad\qquad (seq\ (A :: L)\ B) \rightarrow (seq\ L\ (A \Rightarrow B))\\
\ \mid sall : \forall B : tau \rightarrow prp.\forall L : list\ atm.\\
\qquad\qquad (\forall x : tau.(seq\ L\ (B\ x))) \rightarrow (seq\ L\ (\bigwedge B))\\
\ \mid ssome : \forall B : tau \rightarrow prp.\forall L : list\ atm.\\
\qquad\qquad \forall x : tau.(seq\ L\ (B\ x)) \rightarrow (seq\ L\ (\bigvee B)).
\end{array}
$$

Fig. 4. Definition of the Specification Logic without Natural Numbers

Note that in the statement of the *sr* theorem, all of the sequents have empty assumption lists and atomic formulas on the right. Using the induction principle for *seq* to prove such properties often requires generalizing the induction hypothesis to handle sequents with a non-empty assumption list and a non-atomic formula on the right. We provide an inductive definition to facilitate proofs that require these extensions and we parameterize this definition with two properties, one which represents the desired property restricted to atomic formulas (denoted here as P), and one which represents the property that must hold of formulas in the assumption list (denoted here as *Phyp*). We require that the property on atomic formulas follows from the property on assumptions, so that for the base case when the *sinit* rule is applied to a sequent of the form $(seq\ L\ A)$, it will follow from the fact that A is in L that the desired property holds. (In many proofs, the two properties are the same, which means that this requirement is trivially satisfied.) The following are the two properties that we use in the proof of *sr*.

$$
\begin{array}{l}
\textsf{Definition } P := \lambda l : list\ atm.\lambda A : atm.\textsf{Cases } A\ of\\
\quad (typeof\ m\ t) \Rightarrow True\\
\quad \mid (p \Downarrow v) \Rightarrow \forall t : tm.(seq\ l\ \langle typeof\ p\ t\rangle) \rightarrow (seq\ l\ \langle typeof\ v\ t\rangle)\ \textsf{end}.\\
\textsf{Definition } Phyp := \lambda l : list\ atm.\lambda A : atm.\exists p, t : tm.A = (typeof\ p\ t) \wedge\\
\quad (seq\ nil\ \langle A\rangle) \wedge (\forall v : tm.(seq\ l\ \langle p \Downarrow v\rangle) \rightarrow (seq\ l\ \langle typeof\ v\ t\rangle)).
\end{array}
$$

The proof of *sr* (in this and in the previous section) uses induction on the height of the proof of the evaluation judgment $\rhd_0\langle p \Downarrow v\rangle$. Thus in defining P, we ignore *typeof* judgments. The clause for $\Downarrow$ simply states a version of the subject reduction property but with assumption list l. The property that we require of assumption lists is that they only contain atomic formulas of the form $(typeof\ p\ t)$ and that each such assumption can be proven from an empty set of assumptions and itself satisfies the subject reduction property.

The inductive definition which handles generalized induction hypotheses (parameterized by P and *Phyp*) is given in Fig. 5. The definition of *mapP* mimics

Variable $P : list\ atm \rightarrow atm \rightarrow Prop.$
Variable $Phyp : list\ atm \rightarrow atm \rightarrow Prop.$
Inductive $mapP : list\ atm \rightarrow prp \rightarrow Prop :=$
  $mbc : \forall A : atm.\forall L : list\ atm.(P\ L\ A) \rightarrow (mapP\ L\ \langle A \rangle)$
  $\mid minit : \forall A : atm.\forall L : list\ atm.(element\ A\ L) \rightarrow (mapP\ L\ \langle A \rangle)$
  $\mid mtrue : \forall L : list\ atm.(mapP\ L\ tt)$
  $\mid mand : \forall B, C : prp.\forall L : list\ atm.$
    $(mapP\ L\ B) \rightarrow (mapP\ L\ C) \rightarrow (mapP\ L\ (B \& C))$
  $\mid mimp : \forall A : atm.\forall B : prp.\forall L : list\ atm.$
    $((Phyp\ L\ A) \rightarrow (mapP\ (A :: L)\ B)) \rightarrow (mapP\ L\ (A \Rightarrow B))$
  $\mid mall : \forall B : tau \rightarrow prp.\forall L : list\ atm.$
    $(\forall x : tau.(mapP\ L\ (B\ x))) \rightarrow (mapP\ L\ (\bigwedge B))$
  $\mid msome : \forall B : tau \rightarrow prp.\forall L : list\ atm.$
    $\forall x : tau.(mapP\ L\ (B\ x)) \rightarrow (mapP\ L\ (\bigvee B)).$

Fig. 5. A Definition for Extending Properties on Atoms to Properties on Propositions

the definition of *seq* except where atomic properties appear. For example, the clause *mand* can be read as: if the generalized property holds for sequents with arbitrary propositions B and C under assumptions L, then it holds for their conjunction under the same set of assumptions. In *mbc*, the general property holds of an atomic proposition under the condition that the property P holds of the atom. In *minit*, the general property holds simply because the atom is in the list of assumptions. The only other clause involving an atom is *mimp*. The general property holds of an implication $(A \Rightarrow B)$ as long as whenever *Phyp* holds of A, the general property holds of B under the list of assumptions extended with A.

Using the definition of *mapP*, we can structure proofs of many properties so that they involve a direct induction on the definition of the specification logic, and a subinduction on *prog* for the atomic formula case. The theorem below takes care of the first induction.

Definition $PhypL := \lambda L : list\ atm.(\forall a : atm.(element\ a\ L) \rightarrow (Phyp\ L\ a)).$
Theorem $seq\_mapP :$
  $(\forall L : list\ atm.\forall A : atm.(PhypL\ L) \rightarrow$
    $(Phyp\ L\ A) \rightarrow \forall A' : atm.(element\ A'\ (A :: L)) \rightarrow (Phyp\ (A :: L)\ A')) \rightarrow$
  $(\forall L : list\ atm.\forall A : atm.\forall b : prp.(PhypL\ L) \rightarrow$
    $(prog\ A\ b) \rightarrow (seq\ L\ b) \rightarrow (mapP\ L\ b) \rightarrow (P\ L\ A)) \rightarrow$
  $\forall l : list\ atm.\forall B : prp.(PhypL\ l) \rightarrow (seq\ l\ B) \rightarrow (mapP\ l\ B).$

The first two lines of the *seq_mapP* statement roughly state that *Phyp* must be preserved as new atomic formulas are added to the list of assumptions. Here, *PhypL* states that *Phyp* holds of all elements of a list. More specifically, these lines state that whenever $(Phyp\ L\ A')$ holds for all A' already in L, and it is also the case that $(Phyp\ L\ A)$ holds for some new A, then $(Phyp\ (A :: L)\ A')$ also holds for every A' in the list L extended with A. The next two lines of the theorem state the base case, which is likely to be proved by a subinduction on *prog*. Under these two conditions, we can conclude that the generalized property holds of l and B whenever *Phyp* holds of all assumptions in l.

For the new proof of *sr*, the definitions of *prp* and *prog* remain exactly as in Sect. 3, as do the definitions of the parameters that represent syntax along with their distinctness and injectivity axioms. Inversion theorems such as *seq_atom_inv* and *eval_nil_inv* are stated and proved similarly as in the previous section, but without the additional natural number arguments.

Now we can state the generalized version of the *sr* property which is just:

$$\begin{array}{l}\textsf{Theorem}\ sr\_mapP : \forall L : list\ atm.\forall B : prp.\\ \quad (seq\ L\ B) \rightarrow (PhypL\ L) \rightarrow (mapP\ L\ B).\end{array}$$

To prove this theorem, we directly apply *seq_mapP*. The proof that *Phyp* is preserved under the addition of new assumptions is straightforward. The base case for atomic formulas is proved by induction on *prog*. This induction gives us four cases. The two cases which instantiate atom A from *seq_mapP* to formulas of the form $(typeof\ m\ t)$ cause $(P\ L\ A)$ to be reduced to $True$. The details of the two cases for $(prog\ (p \Downarrow v)\ b)$ are similar to the corresponding cases in the proof in Sect. 3.

Finally, we can show that the desired *sr* theorem is a fairly direct consequence of *sr_mapP*.

$$\begin{array}{l}\textsf{Theorem}\ sr : \forall p, v : tm.(seq\ nil\ \langle p \Downarrow v\rangle) \rightarrow\\ \quad \forall t : tm.(seq\ nil\ \langle typeof\ p\ t\rangle) \rightarrow (seq\ nil\ \langle typeof\ v\ t\rangle).\end{array}$$

The proof begins with an application of the *sr_mapP* theorem to conclude that $(mapP\ nil\ \langle p \Downarrow v\rangle)$ holds. Now note that the only way such a formula can hold is by the first clause of the definition of *mapP* since the assumption list is empty and the formula is atomic. This fact is captured by the following inversion theorem.

$$\textsf{Theorem}\ mapP\_nil\_atom\_inv : \forall A : atm.(mapP\ nil\ \langle A\rangle) \rightarrow (P\ nil\ A).$$

Applying this theorem and expanding P, we can conclude that

$$\forall t : tm.(seq\ nil\ \langle typeof\ p\ t\rangle) \rightarrow (seq\ nil\ \langle typeof\ p\ v\rangle)$$

which is exactly what is needed to complete the proof.

We have also used the *mapP* definition to prove the following theorem for a functional language which includes *app* and *abs* as well as many other primitives such as booleans, natural numbers, a conditional statement, and a recursion operator:

$$\begin{array}{l}\textsf{Theorem}\ type\_unicity : \forall M, t : tm(seq\ nil\ \langle typeof\ M\ t\rangle) \rightarrow\\ \quad \forall t' : tm(seq\ nil\ \langle typeof\ M\ t'\rangle) \rightarrow (equiv\ t\ t').\end{array}$$

where the *equiv* predicate is defined simply as

$$\textsf{Inductive}\ equiv : tm \rightarrow tm \rightarrow Prop := refl : \forall t : tm(equiv\ t\ t).$$

To do this proof, we of course had to first define P and *Phyp* specialized to this theorem. We do not discuss the details here.

6 Conclusion, Related Work, and Future Work

We have described a methodology for proving properties of objects expressed using higher-order syntax in Coq. Because of the similarity of reasoning in Coq and reasoning in the object logic we use in our PCC system, we hope to be able to carry over this methodology to the PCC setting. In particular, in both our λProlog and Twelf prototypes, we use an object logic that is currently specified directly, but could be specified as *prog* clauses, allowing the kind of reasoning described here.

In addition to our practical goal of building proofs in the PCC domain as well as other domains which use meta-theoretical reasoning about logics and programming languages, another goal of this paper was to provide insight into how a class of proofs in the relatively new logic $FO\lambda^{\Delta\mathbb{N}}$ correspond to proofs in a class of logics that have been around somewhat longer, namely logics that contain dependent types and inductive definitions.

Certainly, many more examples are needed to illustrate that our approach scales to prove all properties that we are interested in. In addition to carrying out more examples, our future work includes providing more flexible support for reasoning in this setting. The *mapP* predicate in Sect. 5 was introduced to provide one kind of support for induction on sequents. Other kinds of support are worth exploring. For example, we have started to investigate the possibility of generating induction principles for various object-level predicates such as *typeof*. One goal is to find induction principles whose proofs would likely use *mapP* and *seq*, but when they are used in proofs, all traces of the middle layer specification logic would be absent.

As in any formal encoding of one system in another, we need to express and prove adequacy theorems for both the specification and object-level logics. Proofs of adequacy of these encodings should follow similarly to the one for the π-calculus in Honsell et al. [11]. Such a proof would require that we do not discharge the type *tm* in Coq, thus preventing it from being instantiated with an inductively defined type, which could violate adequacy.

In related work, there are several other syntactic approaches to using higher-order syntax and induction in proofs, which either do not scale well, or more work is needed to show that they can. For example, Coq was used by Despeyroux et al. [5] to do a proof of the subject reduction property for the untyped λ-calculus. There, the problem of negative occurrences in definitions used for syntax encodings was handled by replacing such occurrences by a new type. As a result, some additional operations were needed to encode and reason about these types, which at times was inconvenient. Miculan uses a similar approach to handling negative occurrences in formalizing meta-theory of both modal μ-calculus [15] and the lazy call-by-name λ-calculus [14] in Coq. These proofs require fairly extensive use of axioms, more complex than those used here, whose soundness are justified intuitively.

Honsell et al. [11] and Despeyroux [4] define a higher-order encoding of the syntax of the π-calculus in Coq and use it formalize various aspects of the metatheory. Although they use higher-order syntax to encode processes, there is no negative occurrence of the type being defined, and so they are able to define processes inductively.

Despeyroux and Hirschowitz [6] studied another approach using Coq. Again, a different type replacing negative occurrences is used, but instead of directly representing the syntax of (closed) terms of the encoded language by terms of types such as *tm*, closed and open terms of the object language are implemented together as functions from lists of arguments (of type *tm*) to terms of type *tm*. Examples are described, but it is not clear how well the approach scales.

Hofmann [10] shows that a particular induction principle for *tm*, which is derived from a straightforward extension of the inductive types in Coq to include negative occurrences, can be justified semantically under certain conditions (though not within Coq). Although he cannot prove the subject reduction property shown here, he shows that it is possible to express a straightforward elegant proof of a different property: that every typed λ-term reduces to a term in weak-head normal form.

In Pfenning and Rohwedder [18], the technique of schema checking is added to the Elf system, a precursor to the Twelf system mentioned earlier. Both systems implement the Logical Framework (LF) [9]. Induction cannot be expressed in LF, so proofs like those shown here cannot be fully formalized inside the system. However, each of the cases of a proof by induction can be represented. The schema checking technique works outside the system and checks that all cases are handled.

Despeyroux et al. [7] present a λ-calculus with a modal operator which allows primitive recursive functionals over encodings with negative occurrences. This work is a first step toward a new type theory that is powerful enough to express and reason about deductive systems, but is not yet powerful enough to handle the kinds of theorems presented here.

Schürmann [20] has developed a logic which extends LF with support for meta-reasoning about object logics expressed in LF. It has been used to prove the Church-Rosser theorem for the simply-typed λ-calculus and many other examples. The design of the component for reasoning by induction does not include induction principles for higher-order encodings. Instead, it is based on a realizability interpretation of proof terms. This logic has been implemented in Twelf, and includes powerful automated support for inductive proofs.

References

1. A.W. Appel and A.P. Felty. Lightweight lemmas in λProlog. In *International Conference on Logic Programming*, Nov. 1999.
2. A.W. Appel and A.P. Felty. A semantic model of types and machine instructions for proof-carrying code. In *The 27th Annual ACM SIGPLAN-SIGACT Symposium on Principles of Programming Languages*, 2000.
3. P. Borras, D. Clément, T. Despeyroux, J. Incerpi, G. Kahn, B. Lang, and V. Pascual. Centaur: The system. In *Proceedings of SIGSOFT'88: Third Annual Symposium on Software Development Environments (SDE3)*, Boston, 1988.

4. J. Despeyroux. A higher-order specification of the π-calculus. In *First IFIP International Conference on Theoretical Computer Science.* Springer-Verlag Lecture Notes in Computer Science, 2000.
5. J. Despeyroux, A. Felty, and A. Hirschowitz. Higher-order abstract syntax in Coq. In *Second International Conference on Typed Lambda Calculi and Applications.* Springer-Verlag Lecture Notes in Computer Science, Apr. 1995.
6. J. Despeyroux and A. Hirschowitz. Higher-order syntax and induction in coq. In *Fifth International Conference on Logic Programming and Automated Reasoning.* Springer-Verlag Lecture Notes in Computer Science, 1994.
7. J. Despeyroux, F. Pfenning, and C. Schürmann. Primitive recursion for higher-order abstract syntax. In *Third International Conference on Typed Lambda Calculi and Applications.* Springer-Verlag Lecture Notes in Computer Science, 1997.
8. L.-H. Eriksson. A finitary version of the calculus of partial inductive definitions. In L.-H. Eriksson, L. Hallnäs, and P. Schroeder-Heister, editors, *Proceedings of the January 1991 Workshop on Extensions to Logic Programming.* Springer-Verlag Lecture Notes in Artificial Intelligence, 1992.
9. R. Harper, F. Honsell, and G. Plotkin. A framework for defining logics. *Journal of the ACM*, 40(1), Jan. 1993.
10. M. Hofmann. Semantical analysis of higher-order abstract syntax. In *Fourteenth Annual Symposium on Logic in Computer Science*, 1999.
11. F. Honsell, M. Miculan, and I. Scagnetto. π-calculus in (co)inductive type theories. *Theoretical Computer Science*, 253(2), 2001.
12. R. McDowell. *Reasoning in a Logic with Definitions and Induction.* PhD thesis, University of Pennsylvania, December 1997.
13. R. McDowell and D. Miller. Reasoning with higher-order abstract syntax in a logical framework. *ACM Transactions on Computational Logic*, 3(1), Jan. 2002.
14. M. Miculan. Developing (meta)theory of λ-calculus in the theory of contexts. *Electronic Notes on Theoretical Computer Science*, 58, 2001.
15. M. Miculan. On the formalization of the modal μ-calculus in the calculus of inductive constructions. *Information and Computation*, 164(1), 2001.
16. G. Nadathur and D. Miller. An overview of λProlog. In K. Bowen and R. Kowalski, editors, *Fifth International Conference and Symposium on Logic Programming.* MIT Press, 1988.
17. G. Necula. Proof-carrying code. In *24th ACM SIGPLAN-SIGACT Symposium on Principles of Programming Languages.* ACM Press, Jan. 1997.
18. F. Pfenning and E. Rohwedder. Implementing the meta-theory of deductive systems. In *Eleventh International Conference on Automated Deduction*, volume 607. Lecture Notes in Computer Science, 1992.
19. F. Pfenning and C. Schürmann. System description: Twelf — A meta-logical framework for deductive systems. In *Sixteenth International Conference on Automated Deduction*, volume 1632 of *Lecture Notes in Artificial Intelligence.* Springer-Verlag, 1999.
20. C. Schürmann. *Automating the Meta Theory of Deductive Systems.* PhD thesis, Carnegie Mellon University, 2000.
21. The Coq Development Team. The Coq Proof Assistant reference manual: Version 7.2. Technical report, INRIA, 2002.

PuzzleTool: An Example of Programming Computation and Deduction

Michael J.C. Gordon

University of Cambridge Computer Laboratory
William Gates Building, JJ Thomson Avenue, Cambridge CB3 0FD, U.K.
mjcg@cl.cam.ac.uk
http://www.cl.cam.ac.uk/~mjcg

Abstract. Systems that integrate user-programmable theorem proving with efficient algorithms for boolean formula manipulation are promising platforms for implementing special-purpose tools that combine computation and deduction. An example tool is presented in this paper in which theorem proving is used to compile a class of problems stated in terms of functions operating on sets of integers to boolean problems that can be solved using a BDD oracle. The boolean solutions obtained via BDD calculations are then converted by theorem proving to the high-level representation. Although the example is rather specialised, our goal is to illustrate methodological principles for programming tools whose operation requires embedded proof.

1 Background and Motivation

There are already commercial Electronic Design Automation (EDA) tools[1] that have embedded model checkers. It seems likely that future tools will also require embedded theorem proving.

The PROSPER EU Esprit project[2] investigated embedding theorem proving inside both hardware and software applications. PROSPER focused on architectural issues in linking systems [4]. The work reported here is narrower and concerns the programming of deduction and computation within a single system, HOL, and describes:

- scripting combinations of theorem proving and symbolic computation;
- the 'logical flow' in compiling from an abstract representation to a boolean representation and lifting the boolean results back to the abstract level;
- a case study of a particular tool implementation.

[1] Examples are Avanti's clock domain and one-hot checkers (`http://www.avanticorp.com`) and 0-In search (`http://www.0in.com`).

[2] See `http://www.dcs.gla.ac.uk/prosper`.

V.A. Carreño, C. Muñoz, S. Tahar (Eds.): TPHOLs 2002, LNCS 2410, pp. 214–229, 2002.

In Sections 3 and 4 we describe the kind of problems our example tool is intended to solve (analyse a class of solitaire-like puzzles). In the subsequent sections the embedded theorem proving is explained and discussed.

The conventional way to implement special purpose symbolic model checking point tools is either to program them directly using a BDD package such as MuDDy [9] or CUDD [15], or to use an off-the-shelf model checker like SMV or SPIN, to provide high level modelling and property languages. The advantage of the approach described here is that the implementer can develop a model within the expressive language of higher order logic and then implement customised functions to efficiently perform exactly the desired computations. The theorem proving infrastructure of tactics and simplifiers applied to a general logic becomes available for compiling problems into the domain of specialised oracles like BDD and SAT.

Our tool is not intended for deployment on critical systems, so the high assurance provided by the LCF-style proof used to implement it is of little value. It is possible that other tools implemented using a similar methodology might gain value from this assurance[3], but the point we want to make in this paper is that even if assurance of soundness is not needed, the implementation style described here is a powerful way of programming combinations of computation and deduction.

2 Related Work

One of the earliest combinations of theorem proving and model checking is the HOL-Voss system of Joyce and Seger [8]. Voss [14] consists of a lazy ML-like functional language, called FL, with BDDs as a built-in data-type. Quantified Boolean formulae can be input and are parsed to BDDs. Algorithms for model checking are easily programmed. Joyce and Seger interfaced an early HOL system (HOL88) to Voss and in a pioneering paper showed how to verify complex systems by a combination of theorem proving deduction and symbolic trajectory evaluation (STE). The HOL-Voss system integrates HOL deduction with BDD computations. BDD tools are programmed in FL and can then be invoked by HOL-Voss tactics, which can make external calls into the Voss system, passing subgoals via a translation between the HOL and Voss term representations. After Seger moved to Intel, the development of Voss evolved into a system called Forte that "is an LCF-style implementation of a higher-order classical logic" [1] and "seamlessly integrates several types of model-checking engines with lightweight theorem proving and extensive debugging capabilities, creating a productive high-capacity formal verification environment". Only partial details of this are in the public domain [11,1], but a key idea is that FL is used both as a specification language and as an LCF-style metalanguage – i.e. there is no separation, as there is in LCF-like systems, between object language (higher order logic) and metalanguage (Standard ML).

[3] The tool creates a theory each time it is invoked which contain various intermediate theorems. It is possible another application might find such theories a useful result.

Another key paper on connecting theorem proving and model checking is *An Integration of Model-Checking with Automated Proof Checking* by Rajan, Shankar and Srivas [13] in which PVS [12] is extended to support symbolic model checking of properties stated in the μ-calculus via a link to an external ROBDD-based μ-calculus checker. Model checking is invoked from PVS via a command that translates higher order logic goals into boolean formulae that can be printed into a format suitable for input to the external tool. The translation is not by programmed proof (which is the approach we describe below) but by Lisp code that operates on the internal representation of PVS formulae. PVS is a powerful shrink-wrapped checker, our approach is complementary: it provides a scripting framework for the user to implement bespoke tools.

PVS adds model checking to a theorem proving platform. The dual is to add theorem proving to a model checker. This has been done by McMillan in Cadence SMV [3], which provides problem decomposition commands that split verification goals into components small enough for model checking. The decomposition is based on deductive rules, for example for compositional refinement [10], and implemented by light-weight theorem proving. SMV doesn't provide a user-programmable scripting facility, rather each new deduction method is hard-wired into the system. It may be possible to use the approach described in this paper to program the kind of algorithms that SMV builds-in as derived rules, but more work is needed to investigate this.

All the work described above uses external oracles (a BDD engine or model checker). John Harrison implemented BDDs inside the HOL system as a derived proof rule without making use of an external oracle [7] and then showed that the BDD algorithms provide a way of implementing tautology-checking that is significantly better than the methods previously used in HOL. He found, however, that performance was about a thousand times slower than with a BDD engine implemented in C.[4] By reimplementing some of HOL's primitive rules, performance could be improved by around ten times. Harrison only provided data for boolean equivalence checking. The approach in this paper aims to get near the performance of C-based model checking (by using a BDD package implemented in C), whilst remaining within the spirit of the fully-expansive LCF style. Harrison's work is 'logically purer' than ours, but less efficient. The trade-off between logical security and efficiency depends on the application, but Harrison's experiments on an internal implementation of BDDs provides a very interesting point between pure fully-expansive theorem proving and the use of an external oracle.

3 A Class of Puzzles Generalising Peg Solitaire

Two example puzzles are first described, followed by a general setting in which they can be formulated. A third very simple example is given in Section 5.3 to illustrate the operation of our puzzle solving tool `PuzzleTool`.

[4] Some examples were much worse than this and some better.

3.1 Classical Peg Solitaire

The Peg Solitaire board is shown below. All the positions, except the one in the middle, are occupied by pegs, denoted by xxxxx. A move consists of 'jumping' a peg over an adjacent peg in the same row or column into a hole, and removing the peg that was jumped over from the board (thereby reducing the number of pegs on the board by one).

```
                    -------------------------
                    | xxxxx | xxxxx | xxxxx |
                    -------------------------
                    | xxxxx | xxxxx | xxxxx |
    -----------------------------------------------------
    | xxxxx | xxxxx | xxxxx | xxxxx | xxxxx | xxxxx | xxxxx |
    -----------------------------------------------------
    | xxxxx | xxxxx | xxxxx |       | xxxxx | xxxxx | xxxxx |
    -----------------------------------------------------
    | xxxxx | xxxxx | xxxxx | xxxxx | xxxxx | xxxxx | xxxxx |
    -----------------------------------------------------
                    | xxxxx | xxxxx | xxxxx |
                    -------------------------
                    | xxxxx | xxxxx | xxxxx |
                    -------------------------
```

The puzzle is to find a sequence of moves, starting from the above configuration, to a configuration consisting of just one peg in the middle.

Peg Solitaire can be formulated as a state exploration problem by assigning a boolean variable to each board position, for example:

```
                    -------------------------
                    |  v00  |  v02  |  v03  |
                    -------------------------
                    |  v04  |  v05  |  v06  |
    -----------------------------------------------------
    |  v07  |  v08  |  v09  |  v10  |  v11  |  v12  |  v13  |
    -----------------------------------------------------
    |  v14  |  v15  |  v16  |  v17  |  v18  |  v19  |  v20  |
    -----------------------------------------------------
    |  v21  |  v22  |  v23  |  v24  |  v25  |  v26  |  v27  |
    -----------------------------------------------------
                    |  v28  |  v29  |  v30  |
                    -------------------------
                    |  v31  |  v32  |  v33  |
                    -------------------------
```

The initial state is represented by
`v01 ∧ v02 ∧ ... ∧ v16 ∧ ¬v17 ∧ v18 ∧ ... ∧ v33`
and the final(goal) state by
`¬v01 ∧ ¬v02 ∧ ... ∧ ¬v16 ∧ v17 ∧ ¬v18 ∧ ... ∧ ¬v33`

The transition relation, say `R`, is then defined to be a disjunctions of terms, with one disjunct per possible move, so that `R((v01,...,v33),(v01',...,v33'))` is true if and only if there exists a move in state `(v01,...,v33)` that results in state `(v01',...,v33')`.

Standard BDD methods can be used to compute a sequence of states starting with the initial state, ending with the final state, and such that adjacent elements in the sequence satisfy the transition relation `R`.

3.2 HexSolitaire

Consider now a variation of Peg Solitaire called HexSolitaire.

```
                 _____                                       _____
                /     \                                     /     \
          _____/ xxxxx \_____                         _____/       \_____
         /     \       /     \                       /     \       /     \
   _____/ xxxxx \_____/ xxxxx \_____           _____/       \_____/       \_____
  /     \       /     \       /     \         /     \       /     \       /     \
 / xxxxx \_____/ xxxxx \_____/ xxxxx \       /       \_____/       \_____/       \
 \       /     \       /     \       /       \       /     \       /     \       /
  \_____/ xxxxx \_____/ xxxxx \_____/         \_____/       \_____/       \_____/
  /     \       /     \       /     \         /     \       /     \       /     \
 / xxxxx \_____/       \_____/ xxxxx \ ====> /       \_____/ xxxxx \_____/       \
 \       /     \       /     \       /       \       /     \       /     \       /
  \_____/ xxxxx \_____/ xxxxx \_____/         \_____/       \_____/       \_____/
  /     \       /     \       /     \         /     \       /     \       /     \
 / xxxxx \_____/ xxxxx \_____/ xxxxx \       /       \_____/       \_____/       \
 \       /     \       /     \       /       \       /     \       /     \       /
  \_____/ xxxxx \_____/ xxxxx \_____/         \_____/       \_____/       \_____/
        \       /     \       /                     \       /     \       /
         \_____/ xxxxx \_____/                       \_____/       \_____/
               \       /                                   \       /
                \_____/                                     \_____/
```

In HexSolitaire the moves are vertical or diagonal jumps, i.e. over abutting edges of hexagons. Our tool, described below, can prove that there are no solutions to HexSolitaire.

4 PuzzleTool: A Tool for Puzzle Designers

PuzzleTool is an aid for puzzle designers. Its input is a description of a board, an initial state, a goal state and a set of moves. The output is a solution if one exists. If there is no solution, then a set of alternative end states that do have solutions can be computed. The puzzle designer can then choose one of these computed reachable end states to create a solvable puzzle. In the rest of this paper it is shown how to program PuzzleTool in HOL. Sections 2 and 7 contain a brief description of related work and some conclusions, respectively.

A board is represented by a finite set of holes located at positions (x,y) in a 2D grid (the co-ordinates x and y can be arbitrary integers). A state of the board is the subset of holes that have pegs in them.

For example, classical Peg Solitaire is represented as follows.[5]

```
SolitaireBoard =   {              (2,6); (3,6); (4,6);
                                  (2,5); (3,5); (4,5);
                    (0,4); (1,4); (2,4); (3,4); (4,4); (5,4); (6,4);
                    (0,3); (1,3); (2,3); (3,3); (4,3); (5,3); (6,3);
                    (0,2); (1,2); (2,2); (3,2); (4,2); (5,2); (6,2);
                                  (2,1); (3,1); (4,1);
                                  (2,0); (3,0); (4,0)}

SolitaireInitialState = SolitaireBoard DIFF {(3,3)}
SolitaireFinalState   = {(3,3)}
```

To specify the moves, eight auxiliary functions are defined which map a hole at (x,y) to the coordinates of the hole n units away in one of eight directions: north, south, east, west, north-west, north-east, south-west, south-east.

The following diagram illustrates the moves and auxiliary functions N, S, E, W, NW, NE, SW and SE. How these eight functions are used to define moves is explained later when the general transition function Trans is defined.

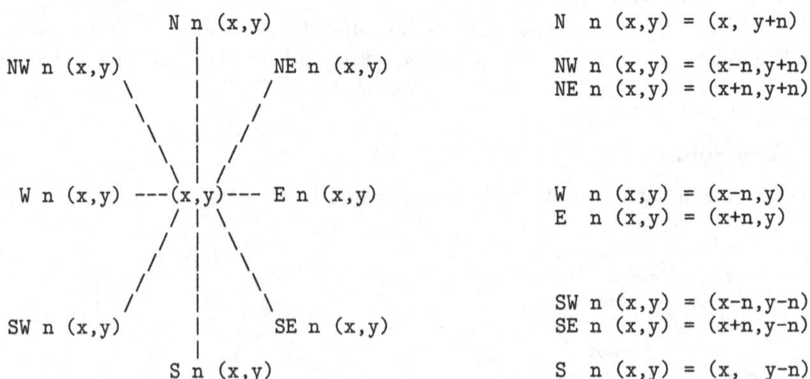

A board and set of moves define a transition relation. For a particular puzzle a transition relation is obtained by applying Trans, defined below, to appropriate arguments.

[5] In HOL notation, a finite set is written $\{e_1;\ldots;e_n\}$, with semi-colon separating the elements. S_1 DIFF S_2 denotes the result of deleting all element in S_2 from S_1.

```
⊢ Trans board moves (state, state') =
    ∃p q r ∈ board. (∃move ∈ moves. (q = move p) ∧ (r = move q)) ∧
                    (p IN state) ∧ (q IN state) ∧ ¬(r IN state) ∧
                    (state' = (state DIFF {p;q}) UNION {r})
```

the arguments `board` and `moves` are, respectively, the set of holes and set of moves of the puzzle. A peg at `p` can jump over a peg at `q` into a hole at `r` in state `state` if they are in a line (i.e. `(q = move p)∧(r = move q)` for some move `move`), and if `p` and `q` have pegs in `state` (i.e. `(p IN state)∧(q IN state)`) and there is a hole at `r` in `state` (i.e. `¬(r IN state)`). If `p` jumps over `q` into `r` then the successor state (i.e. `state'`) is obtained by removing `p` and `q` and adding `r` (i.e. `(state' = (state DIFF {p;q}) UNION {r})`, where S_1 `UNION` S_2 denotes the union of S_1 and S_2.

For classical Peg Solitaire the holes are adjacent, so `n=1`, and the set of moves are defined by `SolitaireMoves = {N 1; S 1; E 1; W 1}`.

The term `Trans SolitaireBoard SolitaireMoves` denotes the transition relation for Peg Solitaire, where `SolitaireBoard` and `SolitaireMove` are as defined above. The low-level representation consisting of the disjunction of 76 conjunctions can be derived by deduction from the definitions of `Trans`, `SolitaireBoard` and `SolitaireMove`. How this is done is explained in the next section.

HexSolitaire is most naturally specified with a board having holes in the same row two horizontal units apart (`n=2`), but holes in the same column one vertical unit apart (`n=1`).

```
HexSolitaireBoard =
{                           (2,8);
                     (1,7);        (3,7);
              (0,6);        (2,6);        (4,6);
                     (1,5);        (3,5);
              (0,4);        (2,4);        (4,4);
                     (1,3);        (3,3);
              (0,2);        (2,2);        (4,2);
                     (1,1);        (3,1);
                            (2,0)}

HexSolitaireInitialState = HexSolitaireBoard DIFF {(2,4)}
HexSolitaireFinalState   = {(2,4)}
HexSolitaireMoves        = {N 2; S 2; NE 1 ; SW 1 ; SE 1; NW 1}
```

In general, a puzzle is specified by four definitions: the board, the set of moves, the initial state and the final (or goal) state.

For `PuzzleTool` these are represented as HOL definitions of constants. For example, HexSolitaire is represented by the definitions of `HexSolitaireBoard`, `HexSolitaireInitialState`, `HexSolitaireFinalState` and `HexSolitaireMoves`.

`PuzzleTool` takes three definitions as arguments and returns a pair of functions. It has ML type `:thm * thm * thm -> (thm -> thm) * (unit -> thm)`. Evaluating

```
PuzzleTool(Board_def,Moves_def,InitialState_def)
```

returns two functions. The first is applied to the definition of a desired goal state to compute a solution. The second is applied to `()` to compute a theorem giving the set of reachable end states (an end state is a state from which no moves are possible). See Section 5.3 for an example.

5 How PuzzleTool Works

First we describe abstractly how **PuzzleTool** works, then this account is specialised to the domain of puzzles and finally a very simple example is worked through.

5.1 Abstract View

Let B and R be polymorphic predicates representing a set of initial states and a transition relation, respectively.

$B : \alpha \rightarrow \texttt{bool}, \qquad R : \alpha \times \alpha \rightarrow \texttt{bool}$

Let `Reachable` $R\ B\ a$ if and only if a is reachable from some value satisfying B via a finite number of transitions of R. `Reachable` is a polymorphic function of type $(\alpha \times \alpha \rightarrow \texttt{bool}) \rightarrow (\alpha \rightarrow \texttt{bool}) \rightarrow \alpha \rightarrow \texttt{bool}$. We will use `Reachable` at two particular types: σ and τ.

Suppose we have a specific type σ, a specific predicate `B`, a specific transition relation `R` and a specific set of goal states specified by a predicate `Q`:

$\texttt{B} : \sigma \rightarrow \texttt{bool}, \qquad \texttt{R} : \sigma \times \sigma \rightarrow \texttt{bool}, \qquad \texttt{Q} : \sigma \rightarrow \texttt{bool}$

Think of σ as the type of states of puzzles (i.e. sets of those coordinates containing pegs). Suppose also that we have a way of encoding the subset of σ satisfying a predicate `D` as values of another type τ. In our application, values of type τ will be vectors of booleans and `D` characterises the states for a particular puzzle. The encoding is defined by functions `Abs` from σ to τ and `Rep` from τ to σ which satisfy the properties below.

$\texttt{D} : \sigma \rightarrow \texttt{bool}, \qquad \texttt{Rep} : \sigma \rightarrow \tau\ , \qquad \texttt{Abs} : \tau \rightarrow \sigma$

$\vdash \forall \texttt{s}{:}\sigma.\ \texttt{D s = (Abs(Rep s) = s)}$
$\vdash \forall \texttt{t}{:}\tau.\ \texttt{Rep(Abs t) = t}$
$\vdash \forall \texttt{s}{:}\sigma.\ \texttt{B s} \Rightarrow \texttt{D s}$
$\vdash \forall \texttt{s s'}{:}\sigma.\ \texttt{D s} \wedge \texttt{R(s,s')} \Rightarrow \texttt{D s'}$

Our first goal is to calculate a sequence of values of type σ starting from a value satisfying `B`, ending with a value satisfying `Q` and such that adjacent values in the sequence are related by `R`. In our application, such a sequence will be a puzzle solution.

We solve our first goal by calculating a sequence $\texttt{t}_1, \ldots, \texttt{t}_p$ of values of τ such that

$\vdash \texttt{B(Abs t}_1\texttt{)},\ \vdash \texttt{R(Abs t}_1\texttt{,Abs t}_2\texttt{)},\ldots,\ \vdash \texttt{R(Abs t}_{p-1}\texttt{,Abs t}_p\texttt{)},\ \vdash \texttt{Q(Abs t}_p\texttt{)}$

In our application, $\texttt{t}_1, \ldots, \texttt{t}_p$ will be vectors of boolean constants computed using BDDs. Then by evaluating the terms $\texttt{Abs t}_1, \ldots, \texttt{Abs t}_p$ one obtains the desired sequence of values of type σ. This is how solutions to puzzles are computed.

Suppose now we want to compute all the reachable states starting from `B` via transitions of `R` that satisfy some predicate `P`. In our example `P` characterises end states. We proceed by computing a term `W t` such that

$\vdash \texttt{Reachable}\ (\lambda\texttt{(t,t')}.\ \texttt{R(Abs t,Abs t')})\ (\lambda\texttt{t}.\ \texttt{B(Abs t)})\ \texttt{t} \wedge \texttt{P(Abs t) = W t}$

In our example `W t` is obtained by using standard fixed-point methods to compute the BDD of `Reachable (λ(t,t'). R(Abs t,Abs t')) (λt. B (Abs t)) t`, then conjoining this with the BDD of `P(Abs t)`, and then converting the resulting BDD to a term [5].

The following are routine theorems:

```
⊢ ∀D R B.  (∀s. B s ⇒ D s) ∧ (∀s s'. D s ∧ R(s,s') ⇒ D s')
            ⇒
            (∀s. Reachable R B s ⇒ D s)

⊢ ∀D abs rep R B.
     (∀s. D s = (abs (rep s) = s)) ∧ (∀s. Reachable R B s ⇒ D s)
     ⇒
     ∀t. Reachable R B (abs t) =
           Reachable (λ(t,t'). R(abs t,abs t')) (λt. B (abs t)) t
```

Hence it follows by elementary reasoning from the properties on the previous page that:

```
⊢ ∀s. Reachable R B s ⇒ D s

⊢ ∀t. Reachable R B (Abs t) =
       Reachable (λ(t,t').R(Abs t,Abs t')) (λt.B (Abs t)) t
```

Simplifying the computed equation for `W t` with the second of these theorems yields:

```
⊢  ∀t. Reachable R B (Abs t) ∧ P(Abs t) = W t
```

hence by specialising `t` to `Rep s` and then generalising `s`

```
⊢  ∀s. Reachable R B (Abs(Rep s)) ∧ P(Abs(Rep s)) = W(Rep s)
```

hence

```
⊢  ∀s. Reachable R B s ∧ P s = D s ∧ W(Rep s)
```

Finally, by suitable logical manipulation one can prove for specific states $s_1,\cdots,s_q$ that:

```
⊢  W(Rep s) = s ∈ {s1,···,sq}
```

See the following section for an example.

5.2 Concrete Instantiation for PuzzleTool

The user of `PuzzleTool` starts by defining three constants to represent the board, the set of moves and the initial states. The names of these constants are chosen by the user, but let us assume here that they are `Board`, `Moves` and `Init` , respectively. Thus the inputs are something like:

```
val Board_def        = Define `Board = ... `;
val Moves_def        = Define `Moves = ... `;
val InitialState_def = Define `Init  = ... `;
```

The operation of `PuzzleTool` is then fully automatic and is split into two phases.

Phase 1 computes a bespoke theory of the puzzle to be analysed and theorems that relate the high level specifications to a boolean encoding. These are then packaged into two functions which, when invoked, perform BDD analysis. Phase 1 is invoked by evaluating something like:

```
val (SolveFun, EndStatesFun) =
 PuzzleTool(Board_def,Moves_def,InitialState_def)
```

Phase 2 uses the HOL library `HolBddLib` [6], to compute a solution (`SolveFun`) and/or the set of reachable end states (`EndStatesFun`). If a solution is computed then the user must first define the desired goal state by something like:

```
val Final_def = Define ‘Final = ... ‘;
```

and then apply `SolveFun` to `Final_def`.

The theorems proved in Phase 1 are used by the functions returned by `PuzzleTool` both to encode the problem for the BDD-based calculation in Phase 2 and also to lift the solutions found in Phase 2 to the original representation in terms of sets of holes.

The ML program implementing `PuzzleTool` starts by making a number of definitions derived from the arguments it is invoked with. First some predicates are defined:

```
⊢ BoardInit  state = (state = Init)
⊢ BoardDom   state = (state SUBSET Board)
```

where S_1 `SUBSET` S_2 if and only if S_1 is included in S_2.

Next a transition relation for the specific puzzle is defined by applying `Trans` to `Board` and `Moves`.

```
⊢ BoardTrans = Trans Board Moves
```

and it is proved that

```
⊢ ∀state state'.
    BoardDom state ∧ BoardTrans(state,state') ⇒ BoardDom state'
```

abstraction and representation functions `BoardAbs` and `BoardRep` are then defined. `BoardAbs` maps states to tuples of booleans (boolean vectors) and `BoardRep` is a partial inverse. A boolean vector `(b1,...,bn)` represents a set of states via an enumeration of the holes in the board. Hole `p` is in the set represented by `(b1,...,bn)` if and only if `bi` is true, where `i` is the position of `p` in the enumeration.

The enumeration of holes is specified by defining a non-repeating list of the board positions:

```
⊢ BoardList = [...]
```

where the elements of the list are computed from `Board`. The following general functions are used to define the abstraction and representation:

```
⊢ REP pl state = MAP (λp. p IN state) pl

⊢ (ABS [] [] = {})
  ∧
  (ABS (p::pl) (b::bl) =
    if b then p INSERT ABS pl bl else ABS pl bl)
```

where MAP maps a function over a list, :: (a infix) is the list 'cons' function and INSERT (an infix) inserts an element into a set. To convert between tuples and lists, functions dependent on the size of the puzzle are automatically defined:

```
⊢ BoardL2T [b1;...;bn] = (b1,...,bn)
⊢ BoardT2L (b1,...,bn) = [b1;...;bn]
```

Using these, PuzzleTool defines

```
⊢ BoardAbs bv    = ABS BoardList (BoardT2L bv)
⊢ BoardRep state = BoardL2T(REP BoardList state)
```

The following theorems are then proved:

```
⊢ ∀state. BoardDom state = (BoardAbs(BoardRep state) = state)
⊢ ∀bv. BoardRep(BoardAbs bv) = bv
⊢ ∀state. BoardInit state ⇒ BoardDom state
```

The variable bv here and below is so named because it ranges over boolean vectors. Next it is proved that:

```
⊢ ∀state. Reachable BoardTrans BoardInit state ⇒ BoardDom state
```

and hence

```
⊢  Reachable BoardTrans BoardInit (BoardAbs(b1,...,bn)) =
     Reachable
      (λ((b1,...,bn),(b1',...,bn')).
        BoardTrans(BoardAbs(b1,...,bn),BoardAbs(b1',...,bn')))
      (λ(b1,...,bn). BoardInit(BoardAbs(b1,...,bn)))
      (b1,...,bn)
```

In order to evaluate the right hand side of this equation using BDDs, the terms BoardTrans(BoardAbs(b1,...,bn),BoardAbs(b1',...,bn')) and also the term BoardInit(BoardAbs(b1,...,bn)) are rewritten using derived equations to get theorems of the form:

```
⊢  BoardTrans(BoardAbs(b1,...,bn),BoardAbs(b1',...,bn')) = ...
⊢  BoardInit(BoardAbs(b1,...,bn)) = ...
```

where the occurrences of "..." on the right hand side are boolean formulae suitable for direct representation as BDDs (see Section 5.3 for examples).

Suitable equations must be assumed or proved that enable the HOL rewriting tools[6] to automatically derive theorems of the form shown above. In the current implementation some of these equations are proved as general theorems that can be instantiated, some are proved on-the-fly each time the tool is invoked, and some are even assumed as axioms (Section 7 discusses this).

Phase 2 now takes over. To search for a solution, a theorem:

```
⊢  BoardFinal(BoardAbs(b1,...,bn)) = ...
```

is also derived. If a solution exists, then standard state enumeration methods in HolBddLib can compute a sequence of theorems:

$\vdash$ BoardInit(BoardAbs(c_{11},...,c_{1n}))
$\vdash$ BoardTrans(BoardAbs(c_{11},...,c_{1n}),BoardAbs(...))

$\vdots$

$\vdash$ BoardTrans(BoardAbs(...),BoardAbs(c_{m1},,...,c_{mn}))
$\vdash$ BoardFinal(BoardAbs(c_{m1},...,c_{mn}))

[6] These tools are in the HOL libraries simpLib and computeLib.

where (c_{11},...,c_{1n}), ... , (c_{m1},...,c_{mn}) are vectors of the boolean constants T (true) and F (false). Evaluating BoardAbs(c_{11},...,c_{1n}), ... , BoardAbs(c_{m1},...,c_{mn}) using computeLib [2] yields a sequence of states representing a possible solution.

To compute the reachable end states, standard HolBddLib methods can compute the BDD representing the term

```
Reachable BoardTrans BoardInit (BoardAbs(b1,...,bn))
```

An end state is defined to be a state from which there are no transitions. The general definition is:

```
⊢ ∀R state. End R state = ¬∃state'. R(state,state')
```

Instantiating R to (λ(bv,bv'). BoardTrans(BoardAbs bv,boardAbs bv')), bv to (b1,...,bn) and then simplifying yields a theorem of the form

```
⊢ End (λ(bv,bv'). BoardTrans(BoardAbs bv,BoardAbs bv')) (b1,...,bn) = ...
```

HolBddLib is then used to compute the BDD of the right hand side of this equation, and hence the BDD representing the left hand side. This BDD can then be combined with the BDD representing the end states and the result converted to a theorem of the form:

```
⊢  Reachable BoardTrans BoardInit (BoardAbs(b1,...,bn)) ∧
     End BoardTrans (BoardAbs(b1,...,bn)) = ...
```

where the right hand side "..." is the term obtained from the BDD. Rewriting this to DNF and also using the equations:

```
⊢ b = (b = T)
⊢ ¬b = (b = F)
⊢ (b1 = c1) ∧ (b2 = c2) == ((b1,b2) = (c1,c2))
```

yields a theorem of the form

```
⊢  ∀b1 ... bn
     Reachable BoardTrans BoardInit (BoardAbs(b1,...,bn)) ∧
      End BoardTrans (BoardAbs(b1,...,bn)) =
      ((b1,...,bn) = (c11,...,c1n)) ∨ ... ∨ ((b1,...,bn) = (cp1,...,cpn))
```

where each constant c_{jk} is T or F. This is equivalent to

```
⊢  ∀bv. Reachable BoardTrans BoardInit (BoardAbs bv) ∧
          End BoardTrans (BoardAbs bv) =
          (bv = (c11,...,c1n)) ∨ ··· ∨ (bv = (cp1,...,cpn))
```

i.e. quantification of n boolean variables is replaced by a single quantification over a tuple. Instantiating bv to BoardRep state and then generalising yields:

```
⊢ ∀state. Reachable BoardTrans BoardInit (BoardAbs(BoardRep state)) ∧
           End BoardTrans (BoardAbs(BoardRep state)) =
           (BoardRep state = (c11,...,c1n))
           ∨
           ⋮
           ∨
           (BoardRep state = (cp1,...,cpn))
```

The following is a theorem

```
⊢ ∀abs rep.
    (∀bv. rep(abs bv) = bv)
    ⇒
    ∀s bv. (rep s = bv) = (abs(rep s) = abs bv)
```

hence as ⊢ ∀bv. BoardRep(BoardAbs bv) = bv

```
⊢ ∀state bv.
    (BoardRep state = bv) = (BoardAbs(BoardRep state) = BoardAbs bv)
```

hence

```
⊢ ∀state.
     Reachable BoardTrans BoardInit (BoardAbs(BoardRep state)) ∧
      End BoardTrans (BoardAbs(BoardRep state)) =
      (BoardAbs(BoardRep state) = BoardAbs(c11,...,c1n))
      ∨
      ⋮
      ∨
      (BoardAbs(BoardRep state) = BoardAbs(cp1,...,cpn))
```

using

```
⊢ ∀state. BoardDom state = (BoardAbs(BoardRep state) = state)
⊢ ∀state. Reachable BoardTrans BoardInit state ⇒ BoardDom state
```

the theorem above simplifies to

```
⊢ ∀state.
     Reachable BoardTrans BoardInit state ∧ End BoardTrans state =
       BoardDom state
       ∧
       ((state=BoardAbs(c11,...,c1n))∨...∨(state=BoardAbs(cp1,...,cpn)))
```

which can be rewritten to

```
⊢ ∀state.
     Reachable BoardTrans BoardInit state ∧ End BoardTrans state =
       state SUBSET Board
       ∧
       state IN {BoardAbs(c11,...,c1n)); ... ; BoardAbs(cp1,...,cpn)}
```

which shows the set of reachable end states to be:

```
{BoardAbs(c11,...,c1n) ; ... ; BoardAbs(cp1,...,cpn)}
```

Each member of this set can be evaluated to state (i.e. a set of hole positions).

5.3 Example: Puzzle Triv

A very simple puzzle called Triv will be used to illustrate the operation of PuzzleTool.

```
TrivBoard = {(0,4);
             (0,3);
             (0,2);
             (0,1);
             (0,0)}

TrivMoves = {N 1; S 1}
TrivInitialState = {(0,0);(0,1);(0,3)}
TrivFinalState = {(0,4)}
```

```
---------          ---------
|       |          | XXXXX |
---------          ---------
| XXXXX |          |       |
---------          ---------
|       |   ===>   |       |
---------          ---------
| XXXXX |          |       |
---------          ---------
| XXXXX |          |       |
---------          ---------
```

The first step in solving Triv is to define constants representing the puzzle:

```
val TrivBoard_def      = Define`TrivBoard={(0,4);(0,3);(0,2);(0,1);(0,0)}`
val TrivMoves_def      = Define`TrivMoves={N 1;S 1}`
val TrivInitialState_def = Define`TrivInitialState={(0,0);(0,1);(0,3)}`
val TrivFinalState_def   = Define`TrivFinalState={(0,4)}`
```

The function puzzleTool performs the Phase 1 analysis and returns a pair of functions for doing the Phase 2 BDD analysis. The first function computes solutions and the second one computes reachable end states. Here is a complete session to solve and analyse Triv (some system output deleted).

```
- val (TrivSolveFn,TrivEndStatesFn) =
    PuzzleTool(TrivBoard_def, TrivMoves_def, TrivInitialState_def);

> val TrivSolveFn = fn : thm -> thm
  val TrivEndStatesFn = fn : unit -> thm

- TrivEndStatesFn();
> val it = |- Reachable TrivTrans TrivInit s /\ End TrivTrans s =
              s SUBSET TrivBoard /\ s IN {{(0,4)}; {(0,1)}} : thm

- val TrivFinalState_def = Define `TrivFinalState = {(0,4)}`;
> val TrivFinalState_def = |- TrivFinalState = {(0,4)} : thm

- TrivSolveFn TrivFinalState_def;
> val it =
        |- TrivInit {(0,3); (0,1); (0,0)} /\
           TrivTrans ({(0,3); (0,1); (0,0)},{(0,3); (0,2)}) /\
           TrivTrans ({(0,3); (0,2)},{(0,4)}) /\ TrivFinal {(0,4)} : thm
```

Thus invoking PuzzleTool on Triv yields the following trace:

```
[{(0,0);(0,1);(0,3)}; {(0,2);(0,3)}; {(0,4)}]
```

and the set {{(0,4)}; {(0,1)}} of reachable end states.

The results of invoking PuzzleTool on the two solitaire games can be found on the web at: http://www.cl.cam.ac.uk/~mjcg/puzzleTool.

Phase 1 begins with a number of definitions being made.

```
⊢ TrivInit state = (state = {(0,0); (0,1); (0,3)})
⊢ TrivAbs (b0,b1,b2,b3,b4)
                   = ABS[(0,4);(0,3);(0,2);(0,1);(0,0)][b0;b1;b2;b3;b4]
⊢ TrivT2L [b0;b1;b2;b3;b4] = (b0,b1,b2,b3,b4)
⊢ TrivRep s = TrivT2L (REP [(0,4);(0,3);(0,2);(0,1);(0,0)] s)
⊢ TrivDom s = s SUBSET {(0,0);(0,1);(0,3)}
⊢ TrivTrans (state,state') = Trans TrivBoard TrivMoves (state,state')
```

From these definitions, some axioms[7] and pre-proved general theorems, the boolean representation of the puzzle is computed:

```
⊢ TrivInit(TrivAbs(b0,b1,b2,b3,b4)) = ¬b0 ∧ b1 ∧ ¬b2 ∧ b3 ∧ b4

⊢ TrivTrans(TrivAbs(b0,b1,b2,b3,b4),TrivAbs(b0',b1',b2',b3',b4')) =
    (b2 ∧ b1 ∧ ¬b0 ∧ b0' ∧ ¬b1' ∧ ¬b2' ∧ (b3' = b3) ∧ (b4' = b4) ∨
     b3 ∧ b2 ∧ ¬b1 ∧ (b0' = b0) ∧ b1' ∧ ¬b2' ∧ ¬b3' ∧ (b4' = b4) ∨
     b4 ∧ b3 ∧ ¬b2 ∧ (b0' = b0) ∧ (b1' = b1) ∧ b2' ∧ ¬b3' ∧ ¬b4') ∨
    b0 ∧ b1 ∧ ¬b2 ∧ ¬b0' ∧ ¬b1' ∧ b2' ∧ (b3' = b3) ∧ (b4' = b4) ∨
    b1 ∧ b2 ∧ ¬b3 ∧ (b0' = b0) ∧ ¬b1' ∧ ¬b2' ∧ b3' ∧ (b4' = b4) ∨
    b2 ∧ b3 ∧ ¬b4 ∧ (b0' = b0) ∧ (b1' = b1) ∧ ¬b2' ∧ ¬b3' ∧ b4'
```

[7] See Section 7 for a discussion of assuming axioms versus proving theorems.

These 'scalarised' equations are built into the functions returned by `PuzzleTool`.

Phase 2 can now start. The BDD state exploration tools in HOL's `HolBddLib` can automatically compute a solution, given a goal state:

```
val FinalState_def = Define 'TrivFinalState = {(0,4)}';
```

from which the function `TrivSolveFn` returned by `PuzzleTool` deduces:

```
⊢ TrivFinal (TrivAbs(b0,b1,b2,b3,b4)) = b0 ∧ ¬b1 ∧ ¬b2 ∧ ¬b3 ∧ ¬b4
```

and then uses `HolBddLib` to compute:

```
⊢ TrivInit(TrivAbs(F,T,F,T,T))
⊢ TrivBoardTrans(TrivAbs(F,T,F,T,T),TrivAbs(F,T,T,F,F))
⊢ TrivBoardTrans(TrivAbs(F,T,T,F,F),TrivAbs(T,F,F,F,F))
⊢ TrivFinal(TrivAbs(T,F,F,F,F))
```

which after evaluation and conjunction yield:

```
⊢ TrivInit {(0,3);(0,1);(0,0)} ∧
  TrivTrans ({(0,3);(0,1);(0,0)},{(0,3);(0,2)}) ∧
  TrivTrans ({(0,3);(0,2)},{(0,4)}) ∧
  TrivFinal {(0,4)}
```

To compute the reachable end states the following is derived.

```
⊢ End (λ(bv,bv'). TrivTrans(TrivAbs bv,TrivAbs bv'))
      (b0,b1,b2,b3,b4) = ((∀b1' b2'. ¬b2 ∨ ¬b1 ∨ b0 ∨ b1' ∨ b2') ∧
                          (∀b2' b3'. ¬b3 ∨ ¬b2 ∨ b1 ∨ b2' ∨ b3') ∧
                          ∀b3' b4'. ¬b4 ∨ ¬b3 ∨ b2 ∨ b3' ∨ b4') ∧
                          (∀b0' b1'. ¬b0 ∨ ¬b1 ∨ b2 ∨ b0' ∨ b1') ∧
                          (∀b1' b2'. ¬b1 ∨ ¬b2 ∨ b3 ∨ b1' ∨ b2') ∧
                          ∀b2' b3'. ¬b2 ∨ ¬b3 ∨ b4 ∨ b2' ∨ b3'
```

The BDD of this is computed. `HolBddLib` is then invoked to compute the BDD of the set of all reachable states and the two BDDs are combined and the resulting BDD converted to a term to obtain:

```
⊢ Reachable TrivTrans TrivInit (TrivAbs(b0,b1,b2,b3,b4)) ∧
   End TrivTrans (TrivAbs(b0,b1,b2,b3,b4)) =
    b0 ∧ ¬b1 ∧ ¬b2 ∧ ¬b3 ∧ ¬b4 ∨ ¬b0 ∧ ¬b1 ∧ ¬b2 ∧ b3 ∧ ¬b4
```

from this it follows by deduction:

```
⊢ ∀b0 b1 b2 b3 b4.
    Reachable TrivTrans TrivInit (TrivAbs(b0,b1,b2,b3,b4)) ∧
    End TrivTrans (TrivAbs(b0,b1,b2,b3,b4)) =
     ((b0,b1,b2,b3,b4) = (T,F,F,F,F)) ∨ ((b0,b1,b2,b3,b4) = (F,F,F,T,F))
```

hence:

```
⊢ Reachable TrivTrans TrivInit (TrivAbs(TrivRep s)) ∧
   End TrivTrans (TrivAbs(TrivRep s)) =
    (TrivRep s = (T,F,F,F,F)) ∨ (TrivRep s = (F,F,F,T,F))
```

and then using theorems relating `TrivDom`, `TrivRep` and `TrivAbs`, and also the easily proved theorem ⊢ `∀s. Reachable TrivTrans TrivInit s ⇒ TrivDom s`

```
⊢ Reachable TrivTrans TrivInit s ∧ End TrivTrans s =
    TrivDom s ∧ ((s = {(0,4)}) ∨ (s = {(0,1)}))
```

and hence

```
⊢ Reachable TrivTrans TrivInit s ∧ End TrivTrans s =
    s SUBSET TrivBoard ∧ s IN {{(0,4)};{(0,1)}}
```

6 Computation and Deduction

The flow from input to output described in the previous section is programmed in the HOL metalanguage, Standard ML. `PuzzleTool` can be viewed logically as two derived inference rules in higher order logic. The rule corresponding to computing solutions is:

```
⊢ Board = ...     ⊢ Moves = ...     ⊢ Init  = ...     ⊢ Final = ...
------------------------------------------------------------------------
⊢ state₁ IN Init
   ∧
   BoardTrans(state₁, ...) ∧     . . .     ∧ BoardTrans(..., stateₙ)
   ∧
   stateₙ IN Final
```

the rule corresponding to computing all reachable end states is:

```
⊢ Board = ...     ⊢ Moves = ...     ⊢ Init  = ...
------------------------------------------------------------------------
⊢ ∀state.
    Reachable BoardTrans BoardInit state ∧ End BoardTrans state =
     state SUBSET Board ∧ state IN { . . . }
```

7 Conclusions

The implementation of efficient special purpose tools in a general theorem prover like HOL is practicable. The implementer can produce a quick prototype using 'brute force' methods and by assuming general results as axioms before going to the effort of proving them. This can establish proof of concept.

The tool can then be tuned by optimising deductions using standard theorem proving and verification methods. The assurance of soundness can be improved by checking the assumed theorems for each problem 'on-the-fly' by symbolic execution, or by proving general theorems that can be instantiated. How much optimisation and assurance enhancement to do requires a cost/benefit analysis.

The needs of a tool implementation platform are wider than the needs of a pure proof assistant. Scripting and encapsulation mechanisms are necessary, and access to external oracles (like BDD and SAT engines) are essential for some applications. There is potential for synergy between the automatic theorem proving, proof assistant and verification tool communities. This paper has explored widening the applicability of a proof assistant, HOL, to be a tool implementation platform. In the long term it might be worthwhile designing such a platform from scratch, but in the short to medium term there is plenty of potential left in existing programmable proof assistants.

Acknowledgements

Thanks to members of the Automated Reasoning Group at Cambridge for feedback on an early presentation of this work. I got the Peg Solitaire example via Tony Hoare and Bill Roscoe of Oxford University. This research is supported by EPSRC Project GR/R27105/01 entitled *Fully Expansive Proof and Algorithmic Verification*.

References

1. Mark D. Aagaard, Robert B. Jones, and Carl-Johan H. Seger. Lifted-FL: A Pragmatic Implementation of Combined Model Checking and Theorem Proving. In *Theorem Proving in Higher Order Logics (TPHOLs'99)*, number 1690 in Lecture Notes in Computer Science, pages 323–340. Springer-Verlag, 1999.
2. Bruno Barras. Programming and computing in HOL. In J. Harrison and M. Aagaard, editors, *Theorem Proving in Higher Order Logics: 13th International Conference, TPHOLs 2000*, volume 1869 of *Lecture Notes in Computer Science*, pages 17–37. Springer-Verlag, 2000.
3. See web page `http://www-cad.eecs.berkeley.edu/~kenmcmil/smv\/`.
4. L. A. Dennis, G. Collins, M. Norrish, R. Boulton, K. Slind, G. Robinson, M. Gordon, and T. Melham. The prosper toolkit. In S. Graf and M. Schwartbach, editors, *Tools and Algorithms for Constructing Systems (TACAS 2000)*, number 1785 in Lecture Notes in Computer Science, pages 78–92. Springer-Verlag, 2000.
5. Michael J.C. Gordon. Reachability programming in HOL using BDDs. In J. Harrison and M. Aagaard, editors, *Theorem Proving in Higher Order Logics: 13th International Conference, TPHOLs 2000*, volume 1869 of *Lecture Notes in Computer Science*, pages 180–197. Springer-Verlag, 2000.
6. Mike Gordon. Reachability programming in HOL98 using BDDs. In *The 13th International Conference on Theorem Proving and Higher Order Logics*. Springer-Verlag, 2000.
7. John Harrison. Binary decision diagrams as a HOL derived rule. *The Computer Journal*, 38:162–170, 1995.
8. J. Joyce and C. Seger. The HOL-Voss System: Model-Checking inside a General-Purpose Theorem-Prover. In J. J. Joyce and C.-J. H. Seger, editors, *Higher Order Logic Theorem Proving and its Applications: 6th International Workshop, HUG'93, Vancouver, B.C., August 11-13 1993*, volume 780 of *Lecture Notes in Computer Science*, pages 185–198. Spinger-Verlag, 1994.
9. Moscow ML interface to BuDDy by Ken Friis Larsen and Jakob Lichtenberg documented at `http://www.it-c.dk/research/muddy/`.
10. K.L. McMillan. A compositional rule for hardware design refinement. In Orna Grumberg, editor, *Computer-Aided Verification, CAV '97*, Lecture Notes in Computer Science, pages 24–35, Haifa, Israel, June 1997. Springer-Verlag.
11. John O'Leary, Xudong Zhao, Robert Gerth, and Carl-Johan H. Seger. Formally verifying IEEE compliance of floating-point hardware. *Intel Technology Journal*, First Quarter 1999. Online at `http://developer.intel.com/technology/itj/`.
12. See web page `http://www.csl.sri.com/pvs.html`.
13. S. Rajan, N. Shankar, and M.K. Srivas. An integration of model-checking with automated proof checking. In Pierre Wolper, editor, *Computer-Aided Verification, CAV'95*, volume 939 of *Lecture Notes in Computer Science*, pages 84–97, Liege, Belgium, June 1995. Springer-Verlag.
14. Carl-Johan H. Seger. Voss - a formal hardware verification system: User's guide. Technical Report UBC TR 93-45, The University of British Columbia, December 1993.
15. Fabio Somenzi's CUDD: CU Decision Diagram Package documented at `http://vlsi.colorado.edu/~fabio/CUDD/`.

A Formal Approach to Probabilistic Termination

Joe Hurd*

Computer Laboratory
University of Cambridge
joe.hurd@cl.cam.ac.uk

Abstract. We present a probabilistic version of the while loop, in the context of our mechanised framework for verifying probabilistic programs. The while loop preserves useful program properties of measurability and independence, provided a certain condition is met. This condition is naturally interpreted as *"from every starting state, the while loop will terminate with probability 1"*, and we compare it to other probabilistic termination conditions in the literature. For illustration, we verify in HOL two example probabilistic algorithms that necessarily rely on probabilistic termination: an algorithm to sample the Bernoulli(p) distribution using coin-flips; and the symmetric simple random walk.

1 Introduction

Probabilistic algorithms are used in many areas, from discrete mathematics to physics. There are many examples of simple probabilistic algorithms that cannot be matched in performance (or sometimes even complexity) by deterministic alternatives [12]. It is our goal to specify and verify probabilistic algorithms in a theorem-prover. Formal verification is particularly attractive for probabilistic algorithms, because black-box testing is limited to statistical error reports of the form: *"With confidence 90%, the algorithm is broken."* Additionally, even small probabilistic algorithms can be difficult to implement correctly. A whole new class of errors becomes possible and one has to be mathematically sophisticated to avoid them.

In Section 2 we show how probabilistic algorithms can be specified and verified in the HOL[1] theorem prover[2] [2], by thinking of them as deterministic functions having access to an infinite sequence of coin-flips. This approach is general enough to verify many probabilistic algorithms in HOL (including the Miller-Rabin primality test [5]), but (in its raw form) it is limited to algorithms that are guaranteed to terminate.

Example 1. A sampling algorithm simulates a probability distribution ρ by generating on demand a value x with probability $\rho(x)$. In this spirit, an algorithm

* Supported by EPSRC project GR/R27105/01.

[1] As will be seen, higher-order logic is essential for our approach, since many of our results rely on quantification over predicates and functions.

[2] hol98 is available from http://www.cl.cam.ac.uk/Research/HVG/FTP/.

V.A. Carreño, C. Muñoz, S. Tahar (Eds.): TPHOLs 2002, LNCS 2410, pp. 230–245, 2002.

to sample the Geometric($\frac{1}{2}$) distribution will return the natural number n with probability $(\frac{1}{2})^{n+1}$. A simple way to implement this is to return the index of the first coin-flip in the sequence that is 'heads'. However, this is not guaranteed to terminate on every possible input sequences of coin-flips, the counter-example being the 'all-tails' sequence.[3] However, the algorithm does satisfies probabilistic termination, meaning that the probability that it terminates is 1 (*"in all practical situations it must terminate"*).

In fact, there is a large class of probabilistic algorithms that cannot be defined without using probabilistic termination. In Section 3 we present our approach to overcoming this limitation: a probabilistic version of the 'while' loop that slots into our HOL framework and supports probabilistic termination. If a certain probabilistic termination condition is satisfied, then an algorithm defined in terms of a probabilistic while loop automatically satisfies useful properties of measurability and independence. In Section 4 we examine the relationship between our probabilistic termination condition and others in the literature.

In Sections 5 and 6 we use probabilistic termination to define two algorithms in HOL. The first uses coin-flips to sample from the Bernoulli(p) distribution, where p can be any real number between 0 and 1. The second is the symmetric simple random walk, a classic example from probability theory that requires a subtle termination argument to even define.

The contributions of this paper are as follows:

- an overview of our formal framework for verifying probabilistic programs in the HOL theorem prover;
- the formal definition of a probabilistic while loop, preserving compositional properties of measurability and independence;
- a comparison of our naturally occurring probabilistic termination condition with others in the literature;
- the verification in HOL of two probabilistic algorithms requiring probabilistic termination: an algorithm to sample the Bernoulli(p) distribution using coin-flips; and the symmetric simple random walk.

2 Verifying Probabilistic Algorithms in HOL

In this section we provide an overview of our framework for verifying probabilistic algorithms in HOL. Although novel this is not the main focus of this paper, and so the section is necessarily brief. For the complete explanation please refer to my Ph.D. thesis [6].

2.1 Modelling Probabilistic Programs in HOL

Probabilistic algorithms can be modelled in HOL by thinking of them as deterministic algorithms with access to an infinite sequence of coin-flips. The infinite

[3] For a literary example of a coin that always lands on the same side, see the beginning of the Tom Stoppard play: *Rosencrantz & Guildenstern Are Dead.*

sequence of coin flips is modelled by an element of the type $\mathbb{B}^\infty$ of infinite boolean sequences, and serves as an 'oracle'. The oracle provides an inexhaustible source of values 'heads' and 'tails', encoded by $\top$ and $\bot$. Every time a coin is flipped, the random result of the coin flip is popped and consumed from the oracle. A probabilistic algorithm takes, besides the usual parameters, another oracle parameter from which it may pop the random values it needs. In addition to its result, it returns the rest of the oracle for consumption by someone else.

A simple example of a probabilistic algorithm is a random variable. Consider the random variable V ranging over values of type α, which we represent in HOL by the function

$$v : \mathbb{B}^\infty \to \alpha \times \mathbb{B}^\infty$$

Since random variables do not take any parameters, the only parameter of v is the oracle: an element of the type $\mathbb{B}^\infty$ of infinite boolean sequences. It returns the value of the random variable (an element of type α) and the rest of the oracle (another element of $\mathbb{B}^\infty$).

Example 2. If shd and stl are the sequence equivalents of the list operations 'head' and 'tail', then the function

$$\vdash \mathsf{bit} = \lambda s.\ (\mathsf{if\ shd}\ s\ \mathsf{then}\ 1\ \mathsf{else}\ 0,\ \mathsf{stl}\ s) \tag{1}$$

models a Bernoulli($\frac{1}{2}$) random variable that returns 1 with probability $\frac{1}{2}$, and 0 with probability $\frac{1}{2}$. For example,

$$\mathsf{bit}\ (\top, \bot, \top, \bot, \ldots) = (1, (\bot, \top, \bot, \ldots))$$

shows the result of applying bit to one particular infinite boolean sequence.

It is possible to combine random variables by 'passing around' the sequence of coin-flips.

Example 3. We define a $\mathsf{bin}\ n$ function that combines several applications of bit to calculate the number of 'heads' in the first n flips.

$$\begin{aligned} &\vdash \mathsf{bin}\ 0\ s = (0, s)\ \wedge \\ &\quad \forall n.\ \mathsf{bin}\ (\mathsf{suc}\ n)\ s = \mathsf{let}\ (x, s') \leftarrow \mathsf{bin}\ n\ s\ \mathsf{in}\ \left(\mathsf{let}\ (y, s'') \leftarrow \mathsf{bit}\ s'\ \mathsf{in}\ (x + y, s'')\right) \end{aligned} \tag{2}$$

The HOL function $\mathsf{bin}\ n$ models a Binomial($n, \frac{1}{2}$) random variable.

Concentrating on an infinite sequence of coin-flips as the only source of randomness for our programs is a boon to formalisation in HOL, since only one probability space needs to be formalised in the logic. It also has a practical significance, since we can extract our HOL implementations of probabilistic programs to ML, and execute them on a sequence of high quality random bits from the operating system. These random bits are derived from system noise, and are so designed that a sequence of them should have the same probability distribution as a sequence of coin-flips. An example of this extraction process is given in Section 6 for the random walk, and a more detailed examination of the issues can be found in a previous case study of the Miller-Rabin primality test [5].

2.2 Monadic Operator Notation

The above representation is also used in Haskell[4] and other pure functional languages to write probabilistic programs [13,9]. In fact, these programs live in the more general state-transforming monad: in this case the state that is transformed is the sequence of coin-flips. The following monadic operators can be used to reduce notational clutter when combining state-transforming programs.

Definition 1. *The state-transformer monadic operators* unit *and* bind.

$$\vdash \forall a, s.\ \mathsf{unit}\ a\ s = (a, s)$$
$$\vdash \forall f, g, s.\ \mathsf{bind}\ f\ g\ s = \mathsf{let}\ (x, s') \leftarrow f(s)\ \mathsf{in}\ g\ x\ s'$$

The unit *operator is used to lift values to the monad, and* bind *is the monadic analogue of function application.*

Example 4. Our bin n function can now be defined more concisely:

$$\begin{aligned} &\vdash \mathsf{bin}\ 0 = \mathsf{unit}\ 0\ \wedge \\ &\quad \forall n.\ \mathsf{bin}\ (\mathsf{suc}\ n) = \mathsf{bind}\ (\mathsf{bin}\ n)\ (\lambda x.\ \mathsf{bind}\ \mathsf{bit}\ (\lambda y.\ \mathsf{unit}\ (x + y))) \end{aligned} \tag{3}$$

Observe that the sequence of coin-flips is never referred to directly, instead the unit and bind operators pass it around behind the scenes.

2.3 Formalised Probability Theory

By formalising some mathematical measure theory in HOL, it is possible to define a probability function

$$\mathbb{P} : \mathcal{P}(\mathbb{B}^\infty) \to \mathbb{R}$$

from sets of sequences to real numbers between 0 and 1.

Since the Banach-Tarski paradox prevents us from assigning a well-defined probability to *every* set of sequences, it is helpful to think of $\mathbb{P}$ as a partial function. The domain of $\mathbb{P}$ is the set

$$\mathcal{E} : \mathcal{P}(\mathcal{P}(\mathbb{B}^\infty))$$

of events of the probability.[5] Our current version of formalised measure theory is powerful enough that any practically occurring set of sequences is an event. Specifically, we define $\mathbb{P}$ and $\mathcal{E}$ using Carathéodory's Extension Theorem, which ensures that $\mathcal{E}$ is a σ-algebra: closed under complements and countable unions.

Once we have formally defined $\mathbb{P}$ and $\mathcal{E}$ in HOL, we can derive the usual laws of probability from their definitions. One such law is the following, which says that the probability of two disjoint events is the sum of their probabilities:

$$\vdash \forall A, B.\ A \in \mathcal{E} \wedge B \in \mathcal{E} \wedge A \cap B = \emptyset\ \Rightarrow\ \mathbb{P}(A \cup B) = \mathbb{P}(A) + \mathbb{P}(B)$$

---

[4] `http://www.haskell.org`.

[5] Of course, $\mathbb{P}$ must be a total function in HOL, but the values of $\mathbb{P}$ outside $\mathcal{E}$ are never logically significant.

Example 5. Our formalised probability theory allows us to prove results such as

$$\vdash \forall n, r.\ \mathbb{P}\{s \mid \mathsf{fst}\ (\mathsf{bin}\ n\ s) = r\} = \binom{n}{r} \left(\tfrac{1}{2}\right)^n \tag{4}$$

making explicit the Binomial($n, \frac{1}{2}$) probability distribution of the bin n function. The fst function selects the first component of a pair, in this case the $\mathbb{N}$ from $\mathbb{N} \times \mathbb{B}^\infty$. The proof proceeds by induction on n, followed by a case split on the first coin-flip in the sequence (a probability weight of $\frac{1}{2}$ is assigned to each case). At this point the goal may be massaged (using real analysis and the laws of probability) to match the inductive hypothesis.

2.4 Probabilistic Quantifiers

In probability textbooks, it is common to find many theorems with the qualifier 'almost surely', 'with probability 1' or just 'w.p. 1'. Intuitively, this means that the set of points for which the theorem is true has probability 1 (which probability space is usually clear from context). We can define probabilistic versions of the $\forall$ and $\exists$ quantifiers that make this notation precise.[6]

Definition 2. *Probabilistic Quantifiers*

$$\vdash \forall \phi.\ (\forall^* s.\ \phi(s)) = \{s \mid \phi(s)\} \in \mathcal{E}\ \wedge\ \mathbb{P}\{s \mid \phi(s)\} = 1$$
$$\vdash \forall \phi.\ (\exists^* s.\ \phi(s)) = \{s \mid \phi(s)\} \in \mathcal{E}\ \wedge\ \mathbb{P}\{s \mid \phi(s)\} \neq 0$$

Observe that these quantifiers come specialised to the probability space $(\mathcal{E}, \mathbb{P})$ of infinite sequences of coin-flips: this cuts down on notational clutter.

2.5 Measurability and Independence

Recall that we model probabilistic programs with HOL functions of type

$$\mathbb{B}^\infty \to \alpha \times \mathbb{B}^\infty$$

However, not all functions f of this HOL type correspond to reasonable probabilistic programs. Some are not measurable, and hence a set of sequences

$$S = \{s \mid P(\mathsf{fst}\ (f(s)))\}$$

that satisfy some property P of the result is not an event of the probability (i.e., $S \notin \mathcal{E}$). Alternatively, f might not be independent, and hence it may use some coin-flips to compute a result and also return those 'used' coin-flips, like this:

$$\mathsf{broken\_bit} = \lambda s.\ (\mathsf{fst}\ (\mathsf{bit}\ s),\ s)$$

We therefore introduce a property indep called strong function independence. If $f \in$ indep, then f will be both measurable and independent. All reasonable probabilistic programs satisfy strong function independence, and the extra properties are a great aid to verification.

[6] We pronounce $\forall^*$ as "probably" and $\exists^*$ as "possibly".

Definition 3. *Strong Function Independence*

$$\begin{array}{l}\vdash \mathsf{indep} = \\ \quad \{f \mid \\ \quad\quad (\mathsf{fst} \circ f) \in \mathsf{measurable}\ \mathcal{E}\ \mathcal{U}\ \wedge\ (\mathsf{snd} \circ f) \in \mathsf{measurable}\ \mathcal{E}\ \mathcal{E}\ \wedge \\ \quad\quad \mathsf{countable}\ (\mathsf{range}\ (\mathsf{fst} \circ f))\ \wedge\ \exists C.\ \mathsf{is\_prefix\_cover}\ f\ C\}\end{array}$$

In Definition 3 we give the HOL definition of indep. Strongly independent functions must be measurable, and satisfy a compositional form of independence that is enforced by their range being countable and having a 'prefix cover' of probability 1.

Strong function independence fits in neatly with the monadic operator notation we introduced earlier, as the following theorem shows.

Theorem 1. *Strong Function Independence is Compositional*

$$\vdash \forall a.\ \mathsf{unit}\ a \in \mathsf{indep}$$
$$\vdash \forall f, g.\ f \in \mathsf{indep} \wedge (\forall a.\ g(a) \in \mathsf{indep}) \Rightarrow \mathsf{bind}\ f\ g \in \mathsf{indep}$$
$$\vdash \forall f, g.\ f \in \mathsf{indep} \wedge g \in \mathsf{indep} \Rightarrow \mathsf{coin\_flip}\ f\ g \in \mathsf{indep}$$

Proof (sketch). The proof of each statement begins by expanding the definition of indep. *The measurability conditions are proved by lifting results from the underlying algebra to $\mathcal{E}$, and the countability condition is easily established. Finally, in each case the required prefix cover is explicitly constructed.*

The coin_flip operator flips a coin to decide whether to execute f or g, and is defined as

$$\vdash \mathsf{coin\_flip}\ f\ g = \lambda s.\ (\mathsf{if}\ \mathsf{shd}\ s\ \mathsf{then}\ f\ \mathsf{else}\ g)\ (\mathsf{stl}\ s) \qquad (5)$$

The compositional nature of strong function independence means that it will be satisfied by any probabilistic program that accesses the underlying sequence of coin-flips using only the operators {unit, bind, coin_flip}.

3 Probabilistic While Loop

In the previous section we laid out our verification framework for probabilistic programs, emphasising the monadic operator style which ensures that strong function independence holds. The programs have access to a source of randomness in the form of an infinite sequence of coin-flips, and this allows us to easily extract programs and execute them. As we saw with the example program bin n that sampled from the Binomial(n, $\frac{1}{2}$) distribution, it is no problem to define probabilistic programs using well-founded recursion.

However, well-founded recursion is limited to probabilistic programs that compute a finite number of values, each having a probability of the form $m/2^n$.[7]

[7] This follows from the fact that any well-founded function of type $\mathbb{B}^\infty \to \alpha \times \mathbb{B}^\infty$ can only read a finite number of booleans from the input sequence.

Example 6. The limitations of well-founded recursion prevent the definition of the following probabilistic programs:

- an algorithm to sample the Uniform(3) distribution (the probability of each result is 1/3, which cannot be expressed in the form $m/2^n$);
- and an algorithm to sample the Geometric($\frac{1}{2}$) distribution (there are an infinite number of possible results).

In this section we go further, and show how to define probabilistic programs that are not strictly well-founded, but terminate with probability 1. Every probabilistic program in the literature falls into this enlarged definition, and so (in principle, at least) can be modelled in HOL.

3.1 Definition of the Probabilistic While Loop

We aim to define a probabilistic version of the while loop, where the body

$$b : \alpha \to \mathbb{B}^\infty \to \alpha \times \mathbb{B}^\infty$$

of the while loop probabilistically advances a state of type α, and the condition

$$c : \alpha \to \mathbb{B}$$

is a deterministic state predicate.

We first define a bounded version of the probabilistic while loop with a cut-off parameter n: if the condition is still true after n iterations, the loop terminates anyway.

Definition 4. *Bounded Probabilistic While Loop*

$$\begin{aligned}
&\vdash \forall c, b, n, a. \\
&\quad \mathsf{while_cut}\ c\ b\ 0\ a = \mathsf{unit}\ a\ \wedge \\
&\quad \mathsf{while_cut}\ c\ b\ (\mathsf{suc}\ n)\ a = \mathsf{if}\ c(a)\ \mathsf{then}\ \mathsf{bind}\ (b(a))\ (\mathsf{while_cut}\ c\ b\ n)\ \mathsf{else}\ \mathsf{unit}\ a
\end{aligned}$$

The bounded version of probabilistic while does not employ probabilistic recursion. Rather it uses standard recursion on the cut-off parameter n, and consequently many useful properties follow by induction on n.

We now use while_cut to make a 'raw definition' of an unbounded probabilistic while loop.

Definition 5. *Probabilistic While Loop*

$$\begin{aligned}
&\vdash \forall c, b, a, s. \\
&\quad \mathsf{while}\ c\ b\ a\ s = \\
&\quad \mathsf{if}\ \exists n.\ \neg c(\mathsf{fst}\ (\mathsf{while_cut}\ c\ b\ n\ a\ s))\ \mathsf{then} \\
&\quad\quad \mathsf{while_cut}\ c\ b\ (\mathsf{minimal}\ (\lambda n.\ \neg c(\mathsf{fst}\ (\mathsf{while_cut}\ c\ b\ n\ a\ s))))\ a\ s \\
&\quad \mathsf{else\ arb}
\end{aligned}$$

where arb *is an arbitrary fixed value, and* minimal ϕ *is specified to be the smallest natural number* n *satisfying* $\phi(n)$.

3.2 Characterisation of the Probabilistic While Loop

There are two characterising theorems that we would like to prove about probabilistic while loops. The first demonstrates that it executes as we might expect: check the condition, if true then perform an iteration and repeat, if false then halt and return the current state. Note the bind and unit in the theorem: this is a probabilistic while loop, not a standard one!

Theorem 2. *Iterating the Probabilistic While Loop*

$$\vdash \forall c, b, a.\ \mathsf{while}\ c\ b\ a = \mathsf{if}\ c(a)\ \mathsf{then}\ \mathsf{bind}\ (b(a))\ (\mathsf{while}\ c\ b)\ \mathsf{else}\ \mathsf{unit}\ a$$

Proof. For a given c, b, a, s, if there is some number of iterations of b (starting in state a with sequence s) that would lead to the condition c becoming false, then while *performs the minimum number of iterations that are necessary for this to occur, otherwise it returns* arb. *The proof now splits into the following cases:*

- *The condition eventually becomes false:*
  - *The condition is false to start with: in this case the minimum number of iterations for the condition to become false will be zero.*
  - *The condition is not false to start with: in this case the minimum number of iterations for the condition to become false will be greater than zero, and so we can safely perform an iteration and then ask the question again.*
- *The condition will always be true: therefore, after performing one iteration the condition will still always be true. So both LHS and RHS are equal to* arb.

Note that up to this point, the definitions and theorems have not mentioned the underlying state, and so generalise to any state-transforming while loop. The second theorem that we would like to prove is specific to probabilistic while loops, and states that while preserves strong function independence. This allows us to add while to our set of monadic operators for safely constructing probabilistic programs. However, for while c b to satisfy strong function independence, the following 'termination' condition is placed on b and c.[8]

Definition 6. *Probabilistic Termination Condition*

$$\vdash \forall c, b.\ \mathsf{while\_terminates}\ c\ b = \forall a.\ \forall^* s.\ \exists n.\ \neg c(\mathsf{fst}\ (\mathsf{while\_cut}\ c\ b\ n\ a\ s))$$

This extra condition says that for every state a, there is an event of probability 1 that leads to the termination of the probabilistic while loop. This additional condition ensures that probabilistic while loops preserve strong function independence.

Theorem 3. *Probabilistic While Loops Preserve Strong Function Independence*

$$\vdash \forall c, b.\ (\forall a.\ b(a) \in \mathsf{indep}) \wedge \mathsf{while\_terminates}\ c\ b \Rightarrow \forall a.\ \mathsf{while}\ c\ b\ a \in \mathsf{indep}$$

[8] The $\forall^*$ in the definition is a probabilistic universal quantifier (see Section 2.4).

Proof (sketch). Countability of range and measurability follow easily from the strong function independence of $b(a)$, for every a. The prefix cover is again explicitly constructed, by stitching together the prefix covers of $b(a)$ for all reachable states a that terminate the while loop (i.e., that satisfy $\neg c(a)$).

At this point the definition of a probabilistic while loop is finished. We export Theorems 2 and 3 and Definition 6, and these totally characterise the while operator: users never need to work with (or even see) the raw definition.

Finally, no formal definition of a new while loop would be complete without a Hoare-style while rule, and the following can be proved from the characterising theorems.

Theorem 4. *Probabilistic While Rule*

$$\begin{aligned}&\vdash \forall \phi, c, b, a.\\ &\qquad (\forall a.\ b(a) \in \mathsf{indep}) \ \wedge\ \mathsf{while\_terminates}\ c\ b\ \wedge\\ &\qquad \phi(a)\ \wedge\ (\forall a.\ \forall^* s.\ \phi(a) \wedge c(a) \Rightarrow \phi(\mathsf{fst}\ (b\ a\ s)))\ \Rightarrow\\ &\qquad \forall^* s.\ \phi(\mathsf{fst}\ (\mathsf{while}\ c\ b\ a\ s))\end{aligned}$$

"For a well-behaved probabilistic while loop, if a property is true of the initial state and with probability 1 is preserved by each iteration, then with probability 1 the property will be true of the final state."

4 Probabilistic Termination Conditions

In the previous section we saw that a probabilistic termination condition was needed to prove that a probabilistic while loop satisfied strong function independence. In this section we take a closer look at this condition, in the context of related work on termination.

Let us begin by observing that our termination condition while_terminates $c\ b$ is both necessary and sufficient for each while $c\ b\ a$ to terminate on a set of probability 1. Therefore, the other termination conditions we survey are either equivalent to ours or logically imply it.

In the context of probabilistic concurrent systems, the following 0-1 law was proved by Hart, Sharir, and Pnueli [3]:[9]

> Let process P be defined over a state space S, and suppose that from every state in some subset S' of S the probability of P's eventual escape from S' is at least p, for some fixed $0 < p$.
>
> Then P's escape from S' is certain, occurring with probability 1.

Identifying P with while $c\ b$ and S' with the set of states a for which $c(a)$ holds, we can formulate the 0-1 law as an equivalent condition for probabilistic termination:

[9] This paraphrasing comes from Morgan [10].

Theorem 5. *The 0-1 Law of Probabilistic Termination*

$$\vdash \forall c, b.\ (\forall a.\ b(a) \in \mathsf{indep}) \Rightarrow (\mathsf{while\_terminates}\ c\ b \iff \exists p.\ 0 < p \land \forall a.\ p \le \mathbb{P}\{s \mid \exists n.\ \neg c(\mathsf{fst}\ (\mathsf{while\_cut}\ c\ b\ n\ a\ s))\})$$

This interesting result implies that over the whole state space, the infimum of all the termination probabilities is either 0 or 1, it cannot lie properly in between. An example of its use for proving probabilistic termination will be seen in our verification of a sampling algorithm for the Bernoulli(p) distribution.

Hart, Sharir, and Pnueli [3] also established a sufficient condition for probabilistic termination called the probabilistic variant rule. We can formalise this as a sufficient condition for termination of our probabilistic while loops; the proof is relatively easy from the 0-1 law.

Theorem 6. *The Probabilistic Variant Condition*

$$\vdash \forall c, b.\ (\forall a.\ b(a) \in \mathsf{indep}) \land (\exists f, N, p.\ 0 < p \land \forall a.\ c(a) \Rightarrow f(a) < N \land p \le \mathbb{P}\{s \mid f(\mathsf{fst}\ (b\ a\ s)) < f(a)\}) \Rightarrow \mathsf{while\_terminates}\ c\ b$$

As its name suggests, the probabilistic variant condition is a probabilistic analogue of the variant method used to prove termination of deterministic while loops. If we can assign to each state a a natural number measure from a finite set, and if each iteration of the loop has probability at least p of decreasing the measure, then probabilistic termination is assured. In addition, when $\{a \mid c(a)\}$ is finite, the probabilistic variant condition has been shown to be necessary as well as sufficient.

5 Example: Sampling the Bernoulli(p) Distribution

The Bernoulli(p) distribution is over the boolean values $\{\top, \bot\}$, and models a test where $\top$ is picked with probability p and $\bot$ with probability $1 - p$. Our sequence of coin-flips can be considered as sampling a Bernoulli($\frac{1}{2}$) distribution, and the present goal is to use these to produce samples from a Bernoulli(p) distribution, where p is any real number between 0 and 1.

The sampling algorithm we use is based on the following simple idea. Suppose the binary expansion of p is $0.p_0p_1p_2\cdots$; consider the coin-flips of the sequence s as forming a binary expansion $0.s_0s_1s_2\cdots$.[10] In this way s can also be regarded

[10] We can conveniently ignore the fact that some numbers have two binary expansions (e.g., $\frac{1}{2} = 0.1000\cdots = 0.0111\cdots$), since the set of these 'dyadic rationals' is countable and therefore has probability 0.

as a real number between 0 and 1. Since the 'number' s is uniformly distributed between 0 and 1, we (informally) have

$$\text{Probability}(s < p = \begin{Bmatrix} \top \\ \bot \end{Bmatrix}) = \begin{Bmatrix} p \\ 1-p \end{Bmatrix}$$

Therefore, an algorithm that evaluates the comparison $s < p$ will be sampling from the Bernoulli(p) distribution, and this comparison can easily be decided by looking at the binary expansions. The matter is further simplified since we can ignore awkward cases (such as $s = p$) that occur with probability 0.

Definition 7. *A Sampling Algorithm for the* Bernoulli(p) *Distribution*

$\vdash \forall p.$
bern_iter $p =$
if $p < \frac{1}{2}$ then coin_flip (unit (inr $\bot$)) (unit (inl ($2p$)))
else coin_flip (unit (inl ($2p - 1$))) (unit (inr $\top$))

$\vdash \forall p.$ bernoulli $p =$ bind (while is_inl (bern_iter $\circ$ outl) (inl p)) (unit $\circ$ outr)

To make the sampling algorithm fit into a probabilistic while loop, the definition makes heavy use of the HOL sum type $\alpha + \beta$, which has constructors inl, inr, destructors outl, outr and predicates is_inl, is_inr. However, the intent of the probabilistic while loop is simply to evaluate $s < p$ by iteration on the bits of s:

- if shd $s = \bot$ and $\frac{1}{2} \leq p$, then return $\top$;
- if shd $s = \top$ and $p < \frac{1}{2}$, then return $\bot$;
- if shd $s = \bot$ and $p < \frac{1}{2}$, then repeat with $s :=$ stl s and $p := 2p$;
- if shd $s = \top$ and $\frac{1}{2} \leq p$, then repeat with $s :=$ stl s and $p := 2p - 1$.

This method of evaluation has two important properties: firstly, it is obviously correct since the scaling operations on p just have the effect of removing its leading bit; secondly, probabilistic termination holds, since every iteration has a probability $\frac{1}{2}$ of terminating the loop. Indeed, Hart's 0-1 law of termination (Theorem 5) provides a convenient method of showing probabilistic termination:

$$\vdash \text{while\_terminates is\_inl (bern\_iter} \circ \text{outl)} \tag{6}$$

From this follows strong function independence

$$\vdash \forall p.\ \text{bernoulli}\ p \in \text{indep} \tag{7}$$

and we can then prove that bernoulli satisfies an alternative definition:

$\vdash \forall p.$ (8)
bernoulli $p =$
if $p < \frac{1}{2}$ then coin_flip (unit $\bot$) (bernoulli ($2p$))
else coin_flip (bernoulli ($2p - 1$)) (unit $\top$)

This definition of bernoulli is more readable, closer to the intuitive version, and easier to use in proofs. We use this to prove the correctness theorem:

$$\vdash \forall p.\ 0 \leq p \wedge p \leq 1 \Rightarrow \mathbb{P}\{s \mid \mathsf{bernoulli}\ p\ s\} = p \tag{9}$$

The proof of this is quite simple, once the right idea is found. The idea is to show that the probability gets within $(\frac{1}{2})^n$ of p, for an arbitrary natural number n. As can be shown by induction, this will occur after n iterations.

It is perhaps surprising that the uncountable set $\{\mathsf{bernoulli}\ p \mid 0 \leq p \leq 1\}$ of programs are all distinct, even though each one examines only a finite number of bits (with probability 1).

6 Example: The Symmetric Simple Random Walk

The (1-dimensional) symmetric simple random walk is a probabilistic process with a compelling intuitive interpretation. A drunk starts at point n (the pub) and is trying to get to point 0 (home). Unfortunately, every step he makes from point i is equally likely to take him to point $i+1$ as it is to take him to point $i-1$. The following program simulates the drunk's passage home, and upon arrival returns the total number of steps taken.

Definition 8. *A Simulation of the Symmetric Simple Random Walk*

$$\vdash \forall n.\ \mathsf{lurch}\ n = \mathsf{coin\_flip}\ (\mathsf{unit}\ (n+1))\ (\mathsf{unit}\ (n-1))$$

$$\vdash \forall f, b, a, k.\ \mathsf{cost}\ f\ b\ (a, k) = \mathsf{bind}\ (b(a))\ (\lambda\, a'.\ \mathsf{unit}\ (a', f(k)))$$

$$\vdash \forall n, k.$$

$$\mathsf{walk}\ n\ k =$$

$$\mathsf{bind}\ (\mathsf{while}\ (\lambda\,(n, \_).\ 0 < n)\ (\mathsf{cost}\ \mathsf{suc}\ \mathsf{lurch})\ (n, k))\ (\lambda\,(\_, k).\ \mathsf{unit}\ k)$$

Theorem 7. *The Random Walk Terminates with Probability 1*

$$\vdash \forall n, k.\ \mathsf{while\_terminates}\ (\lambda\,(n, \_).\ 0 < n)\ (\mathsf{cost}\ \mathsf{suc}\ \mathsf{lurch})$$

Proof. *Let $\pi_{i,j}$ be the probability that starting at point i, the drunk will eventually reach point j.*

We first formalise the two lemmas $\pi_{p+i,p} = \pi_{i,0}$ and $\pi_{i,0} = \pi_{1,0}^i$. Therefore, if with probability 1 the drunk eventually gets home from a pub at point 1, with probability 1 he will eventually get home from a pub at any point.

By examining a single iteration of the random walk we have

$$\pi_{1,0} = \tfrac{1}{2}\pi_{2,0} + \tfrac{1}{2} = \tfrac{1}{2}\pi_{1,0}^2 + \tfrac{1}{2}$$

which rewrites to

$$(\pi_{1,0} - 1)^2 = 0$$

and therefore

$$\pi_{1,0} = 1$$

Once probabilistic termination is established, strong independence easily follows:

$$\vdash \forall n, k.\ \mathsf{walk}\ n\ k \in \mathsf{indep} \tag{10}$$

At this point, we may formulate the definition of walk in a more natural way:

$$\begin{aligned} \vdash \forall n, k.& \\ & \mathsf{walk}\ n\ k = \\ & \mathsf{if}\ n = 0\ \mathsf{then}\ \mathsf{unit}\ k\ \mathsf{else} \\ & \quad \mathsf{coin\_flip}\ (\mathsf{walk}\ (n{+}1)\ (k{+}1))\ (\mathsf{walk}\ (n{-}1)\ (k{+}1)) \end{aligned} \tag{11}$$

We have now finished the hard work of defining the random walk as a probabilistically terminating program. To demonstrate that once defined it is just as easy to reason about as any of our probabilistic programs, we prove the following basic property of the random walk:[11]

$$\vdash \forall n, k.\ \forall^* s.\ \mathsf{even}\ (\mathsf{fst}\ (\mathsf{walk}\ n\ k\ s)) = \mathsf{even}\ (n + k) \tag{12}$$

For a pub at point 1001, the drunk must get home eventually, but he will take an odd number of steps to do so!

It is possible to extract this probabilistic program to ML, and repeatedly simulate it using high-quality random bits from the operating system. Here is a typical sequence of results from random walks starting at level 1:

$$57, 1, 7, 173, 5, 49, 1, 3, 1, 11, 9, 9, 1, 1, 1547, 27, 3, 1, 1, 1, \ldots$$

As can be seen, the number of steps that are required for the random walk to hit zero is usually less than 100. But sometimes, the number can be much larger. Continuing the above sequence of simulations, the 34th simulation sets a new record of 2645 steps, and the next record-breakers are the 135th simulation with 603787 steps and the 664th simulation with 1605511 steps. Such large records early on are understandable, since the theoretical expected number of steps for the random walk is actually infinite!

In case it is difficult to see how an algorithm could have infinite expected running time but terminate with probability 1, consider an algorithm where the probability of termination after n steps is $\frac{6}{\pi^2 n^2}$. The probability of termination is then

$$\sum_n \frac{6}{\pi^2 n^2} = \frac{6}{\pi^2} \sum_n \frac{1}{n^2} = \frac{6}{\pi^2} \cdot \frac{\pi^2}{6} = 1$$

and the expected running time is

$$\sum_n n \frac{6}{\pi^2 n^2} = \frac{6}{\pi^2} \sum_n \frac{1}{n} = \infty$$

[11] Note the use of the probabilistic universal quantifier $\forall^* s$. This allows us to ignore the set of sequences that cause the drunk to walk forever, since it has probability 0.

7 Conclusions

In this paper we have described how probabilistic programs can be verified in the HOL theorem prover, and then shown how programs that terminate with probability 1 can be defined in the model. Finally, we applied the technology to verify two example programs that necessarily rely on probabilistic termination. In principle, our HOL framework is powerful enough to verify any probabilistic program that terminates with probability 1. However, the labour-intensive nature of theorem proving means that it is only practical to verify particularly important probabilistic algorithms.

Fixing a sequence of coin-flips as the primitive source of randomness creates a distinction between probabilistic programs that are guaranteed to terminate on every possible sequence of coin-flips, and programs that terminate on a set of sequences having probability 1. Probabilistic programs that are guaranteed to terminate can place an upper bound on the number of random bits they will require for a computation, but programs defined using probabilistic termination may consume an unbounded number of bits. In application areas where random bits are expensive to generate, or where tight bounds are required on execution time, probabilistic termination must be viewed with a certain amount of suspicion.

There is also a logical distinction between guaranteed termination and probabilistic termination. Typically, a program p defined using probabilistic termination generally has properties that are quantified by $\forall^*$ instead of the stronger $\forall$. This is because 'all bets are off' on the set of sequences where p doesn't terminate, and so universal quantification over all sequences usually results in an unprovable property. In our verification of the Miller-Rabin primality test [5], we deliberately avoided using probabilistic termination to get stronger theorems, and the added power meant that we were able to implement a 'composite prover' for natural numbers.

In our random walk example, the proof of probabilistic termination is quite subtle. The random walk is therefore not likely to fit into a standard scheme of programs satisfying probabilistic termination. For this reason it is important that the definition of probabilistic programs in our formal framework is not tied to any particular program scheme. Instead, we can define an arbitrary probabilistic program, then prove it satisfies probabilistic termination, and finally go on to verify it. In the future, it may be useful to define program schemes that automatically satisfy probabilistic termination: these can be implemented by reduction to our current method followed by an automatic termination proof. However, it is important to retain the general method, or unusual programs such as the random walk could not be modelled.

Finally, in both the Bernoulli(p) and random walk examples, we defined a function in terms of probabilistic while loops and then went to great pains to prove that it was equivalent to a simpler version using straightforward recursion. It might reasonably be asked why we don't directly support recursive definitions of probabilistic programs, and the answer is that it's harder to extract the probabilistic termination condition. One possible approach to this, building on the present work, would be to reduce the definition to a probabilistic while loop and then read off the termination condition from that.

8 Related Work

The semantics of probabilistic programs was first tackled by Kozen [8], and developed by Jones [7], He et al. [4] and Morgan et al. [11]. This line of research extends the predicate transformer idea of [1] in which programs are regarded as functions: they take a set of desired end results to the set of initial states from which the program is guaranteed to produce one of these final states. With the addition of probabilistic choice, the 'sets of states' must be generalised to functions from states to the real interval $[0, 1]$.

Jones defines a Hoare-style logic for total correctness, in which termination with probability 1 is covered by using upper continuous functions as pre- and post-conditions. In this model there is no distinction between guaranteed termination and probabilistic termination. The verification of a sampling algorithm for the Geometric($\frac{1}{2}$) distribution provides an instructive proof of probabilistic termination, but (from the perspective of mechanisation) the method appears to be more complicated than the approach presented in this paper. Also in the context of probabilistic predicate transformers, Morgan [10] explicitly looks at "proof rules for probabilistic loops", applying the probabilistic variant condition of Hart, Sharir, and Pnueli [3] to verify a probabilistic self-stabilisation algorithm.

Our semantics of probabilistic programs is very different from the predicate transformer framework. Being concerned with mechanisation, our aim was to minimise the amount of necessary formalisation. This led to a simple view of probabilistic programs in terms of a sequence of coin-flips, and this bears no obvious correspondence to the predicate transformer view. Proofs in the two settings plainly use the same high-level arguments, but soon diverge to match the low-level details of the semantics. However, it may be that our 'shallow embedding' of probabilistic programs as HOL functions is obscuring similarities. An interesting direction for future work would be to formalise the syntax of a simple while language including a probabilistic choice operator, and then derive the rules of the predicate transformer semantics in terms of our own.

Acknowledgements

The research in this paper is drawn from my Ph.D. thesis, which was supervised by Mike Gordon. The work has greatly benefitted from his guidance, as well as numerous discussions with Konrad Slind, Michael Norrish, Mark Staples, Judita Preiss, and numerous past and present members of the Cambridge Automated Reasoning Group. Finally, I am grateful to the TPHOLs referees for many suggestions on how to improve the presentation.

References

1. E.W. Dijkstra. *A Discipline of Programming.* Prentice-Hall, 1976.
2. M.J.C. Gordon and T.F. Melham. *Introduction to HOL (A theorem-proving environment for higher order logic).* Cambridge University Press, 1993.
3. Sergiu Hart, Micha Sharir, and Amir Pnueli. Termination of probabilistic concurrent programs. *ACM Transactions on Programming Languages and Systems (TOPLAS)*, 5(3):356–380, July 1983.
4. Jifeng He, K. Seidel, and A. McIver. Probabilistic models for the guarded command language. *Science of Computer Programming*, 28(2–3):171–192, April 1997.
5. Joe Hurd. Verification of the Miller-Rabin probabilistic primality test. In Richard J. Boulton and Paul B. Jackson, editors, *TPHOLs 2001: Supplemental Proceedings*, number EDI-INF-RR-0046 in University of Edinburgh Informatics Report Series, pages 223–238, September 2001.
6. Joe Hurd. *Formal Verification of Probabilistic Algorithms.* PhD thesis, University of Cambridge, 2002.
7. Claire Jones. *Probabilistic Non-Determinism.* PhD thesis, University of Edinburgh, 1990.
8. Dexter Kozen. Semantics of probabilistic programs. In *20th Annual Symposium on Foundations of Computer Science*, pages 101–114, Long Beach, Ca., USA, October 1979. IEEE Computer Society Press.
9. John Launchbury and Simon L. Peyton Jones. Lazy functional state threads. In *SIGPLAN Symposium on Programming Language Design and Implementation (PLDI'94), Orlando*, pages 24–35, June 1994.
10. Carroll Morgan. Proof rules for probabilistic loops. In *Proceedings of the BCS-FACS 7th Refinement Workshop*, 1996.
11. Carroll Morgan, Annabelle McIver, Karen Seidel, and J. W. Sanders. Probabilistic predicate transformers. Technical Report TR-4-95, Oxford University Computing Laboratory Programming Research Group, February 1995.
12. Rajeev Motwani and Prabhakar Raghavan. *Randomized Algorithms.* Cambridge University Press, Cambridge, England, June 1995.
13. Philip Wadler. The essence of functional programming. In *19th Symposium on Principles of Programming Languages.* ACM Press, January 1992.

Using Theorem Proving for Numerical Analysis

Correctness Proof of an Automatic Differentiation Algorithm

Micaela Mayero

Programming Logic Group
Chalmers University of Technology
Department of Computing Science, S-412 96 Göteborg, Sweden
mayero@cs.chalmers.se
http://www.cs.chalmers.se/~mayero

Abstract. In this paper, we present a formal proof, developed in the Coq system, of the correctness of an automatic differentiation algorithm. This is an example of interaction between formal methods and numerical analysis (involving, in particular, real numbers). We study the automatic differentiation tool, called O∂yssée, which deals with FORTRAN programs, and using Coq we formalize the correctness proof of the algorithm used by O∂yssée for a subset of programs. To do so, we briefly describe the library of real numbers in Coq including real analysis, which was originally developed for this purpose, and we formalize a semantics for a subset of FORTRAN programs. We also discuss the relevance of such a proof.

1 Introduction

In this work[1], we consider an example of applying formal methods to prove the correctness of numerical analysis programs and more generally programs dealing with real numbers. The correctness of such programs is often critical; for example the code on-board planes and trains, real time programs used in medicine, etc. Many of these numerical analysis programs use a notion of differentiation, for example a *gradient*. All these algorithms are well known, but they are also very tedious to implement because, in particular, mistakes may easily arise. Detection and identification of these mistakes is difficult, which means great additional work of testing is needed. Thus, some `automatic differentiation (AD)` tools (like[2] are used by numericians for the programming of a gradients or another similar differentiation. These tools offer certain advantages: speed of development, genericity of the code, and, in particular, the absence of bugs which might

[1] This work has been realised within the LogiCal project (INRIA-Rocquencourt, France. `http://logical.inria.fr/`)

[2] A list of automatic differentiation systems has been presented in ICIAM 1995 in Hamburg and is available at `http://www-unix.mcs.anl.gov/autodiff/AD_Tools/` `ADIFOR` [3], `PADRE2` [2], `ADIC` [1], O∂yssée [5].

V.A. Carreño, C. Muñoz, S. Tahar (Eds.): TPHOLs 2002, LNCS 2410, pp. 246–262, 2002.

be introduced by the programmer. Regarding this last point, this does not mean that the final program is totally safe because implementing the algorithms in those tools may potentially contain some bugs.

There are two main ways of finding bugs in programs: making tests (as usual in the numerical analysis field) or proving programs (not really used in the numerical domain yet). In our case, we prefer the second method which is exhaustive and thus completely secure. One possible solution, to formally prove the correctness of a program, consists in writing the program and proving some properties regarding this code. In the context of Coq [13], this could be done in a functional way using the specification language, but also in an imperative way by means of the tactic `Correctness` [6]. Here, we will use the specification language to implement a functional version of the algorithm, which deals with imperative programs, but does not need to be coded in an imperative way (moreover, this will be closer to the tool that we propose to study and which provides a functional implementation).

Subsequently, we focus on the particular case of O∂yssée, which presents the advantage of having, in addition to a usual differentiation algorithm (direct mode), an inverse mode, which is, in fact, a gradient (even if this work consists only here in studying the direct mode). With the goal of proving the correctness of the algorithm, we propose using the Coq proof assistance system in order to prove the commutation of the diagram in figure 1. This diagram means the derivative of the semantical interpretation of the input program (in FORTRAN) is equal to the semantical interpretation of the output program (in FORTRAN) given by O∂yssée.

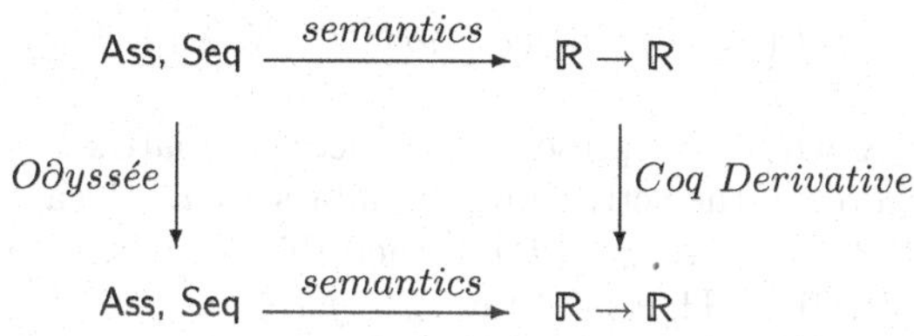

Fig. 1. Commutation diagram

First, we present O∂yssée and the development which has been carried out in Coq regarding real numbers and, more particularly, regarding the derivative. Then, we give a semantics for a subset of FORTRAN programs, a formalization of the differentiation algorithm used by O∂yssée and a correctness proof for this subset. We also discuss, in section 6, the possibility of dealing with a larger set of programs.

2 Presentation of O∂yssée

O∂yssée is a software package for automatic differentiation developed at INRIA Sophia Antipolis in the former `SAFIR` project, now the `TROPICS` project. It implements both derivations: direct mode and inverse mode. It takes as input a program written in Fortran-77, computing a differentiable piecewise function, and returns a new program, which computes a tangent or cotangent derivative. The tangent derivative, called direct mode, corresponds to a usual differentiation, and the cotangent derivative, called inverse mode, is a gradient. Here, we will mainly deal with the direct mode, because it is easier to start.

A preliminary step to both modes is that which marks the *active variables.* A variable v (or value) is called active:

- if v is an input value of REAL type and if the user specifies it as being an active one
  or
- if v is computed from, at least, one active variable.

In general, the generation of derived code follows, for both modes, the following rule: the structure of the program remains unchanged. In other words, the derived code of a subroutine will be a subroutine, that of an "if" will be an "if", and so on for the loops and other (imperative) structures which form the imperative programs.

The direct mode may be summarized as follows: let P be a program, and X and Y the vectors which respectively represent the input and output variables. Then we have:

$$P(X \rightarrow Y) \; \underrightarrow{\textit{direct mode}} \; \bar{P}((X,\bar{X}) \rightarrow (Y,\bar{Y}))$$

with $\bar{Y} = J \cdot \bar{X}$ where J represents the Jacobian matrix.

In other words, if the active input variables are v_i, the active output variables are v_o and if P computes $v_o = f(v_i)$, then this mode will compute v_o and $\bar{v}_o$ in v_i in the $\bar{v}_i$ direction. That is to say $\bar{v}_o = f'(v_i) \cdot \bar{v}_i$.

Let us consider the following small example.

```
subroutine cube (x,z)
real y
  y=x*x
  z=y*x
end
```

We will differentiate this routine with respect to `x`. Therefore, `y` and `z` also become active variables. The following command of O∂yssée differentiates `cube` with respect to `x` in the direct mode (`tl`):

```
diff -head cube -vars x -tl -o std
```

What was previously written $\bar{v}$ is now written VTTL by O∂yssée. Then we have:

```
COD Compilation unit : cubetl
COD Derivative of unit :  cube
COD Dummys:  x z
COD Active IN   dummys:  x
COD Active OUT  dummys:  z
COD Dependencies between IN and OUT:
COD z <--   x

      SUBROUTINE CUBETL (X, Z, XTTL, ZTTL)
      REAL Y
      REAL YTTL
      YTTL = XTTL*X+X*XTTL
      Y = X*X
      ZTTL = XTTL*Y+X*YTTL
      Z = Y*X
      END
```

For more details, we can refer to [5].

3 Presentation of Real Analysis in Coq

As the formalization of this kind of proof requires notions of real analysis, a first step was to develop a real number library (including basic analysis properties) in Coq. Seeing that our purpose is not a construction of real numbers[3], the feasibility to formalize and prove the standard analysis properties seems to us to be sufficient in this case. Thus, our entire formalization is based on 17 axioms, which can be proved if necessary by a similar construction to that made in HOL [10]. Our axioms express the fact that $\mathbb{R}$ is a commutative, ordered, Archimedian and complete field. From these axioms and some other basic properties, we defined the notions of limit and derivative. We present them briefly here, seeing that the rest of the proof depends on the choices taken during these formalizations. This development is available in the V6.3 and V7 versions of Coq.

First, the limit is defined in a metric space which is defined in the following way[4]:

```
Record Metric_Space:Type:= {
  Base:Type;
  dist:Base->Base->R;
  dist_pos:(x,y:Base)``(dist x y) >= 0``;
```

[3] There are some other formalisations of real numbers in Coq based on constructions [4] or intuitionnistic logic [7]

[4] The double quote (``) are used to build a real number expression with a concrete syntax.

```
dist_sym:(x,y:Base)``(dist x y) == (dist y x)``;
dist_refl:(x,y:Base)``(dist x y) == 0``<->x==y;
dist_tri:(x,y,z:Base)``(dist x y) <= (dist x z)+(dist z y)``}.
```

The limit definition we used is:

$$\forall \varepsilon > 0\ \exists \alpha > 0\ s.t.\ \forall x \in D,\ |\ x - x_0\ | < \alpha\ \rightarrow\ |\ f(x) - l\ | < \varepsilon$$

that is to say l is the limit of $f(x)$ when x tends to x_0 in the domain D.

Using the Coq syntax, the limit is expressed as follows:

```
Definition limit_in :=
  [X,X':Metric_Space;f:(Base X)->(Base X');D:(Base X)->Prop;
   x0:(Base X);l:(Base X')]
   (eps:R)``eps > 0`` ->
   (EXT alp:R |``alp > 0``/\(x:(Base X))(D x)/\
               ``(dist X x x0) < alp``->
               ``(dist X' (f x) l) < eps``).
```

Once we have shown that $\mathbb{R}$ is really a metric space, the limit for the functions of type $\mathbb{R} \rightarrow \mathbb{R}$ can be defined in the following way:

```
Definition limit1_in:(R->R)->(R->Prop)->R->R->Prop:=
  [f:R->R; D:R->Prop; l,x0:R](limit_in R_met R_met f D x0 l).
```

Before formalizing the derivative, it is interesting to note that this definition of the limit does not absolutely prescribe $x \neq x_0$. Hence, in order to get a correct definition of the derivative, we must take this first choice[5] into consideration. To formalize the derivative, we chose the following definition:

$$f'(x_0) = \lim_{x \rightarrow x_0, x \neq x_0} \frac{f(x) - f(x_0)}{x - x_0}$$

Considering our definition of limit, we must express that $x \neq x_0$ using the definition domain of the function. The limit which will allow us to build the derivative uses the following notion of domain:

```
Definition D_x: (R->Prop)->R->R->Prop :=
   [D:(R->Prop); y,x:R](D x)/\``y <> x``.
```

In particular, x∈ (D_x D x0) means x∈ D\x0.

Now, we can give the definition of derivative of a function of type $\mathbb{R} \rightarrow \mathbb{R}$. Here as for the limit, we do not compute the derivative d of f, but we show that d is really the derivative of f in x_0 in a domain D.

```
Definition D_in:(R->R)->(R->R)->(R->Prop)->R->Prop:=
  [f,d:(R->R); D:(R->Prop); x0:R]
  (limit1_in [x:R]``((f x)-(f x0))/(x-x0)`` (D_x D x0) (d x0) x0).
```

[5] This choice allows us to prove some lemmas of standard analysis more easily.

This definition allows us to show the basic properties which will be useful to prove our correctness lemma, namely, properties about the derivative of sum, product, composition of functions, etc. Some of these properties can be found in appendix B.

For more details regarding the real library of Coq, we can refer to [11,13].

4 Formalization of the Algorithm in Direct Mode

In this part, we give a formalization of the algorithm in direct mode for a subset of FORTRAN programs, and we prove the correctness of this algorithm. We consider the issues raised, some of which will be answered during the discussion in section 6.

4.1 A Semantics for FORTRAN

Our semantics kernel will have, among other aspects, an environment, an abstract syntax of a subset of FORTRAN expressions, the interpretation of this abstract syntax in Coq, and an interpretation of the differentiation algorithm.

Environment

In this example, our environment is an associated list between the real type variables, indexed by natural numbers, and their associated values. Therefore we have:

```
Inductive rvar:Type:=varreal:nat->rvar.
Definition Env:=(listT (prodT rvar R)).
```

where `listT` and `prodT` represent respectively the type of lists and the type of couples.

Expressions

Now we define the type of the arithmetic expressions, followed by the type of some imperative structures. The arithmetic expressions will include constants, real variables and the usual operators of addition, multiplication, minus and power. In this example, we do not consider the transcendental functions because their implementation is still incomplete in Coq. For our example, the minimal type of the arithmetic expressions is simply expressed as follows:

```
Inductive expr:Type:=
         numR:R->expr
        |var:rvar->expr
        |Fplus:expr->expr->expr
```

```
        |Fmult:expr->expr->expr
        |Fminus:expr->expr->expr
        |Fopp:expr->expr
        |Fpow:expr->nat->expr.
```

With regard to the structures of the FORTRAN language, we only consider here the assignment and the sequence. We will explain this choice during the discussion.

```
Inductive For:Type:= Aff:rvar->expr->For
                    |Seq:For->For->For.
```

Remark 1. It is interesting to note that, even if our abstract syntax represents that of FORTRAN (to deal with O∂yssée programs), it also represents every subset of imperative programs and the correctness proof could be applied to other automatic differentiation tools (which deal with programs of another imperative language).

The Semantics Itself

We chose to give a natural semantics [9], that is to say *big step* semantics, which interprets our language towards some functions of the $\mathbb{R} \rightarrow \mathbb{R}$ type. This semantics is quite well adapted because there are no recursive functions involved. This choice allows our definitions and derivative properties in Coq to be used almost immediately. First, we give a semantics to the variables, the arithmetic expressions, then to our mini language. The evaluation of a variable will be its associated value in the environment. The function myassoc, a function written in Coq using a Fixpoint, looks for the value associated to a given variable in the environment.

```
Definition sem_var [v:rvar;l:Env]:R:=(myassoc v l).
```

The evaluation of the mathematical expressions is as usual:

```
Fixpoint sem_arith [l:Env;arith:expr]:R:=
  Cases arith of
     (numR x)=> x
    |(var r) =>  (sem_var r l)
    |(Fplus e1 e2)=>``(sem_arith l e1)+(sem_arith l e2)``
    |(Fmult e1 e2)=>``(sem_arith l e1)*(sem_arith l e2)``
    |(Fminus e1 e2)=>``(sem_arith l e1)-(sem_arith l e2)``
    |(Fopp e)=>``-(sem_arith l e)``
    |(Fpow e i)=>(pow (sem_arith l e) i)
  end.
```

The interpretation of assignments and sequences modifies the environment. Regarding the assignment, we create a new assignment with the interpretation of the right part and regarding the sequence f1;f2 we give an interpretation of f2 in the environment which has been modified by the interpretation of f1:

```
Fixpoint sem_For [l:Env;for:For]:Env:=
  Cases for of
   (Aff v e)=>(replace v (sem_arith l e) l)
  |(Seq f1 f2) => (sem_For (sem_For l f1) f2)
  end.
```

where `replace` is a function which replaces or assigns a value associated to a variable of the environment by another one, which makes the successive assignments possible.

Now, we need an interpretation towards a real function. For this, we can, for example, recover the value associated to an output variable after the evaluation of the program:

```
Definition sem [for:For;l:Env;vi,vo:rvar]:R->R:=
 [r:R](myassoc vo (sem_For (consT (pairT rvar R vi r) l) for)).
```

where `vi` and `vo` are respectively the significant input variable (given active by the user) and the output variable.

4.2 Derivation Algorithm

First of all, to help us understand the algorithm, we start giving a mathematical transcription of the previous example (`cube`) of section 2.

The instruction `y=x*x` is mathematically defined by the function f:
$f : \mathbb{R}^3 \rightarrow \mathbb{R}^3$
$f : (x, y, z) \mapsto (f_1(x, y, z), f_2(x, y, z), f_3(x, y, z)) = (x, x \times x, z)$

The instruction `z=y*x` is defined by the function g:
$g : \mathbb{R}^3 \rightarrow \mathbb{R}^3$
$g : (x, y, z) \mapsto (g_1(x, y, z), g_2(x, y, z), g_3(x, y, z)) = (x, y, y \times x)$

The sequence is the *composition* of these two functions: $g \circ f$. The differentiation of $g \circ f$ in x_0 is $D_{x_0}(g \circ f) = D_{f(x_0)} g \circ D_{x_0} f$. We have the tangent derivative applying this linear function from $\mathbb{R}^3$ to $\mathbb{R}^3$ with the vector $(\bar{x}, \bar{y}, \bar{z})$ given by the user. To do so, we begin computing the Jacobi matrices of f in (x, y, z) and g in $f(x, y, z)$:

$$J_f(x,y,z) = \begin{pmatrix} \frac{\partial f_1}{\partial x} & \frac{\partial f_1}{\partial y} & \frac{\partial f_1}{\partial z} \\ \frac{\partial f_2}{\partial x} & \frac{\partial f_2}{\partial y} & \frac{\partial f_2}{\partial z} \\ \frac{\partial f_3}{\partial x} & \frac{\partial f_3}{\partial y} & \frac{\partial f_3}{\partial z} \end{pmatrix} = \begin{pmatrix} 1 & 0 & 0 \\ 2x & 0 & 0 \\ 0 & 0 & 1 \end{pmatrix} \text{ and } J_g(x,y,z) = \begin{pmatrix} 1 & 0 & 0 \\ 0 & 1 & 0 \\ y & x & 0 \end{pmatrix}$$

Thus,

$$J_g(x, x^2, z) \begin{pmatrix} 1 & 0 & 0 \\ 0 & 1 & 0 \\ x^2 & x & 0 \end{pmatrix}$$

The Jacobi matrice $g \circ f$ is the following:

$$J_{gof}(x,y,z) = \begin{pmatrix} 1 & 0 & 0 \\ 0 & 1 & 0 \\ x^2 & x & 0 \end{pmatrix} \begin{pmatrix} 1 & 0 & 0 \\ 2x & 0 & 0 \\ 0 & 0 & 1 \end{pmatrix} = \begin{pmatrix} 1 & 0 & 0 \\ 2x & 0 & 0 \\ x^2 + 2x^2 & 0 & 0 \end{pmatrix}$$

Making the projection on $(\bar{x}, 0, 0)$, which is the active vector at the beginning (noted `IN` by O∂yssée):

$$\begin{pmatrix} 1 & 0 & 0 \\ 2x & 0 & 0 \\ 3x^2 & 0 & 0 \end{pmatrix} \begin{pmatrix} \bar{x} \\ 0 \\ 0 \end{pmatrix} = \begin{pmatrix} \bar{x} \\ 2x\bar{x} \\ 3x^2\bar{x} \end{pmatrix}$$

We can now check that this corresponds to the O∂yssée program (`CUBETL`): `YTTL` is exactly the result given by our mathematical formalization.
Regarding `ZTTL` we have:
`ZTTL=XTTL*Y+X*YTTL=XTTL*X*X+X*(XTTL*X+X*XTTL)=3*X*X*XTTL`
which is also the expected result.

We have seen that the direct mode is close to a symbolic derivative. In particular, in O∂yssée, to derive a real mathematical expression consists in calling a function, written in `Ocaml`, which makes this "symbolic" derivative. Therefore, we have a similar function in Coq, but which only contains what corresponds to our language. In order to add the transcendental functions, for example, adding the corresponding derivation rules is sufficient. Moreover, although the notion of derivative is enough to deal with our subset of programs, sometimes we will use the notion of differentiation. In particular, it is the case for the variables. If we only derive variables, the sequence of instructions are independent and this solution does not give the expected result. To formalize the O∂yssée algorithm, we must define an association list $(var, dvar)$ of type `lvar` (which will be built later):

```
Definition lvar:=(listT (prodT rvar rvar)).
```

The derivative of an expression is another expression which has been symbolically derived. The derivative of a variable is its corresponding derived variable in the list `l` of type `lvar`, given by the `myassocV` function. The following Coq function gives the derivative of our expressions:

```
Fixpoint deriv_expr [l:lvar;term:expr]:expr:=
  Cases term of
    (numR r)=>(numR R0)
   |(var x) => (var (myassocV x l))
   |(Fplus a1 a2)=>
     (Fplus (deriv_expr l a1) (deriv_expr l a2))
   |(Fmult a1 a2)=>
     (Fplus (Fmult (deriv_expr l a1) a2)
            (Fmult a1 (deriv_expr l a2)))
```

```
  |(Fminus a1 a2)=>
     (Fminus (deriv_expr l a1) (deriv_expr l a2))
  |(Fopp a)=>(Fopp (deriv_expr l a))
  |(Fpow a i)=>
      (Fmult (Fmult (numR (INR i)) (Fpow a (minus i (S O))))
             (deriv_expr l a))
 end.
```

As can be seen previously in the example, the derivative algorithm of our FORTRAN subset can be summarized as follow:

- the derivative of an assignment is a sequence between the assignment to a new variable of the right-hand side expression derivative and the initial assignment
- the derivative of a sequence is the sequence of the expression derivatives of the initial sequence

In Coq, the O∂yssée differentiation algorithm can be formalized as follows:

```
Fixpoint deriv_for_aux [l:lvar; termfor:For]:For:=
  Cases termfor of
    (Aff v e) =>
      (Seq (Aff (myassocV v l) (deriv_expr l e)) (Aff v e))
   |(Seq f1 f2) => (Seq (deriv_for_aux l f1)
                        (deriv_for_aux l f2))
  end.

Definition deriv_for [termfor:For]:For:=
       (deriv_for_aux (make_list termfor) termfor).
```

where make_list builds the association list $(var, dvar)$.

An example of application of this function can be found in [11] (chapter 5).

5 Proof of Correctness

Now, we can give and prove the correctness lemma regarding the algorithm used by O∂yssée to differentiate FORTRAN programs involving sequences and assignments. As said previously, this is equivalent to show that the interpretation in Coq of such a program is derived (in Coq) into the interpretation (in Coq) of the program differentiated by O∂yssée (which is expressed by the diagram of figure 1).

Before giving the mathematical statement and the formal version (Coq), let us remember that the semantical interpretation depends on the variable which is chosen as active, as well as the variable which is chosen as output value. Those two variables must be in the environment (which is a list of variables with their associated values). This method allows us to deal only with functions of type $\mathbb{R} \rightarrow \mathbb{R}$, which can be derived in Coq. Thus, the theorem can be expressed as follows:

Theorem 1 (Correctness). *Given a function sem, the semantics of a* FORTRAN *program involving sequences and affectations. Given v_i, the input active variable, and v_o, the output variable. Given env, the environment.*
For every program p, if $v_i \in env$ and $v_o \in env$ then
$sem'(p\ env\ v_i\ v_o) = (sem\ \bar{p}\ env\ v_i\ v_o)$ where $\bar{p}$ is the differentiated program returned by O∂yssée.

In Coq, this theorem is expressed in the following way:

```
Theorem corr_deriv:(l:Env)(p:For)(vi,vo:rvar)(D:R->Prop)(x0:R)
    (memenv vi l)->(memenv vo l)->
    (D_in (sem p l vi vo) (sem (deriv_for p) l vi vo) D x0).
```

To prove this theorem, we start proving two similar (to theorem 1) auxiliary lemmas regarding the affectation and the assignment:

```
Lemma corr_deriv_Aff:(l:Env)(e:expr)(r,vi,vo:rvar)(D:R->Prop)(x0:R)
  (memenv vi l)->(memenv vo l)->
  (D_in (sem (Aff r e) l vi vo)
    (sem (deriv_for (Aff r e)) l vi vo) D x0).

lemma corr_deriv_Seq:(l:Env)(f,f0:For)(vi,vo:rvar)(D:R->Prop)(x0:R)
      (memenv vi l)->(memenv vo l)->
  (D_in (sem (Seq f f0) l vi vo)
    (sem (deriv_for (Seq f f0)) l vi vo) D x0).
```

In this section, we give only the intuition of the proof. The formal proof can be found in appendix A.

Proof. of `corr_deriv_Aff`
By induction on the expression e.
Proving this property regarding the assignment consists in proving the correctness of the right-and-side expression derivative.
For each case, we have 2 subcases: either $v_i = r$ and $v_o = dr$ (i.e. v_i and v_o are respectively the input and the output variables) or $v_i \neq r$ or $v_o \neq dr$. The second case is always proved by contradiction using the hypotheses that v_i and v_o are in the environment. Thus, we give the intuition for the relevant case ($v_i = r$ and $v_o = dr$).

- Constant case: $e = c$. We have to prove that the semantics of the derivative of e is the derivative of the semantics of e, i.e. $c' = 0$. We only have to apply the appropriate lemma proved in the real number library.
- The other cases: variable, addition, multiplication,... have a similar proof, using the composition property (for power case) and the induction hypotheses (for addition,...)

Proof. of `corr_deriv_Seq`
Double induction on the programs f and f_0.

1. Assignment-assignment: we must deal with different possible cases for the input and the output variables. In the cases not proved by contradiction, after having performed some substitutions in the assignment which contains the output variable, we can apply the previous lemma (we have only one assignment).
2. Assignment-sequence and sequence-assignment: using the induction hypotheses on sequence, we are in the previous case.
3. Sequence-sequence: the idea is similar to the previous case, but here we have 4 induction hypotheses.

Now, the main theorem can be directly proved:

Proof. of theorem 1
By case analysis on the program p.

1. Assignment case: we apply directly the lemma `corr_deriv_Aff`.
2. Sequence case: we apply directly the lemma `corr_deriv_Seq`.

6 Discussion

In the previous section, we have shown the correctness of programs composed of variables, real arithmetic expressions, assignments and sequences. In other words, straight line real programs. Now, we can wonder whether we can prove the correctness of the whole algorithm, that is to say for every program. There are several difficulties:

What Happens with Control Structures? Let us study the conditional case. This is an interesting case, as it introduces a notion of discontinuity. The correctness of the algorithm is not ensured near the discontinuity points. In particular, the correctness cannot be proved for programs which compute functions by gaps, or functions of Dirac[6]. On the other hand, it is possible, if we are able to choose our branch (either the `then` or the `else`), to show the correctness. This remark essentially expresses the fact that the correctness cannot be proved for all imperative programs, which is represented by the diagram[7] in figure 2. We will denote S_i the sequence number i, p a program and $\bar{A}$ the result given by O∂yssée.

The conditional is not the only cause of discontinuity in the programs. The `goto, dowhile` are also potential causes. This problem could be solved either by giving a straight line programs from the original program [14] or defining a valid space for the program.

[6] Let us note that the differentiated programs expressed by O∂yssée are not absurd in these cases, since they return zero values almost everywhere

[7] provided by Laurent Hascoët INRIA-SOPHIA-antipolis, TROPICS project

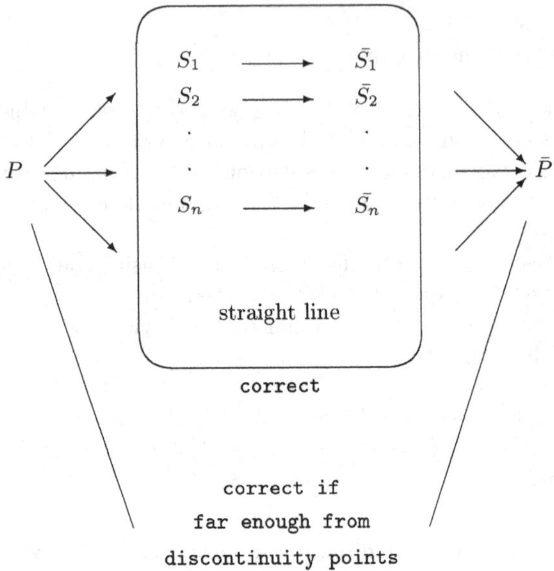

Fig. 2. Correctness field of the algorithm

What Happens with Several Active Variables? We have only dealt with programs having a functional interpretation which corresponds to the diagram in the figure1.

Actually, the semantics we gave is a functional interpretation ($\mathbb{R} \to \mathbb{R}$). From the definition of our current derivative, we cannot give an interpretation towards multidimensional spaces like $\mathbb{R}^n$. In order to deal with these cases, it would be necessary to begin by extending our development of the real analysis by notions of differentiation and Jacobian matrix. This will have to be done if we want to extend this experiment to the computation of the gradient.

What Happens with Other Data Types? Currently, we do not know whether it is possible to find a general way of proceeding[8]. For example, another problem, which we do not deal with here, (nor is it considered in O∂yssée), is a code in which there is an array of variable size. A notion of derivative must be found for that kind of object.
Moreover, the other basic data types (integer or boolean) of FORTRAN can also be considered by the program semantics. To do so, we have to define specific types of variables for each data type. This more complete semantics has also been formalized in Coq.

[8] By the way, automatic differentiation tools consider only certain data types.

7 Conclusion

The application of formal methods to numerical analysis problems is quite new. The formalization and the correctness proof for our subset of programs derived by the O∂yssée system has shown that formal proof systems can be appropriate to deal with that kind of problem. This proof could also have been built in some other proof assistants (like HOL[8] or PVS[12]), as well as for some other automatic differentiation tools for imperative programs. The use of a proof assistance tool allows us to isolate the different cases which must be considered in order to prove the correctness of such algorithms. This makes it possible to emphasize the particular case of programs which compute discontinuous functions.
However, although we are convinced that formal methods are suitable for such a critical domain, we have noticed that we are still rather far from dealing with significant problems and there are two main explanations for this. The first one is that the real analysis formalization is not immediate and that such a library is not developed in every system. The second one is due to numerical analysis itself. Most of the programs written for numerical analysis use approximation principles, which are either controlled or not. There is a kind of conventional *probabilist tolerance*, in the sense that, as can be seen in figure 2 for instance, problems arise rarely, if ever. This view is obviously opposed to our aims, which are to prove formally that no problem can arise. In the particular case we have studied, we can see that the problem of discontinuity points is precisely one example of what we would like to avoid.

Regarding the application of formal proofs to the numerical analysis domain, the research activity is still quite open. However, this formalization, on a subset of imperative straight line programs, is a good start for a more advanced formalization, which would the differentiation more into consideration than the derivative. A next step could be to study algorithms for gradient computation, which are considered as critical in their implementation but which might be dealt with using our methods.

References

1. The ADIC Home Page. `http://www-fp.mcs.anl.gov/adic/`.
2. The PADRE2 Home Page. `http://warbler.ise.chuo-u.ac.jp/Padre2/`.
3. C. Bischof, A. Carle, P. Khademi, A. Mauer, and P. Hovland. ADIFOR2.0 user's guide. Technical Report ANL/MCS-TM-192,CRPC-TR95516-S, Argonne National Laboratory Technical Memorandum, June 1998.
4. Pietro Di Giannantonio and Alberto Ciaffaglione. A co-inductive approach to real numbers. In *Proc. of TYPES'99*, volume 1956, pages 114–130. Springer-Verlag LNCS, 1999.
5. Christèle Faure and Yves Papegay. Odyssée user's guide version 1.7. Technical Report 0224, INRIA Sophia-Antipolis, September 1998.
6. Jean-Christophe Filliâtre. *Preuves de programmes impératifs en théories des types.* PhD thesis, Université Paris-Sud, Juillet 1999.

7. Herman Geuvers, Freek Wiedijk, and Jan Zwanenburg. Equational Reasoning via Partial Reflection. In *Proceedings of TPHOL*. Springer-Verlag, August 2000.
8. M.J.C. Gordon and T.F. Melham. *Introduction to HOL: A Theorem Proving Environment for Higher Order Logic.* Cambridge University Press, 1993.
9. Carl A. Gunter. *Semantics of Programming Languages: Structures and Techniques.* Foundations of Computing. MIT Press, 1992.
10. John Harrison. *Theorem Proving with the Real Numbers.* Springer-Verlag, 1998.
11. Micaela Mayero. *Formalisation et automatisation de preuves en analyses réelle et numérique.* PhD thesis, Université Paris 6, Décembre 2001. `http://logical.inria.fr/~mayero/`.
12. Sam Owre, Natarajan Shankar, and John Rushby. PVS: A prototype verification system. In *Proceedings of CADE 11, Saratoga Springs, New York*, June 1992.
13. LogiCal Project The Coq Development Team. *The Coq Proof Assistant Reference Manual Version 7.2.* INRIA-Rocquencourt, December 2001. `http://coq.inria.fr/doc-eng.html`.
14. R. Towle. *Control and data dependence for program transformations.* PhD thesis, Univ. of Illinois, Urbana, March 1976.

A Complete Formal Correctness Proofs

In the following proofs, the assignment will be noted ":=" and the sequence ";". These proofs are described in such a way that they are rather close to the formal proofs built in Coq. The specific lemmas regarding derivative in Coq used in these proofs, can be found in appendix B.

Proof. of `corr_deriv_Aff`
By induction on the expression e.
Let H0 and H1 be the hypotheses $v_i \in l$ and $v_o \in l$.

1. Constant case: we must prove, using the previous hypotheses, that:
   $\forall c : \mathbb{R}, sem'((r := c)\ l\ v_i\ v_o) = (sem\ \overline{(r := c)}\ l\ v_i\ v_o)$ i.e. that
   $\forall c : \mathbb{R}, sem'((r := c)\ l\ v_i\ v_o) = (sem\ (dr := 0; r := c)\ l\ v_i\ v_o)$ where c is a real constant.
   In the case of $v_i = r$ and $v_o = dr$, it is equivalent to prove, after simplifications and using H0, H1, that $c' = 0$. We only have to apply the lemma `Dconst` of Coq.
   In the case of $v_i \neq r$ or $v_o \neq dr$, we have a contradiction with H0 or H1, because in these cases $v_i \notin l$ or $v_o \notin l$.
2. Variable case: we must prove, using the previous hypotheses, that:
   $\forall c : \mathbb{R}, sem'((r := x)\ l\ v_i\ v_o) = (sem\ \overline{(r := x)}\ l\ v_i\ v_o)$ i.e. that
   $\forall c : \mathbb{R}, sem'((r := x)\ l\ v_i\ v_o) = (sem\ (dr := 1; r := x)\ l\ v_i\ v_o)$ where x is a real variable.
   In the case of $v_i = r$ and $v_o = dr$, it is equivalent to prove, after simplifications and using H0, H1, that $x' = 1$. We only have to apply the lemma `Dx` of Coq.
   In the case of $v_i \neq r$ or $v_o \neq dr$, we have a contradiction with H0 or H1, because in these cases $v_i \notin l$ or $v_o \notin l$. $v_o \notin l$.
3. Addition case: we have also two induction hypotheses:
   - H2: $\forall v_i, v_o, r, sem'((r := e_0)\ l\ v_i\ v_o) = (sem\ \overline{(r := e_0)}\ l\ v_i\ v_o)$ where e_0 is an expression
   - H3: $\forall v_i, v_o, r, sem'((r := e_1)\ l\ v_i\ v_o) = (sem\ \overline{(r := e_1)}\ l\ v_i\ v_o)$ where e_1 is an expression

We must prove that
$sem'((r := e_0 + e_1)\ l\ v_i\ v_o) = (sem\ \overline{(r := e_0 + e_1)}\ l\ v_i\ v_o)$ i.e. that
$sem'((r := e_0 + e_1)\ l\ v_i\ v_o) = (sem\ (dr := e_0' + e_1'; r := e_0 + e_1)\ l\ v_i\ v_o)$.
After simplifications of H2 and H3 we have:
- H2': $\forall v_i, v_o, r, sem'((r := e_0)\ l\ v_i\ v_o) = (sem\ (dr := e_0'; r := e_0)\ l\ v_i\ v_o)$
- H3': $\forall v_i, v_o, r, sem'((r := e_1)\ l\ v_i\ v_o) = (sem\ (dr := e_1'; r := e_1)\ l\ v_i\ v_o)$
In the case of $v_i = r$ and $v_o = dr$, it is equivalent to prove, after simplifications and using H0, H1, H2', H3', that $(e_0 + e_1)' = e_0' + e_1'$. We only have to apply the lemma `Dadd` of Coq.
In the case of $v_i \neq r$ or $v_o \neq dr$, we have a contradiction with H0 or H1, because in these cases $v_i \notin l$ or $v_o \notin l$.
4. Multiplication case: similar to the addition case.
5. Minus case: similar to the addition case.
6. Opposite case: we have also an induction hypothesis:
- H2: $\forall v_i, v_o, r, sem'((r := e_0)\ l\ v_i\ v_o) = (sem\ \overline{(r := e_0)}\ l\ v_i\ v_o)$ where e_0 is an expression
We must prove that $sem'((r := -e_0)\ l\ v_i\ v_o) = (sem\ \overline{(r := -e_0)}\ l\ v_i\ v_o)$ i.e. that $sem'((r := -e_0)\ l\ v_i\ v_o) = (sem\ (dr := -e_0'; r := -e_0)\ l\ v_i\ v_o)$.
After simplifications of H2 we have:
- H2': $\forall v_i, v_o, r, sem'((r := e_0)\ l\ v_i\ v_o) = (sem\ (dr := e_0'; r := e_0)\ l\ v_i\ v_o)$
In the case of $v_i = r$ and $v_o = dr$, it is equivalent to prove, after simplifications and using H0, H1, H2', that $(-e_0)' = -e_0'$. We only have to apply the lemma `Dopp` of `Coq`.
In the case of $v_i \neq r$ or $v_o \neq dr$, we have a contradiction with H0 or H1, because in these cases $v_i \notin l$ or $v_o \notin l$.
7. Power case: similar to the opposite case, using the composition and power lemmas.

Proof. of `corr_deriv_Seq`
Double induction on the programs f and f_0.
Let H0 and H1 be the hypotheses $v_i \in l$ and $v_o \in l$.

1. Assignment-assignment case: we must prove, using the previous hypotheses, that $sem'((r_0 := e_0; r := e)\ l\ v_i\ v_o) = (sem\ \overline{(r_0 := e_0; r := e)}\ l\ v_i\ v_o)$ i.e. that $sem'((r_0 := e_0; r := e)\ l\ v_i\ v_o) = (sem\ ((dr_0 := e_0'; r_0 := e_0); (dr := e'; r := e))\ l\ v_i\ v_o)$ where e_0 and e are expressions.
In the case of $v_i = r$ and $v_o = dr$ or $v_i = r_0$ and $v_o = dr_0$ or $v_i = r$ and $v_o = dr_0$ or $v_i = r_0$ and $v_o = dr$, after some substitutions from the environment and some simplifications, we can apply the previous lemma.
In the other cases, we have either $v_i \notin l$ or $v_o \notin l$, which is a contradiction with respect to the hypotheses.
2. Assignment-sequence case: we have also two induction hypotheses:
- H2: $\forall e, v_i, v_o, r, sem'((r := e; f_1)\ l\ v_i\ v_o) = (sem\ \overline{(r := e; f_1)}\ l\ v_i\ v_o)$ where f_1 is a program
- H3: $\forall e, v_i, v_o, r, sem'((r := e; f_2)\ l\ v_i\ v_o) = (sem\ \overline{(r := e; f_2)}\ l\ v_i\ v_o)$ where f_2 is a program
we must prove, using the previous hypotheses, that
$sem'((r := e; (f_1; f_2))\ l\ v_i\ v_o) = (sem\ \overline{(r := e; (f_1; f_2))}\ l\ v_i\ v_o)$ i.e. that
$sem'((r := e; (f_1; f_2))\ l\ v_i\ v_o) = (sem\ ((dr := e'; r := e); \overline{(f_1; f_2)})\ l\ v_i\ v_o)$ where f_1 and f_2 are programs.
After simplifications of H2 and H3 we have:

- H2': $\forall e, v_i, v_o, r, sem'((r := e; f_1)\ l\ v_i\ v_o) = (sem\ (dr := e'; r := e; \overline{f_1})\ l\ v_i\ v_o)$
- H3': $\forall e, v_i, v_o, r, sem'((r := e; f_2)\ l\ v_i\ v_o) = (sem\ (dr := e'; r := e; \overline{f_2})\ l\ v_i\ v_o)$

Using H2' and H3', we are in the previous case.

3. Sequence-assignment case: symmetric w.r.t. the previous case.
4. Sequence-sequence case: we have 4 induction hypotheses:
- H2: $\forall v_i, v_o, sem'(((f_4; f_3); f_0)\ l\ v_i\ v_o) = (sem\ \overline{((f_4; f_3); f_0)}\ l\ v_i\ v_o)$ where f_4, f_3 and f_0 are programs
- H3: $\forall v_i, v_o, sem'(((f_4; f_3); f_1)\ l\ v_i\ mbv_o) = (sem\ \overline{((f_4; f_3); f_1)}\ l\ v_i\ v_o)$ where f_4, f_3 and f_1 are programs
- H4: si $\forall f_1, sem'((f_1; f_0))\ l\ v_i\ v_o) = (sem\ \overline{(f_1; f_0)}\ l\ v_i\ v_o)$ and if $\forall f_2, sem'((f_2; f_0))\ l\ v_i\ v_o) = (sem\ \overline{(f_2; f_0)}\ l\ v_i\ v_o)$ alors $\forall v_i, v_o, sem'(((f_2; f_1); f_0))\ l\ v_i\ v_o) = (sem\ \overline{((f_2; f_1); f_0)}\ l\ v_i\ v_o)$
- H5: si $\forall f_2, sem'((f_2; f_0))\ l\ v_i\ v_o) = (sem\ \overline{(f_2; f_0)}\ l\ v_i\ v_o)$ and if $\forall f_0, sem'((f_0; f_1))\ l\ v_i\ v_o) = (sem\ \overline{(f_0; f_1)}\ l\ v_i\ v_o)$ alors $\forall v_i, v_o, sem'(((f_0; f_2); f_1))\ l\ v_i\ v_o) = (sem\ \overline{((f_0; f_2); f_1)}\ l\ v_i\ v_o)$

Applying H4 and H5 with f_3 H2 f_4 H3, we have these two following hypotheses:
- H4': $\forall v_i, v_o, sem'(((f_4; f_3); f_0))\ l\ v_i\ v_o) = (sem\ \overline{((f_4; f_3); f_0)}\ l\ v_i\ v_o)$
- H5': $\forall v_i, v_o, sem'(((f_4; f_3); f_1))\ l\ v_i\ v_o) = (sem\ \overline{((f_4; f_3); f_1)}\ l\ v_i\ v_o)$

Thus, we must prove that $\forall v_i, v_o, sem'(((f_4; f_3); (f_0; f_1))\ l\ v_i\ v_o) = (sem\ \overline{(\overline{(f_4; f_3)}; \overline{(f_0; f_1)})}\ l\ v_i\ v_o)$ i.e. that $\forall v_i, v_o, sem'(((f_4; f_3); (f_0; f_1))\ l\ v_i\ v_o) = (sem\ ((f_4; f_3); (f_0; f_1)\ l\ v_i\ v_o)$, and using H4' and H5' we obtain the previous cases.

B Main Derivative Properties (in Coq) Used in the Proofs

```
Lemma Dconst:(D:R->Prop)(y:R)(x0:R)(D_in [x:R]y [x:R]``0`` D x0).
Lemma Dx:(D:R->Prop)(x0:R)(D_in [x:R]x [x:R]``1`` D x0).
Lemma Dadd:(D:R->Prop)(df,dg:R->R)(f,g:R->R)(x0:R)
  (D_in f df D x0)->(D_in g dg D x0)->
  (D_in [x:R]``(f x)+(g x)`` [x:R]``(df x)+(dg x)`` D x0).
Lemma Dmult:(D:R->Prop)(df,dg:R->R)(f,g:R->R)(x0:R)
  (D_in f df D x0)->(D_in g dg D x0)->
  (D_in [x:R]``(f x)*(g x)``
        [x:R]``(df x)*(g x)+(f x)*(dg x)`` D x0).
Lemma Dopp:(D:R->Prop)(f,df:R->R)(x0:R)(D_in f df D x0)->
  (D_in [x:R](Ropp (f x)) [x:R]``-(df x)`` D x0).
Lemma Dminus:(D:R->Prop)(df,dg:R->R)(f,g:R->R)(x0:R)
  (D_in f df D x0)->(D_in g dg D x0)->
  (D_in [x:R](Rminus (f x) (g x)) [x:R]``(df x)-(dg x)`` D x0).
Lemma Dcomp:(Df,Dg:R->Prop)(df,dg:R->R)(f,g:R->R)(x0:R)
  (D_in f df Df x0)->(D_in g dg Dg (f x0))->
  (D_in [x:R](g (f x)) [x:R]``(df x)*(dg (f x))`` (Dgf Df Dg f) x0).
Lemma D_pow_n:(n:nat)(D:R->Prop)(x0:R)(expr,dexpr:R->R)
 (D_in expr dexpr D x0)-> (D_in [x:R](pow (expr x) n)
   [x:R]``(INR n)*(pow (expr x) (minus n (S O)))*(dexpr x)``
        (Dgf D D expr) x0).
```

Quotient Types: A Modular Approach*

Aleksey Nogin

Department of Computer Science
Cornell University, Ithaca, NY 14853
nogin@cs.cornell.edu

Abstract. In this paper we introduce a new approach to axiomatizing quotient types in type theory. We suggest replacing the existing monolithic rule set by a modular set of rules for a specially chosen set of primitive operations. This modular formalization of quotient types turns out to be much easier to use and free of many limitations of the traditional monolithic formalization. To illustrate the advantages of the new approach, we show how the type of collections (that is known to be very hard to formalize using traditional quotient types) can be naturally formalized using the new primitives. We also show how modularity allows us to reuse one of the new primitives to simplify and enhance the rules for the set types.

1 Introduction

NuPRL type theory differs from most other type theories used in theorem provers in its treatment of equality. In Coq's Calculus of Constructions, for example, there is a single global equality relation which is not the desired one for many types (e.g. function types). The desired equalities have to be handled explicitly, which is quite burdensome. As in Martin-Löf type theory [20] (of which NuPRL type theory is an extension), in NuPRL each type comes with its own equality relation (the extensional one in the case of functions), and the typing rules guarantee that well–typed terms respect these equalities. Semantically, a quotient in NuPRL is trivial to define: it is simply a type with a new equality.

Such quotient types have proved to be an extremely useful mechanism for natural formalization of various notions in type theory. For example, rational numbers can be naturally formalized as a quotient type of the type of pairs of integer numbers (which would represent the numerator and the denominator of a fraction) with the appropriate equality predicate.

Somewhat surprisingly, it turns out that formulating rules for these quotient types is far from being trivial and numerous applications of NuPRL [6,19] have run into difficulties. Often a definition involving a quotient will look plausible, but after some (sometimes substantial) work it turns out that a key property is unprovable, or false.

* This work was partially supported by AFRL grant F49620-00-1-0209 and ONR grant N00014-01-1-0765.

V.A. Carreño, C. Muñoz, S. Tahar (Eds.): TPHOLs 2002, LNCS 2410, pp. 263–280, 2002.

A common source of problems is that in NuPRL type theory all true equality predicates are uniformly witnessed by a single canonical constant. This means that even when we know that two elements are equal in a quotient type, we can not in general recover the witness of the equality predicate. In other words, $a = b \in (A//E)$ (where "$A//E$" is a quotient of type A with equivalence relation E) does not always imply $E[a; b]$ (however it does imply $\neg\neg E[a; b]$).

Another common class of problems occurs when we consider some predicate P on type A such that we can show that $P[a] \Leftrightarrow P[b]$ for any $a, b \in A$ such that $E[a; b]$. Since $P[a] \Leftrightarrow P[b]$ does not necessary imply that $P[a] = P[b]$, P may still turn out not to be a well–formed predicate on the quotient type $A//E$ [1].

These problems suggest that there is more to concept of quotient types, than just the idea of changing the equality relation of a type. In this paper we show how we can decompose the concept of quotient type into several simpler concepts and to formalize quotient types based on formalization of those simpler concepts.

We claim that such a "decomposed" theory makes operating with quotient types significantly easier. In particular we show how the new type constructors can be used to formalize the notion of indexed collections of objects. We also claim that the "decomposition" process makes the theory more modular. In particular, we show how to reuse one of the new type constructors to improve and simplify the rules for the set type.

For each of the new (or modified) type constructors, we present a set of derivation rules for this type — both the axioms to be added to the type theory and the rules that can be derived from these axioms. As we will explain in Section 3, the particular axioms we use were carefully chosen to make the theory as modular as possible and to make them as usable as possible in a tactic–based interactive prover. All the new rules were checked and found valid in S. Allen's semantics of type theory [1,2]; these proofs are rather straightforward, so we omit them here. Proofs of all the derived rules were developed and checked in the MetaPRL system [12,14,13].

Although this paper focuses on NuPRL type theory, the author believes that many ideas presented here are relevant to managing witnessing and functionality information in a constructive setting in general.

The paper is organized as follows. First, in Sections 1.1 through 1.3 we will describe some features of NuPRL type theory that are necessary for understanding this work. Sections 2, 4, 5, 6 and 7 present a few new primitive type constructors and show how they help to formulate the rules for quotient and set types; Section 3 explains our approach to choosing particular axioms; Section 8 shows how to use the new type constructors to formalize the notion of collections.

1.1 Propositions-as-Types

NuPRL type theory adheres to the *propositions–as–types principle*. This principle means that a proposition is identified with the type of all its witnesses. A

[1] It will be a function from $A//E$ to $\mathbf{Prop}// \Leftrightarrow$, rather than a predicate (a function from $A//E$ to $\mathbf{Prop}$), where $\mathbf{Prop}$ is a type (universe) of propositions.

proposition is considered true if the corresponding type is inhabited and is considered false otherwise. In this paper we will use words "*type*" and "*proposition*" interchangeably; same with "*witness*" and "*member*".

1.2 Partial Equivalence Relations Semantics

The key to understanding the idea of quotient types is understanding the most commonly used semantics of the NuPRL type theory (and some other type theories as well) — the PER (partial equivalence relations) semantics [28,1,2]. In PER semantics each type is identified with a set of objects and an equivalence relation on that set that serves as an *equality relation* for objects of that type. This causes the equality predicate to be three–place: "$a = b \in C$" stands for "a and b are equal elements of type C", or, semantically, "a and b are related by the equality relation of type C".

Remark 1.1. Note that in this approach an object is an element of a type *iff* it is equal to itself in that type. This allows us to identify $a \in A$ with $a = a \in A$.

According to PER approach, whenever something ranges over a certain type, it not only has to span the whole type, it also has to respect the equality of that type.

Example 1.2. In order for a function f to be considered a function from type A to type B, not only for every $a \in A$, $f(a)$ has to be B, but also whenever a and a' are equal in the type A, $f(a)$ should be equal to $f(a')$ in the type B. Note that in this example the second condition is sufficient since it actually implies the first one. However it is often useful to consider the first condition separately.

Example 1.3. Now consider a set type $T := \{x : A \mid B[x]\}$ (cf. Section 4). Similarly to Example 1.2 above, in order for T to be a well–formed type, not only $B[a]$ has to be a well–formed type for any $a \in A$, but also for any $a = a' \in A$ it should be the case that $B[a]$ and $B[a']$ are equal types.

1.3 Extensional and Intensional Approaches

In this paper we devote significant amount of attention to discussion of choices between what we call *intensional* and *extensional* approaches to certain type operators. The difference between these approaches is in deciding when two objects should be considered equal. In general, in the intensional approach two objects would be considered equal if their *internal* structure is the same, while in the extensional approach two objects would be considered equal if they exhibit the same external behavior.

Example 1.4. In NuPRL type theory the function equality is extensional. Namely, we say that $f = f' \in (A \to B)$ *iff* they both are in $A \to B$ and for all $a \in A$ $f(a) = f'(a) \in B$.

Example 1.5. It is easy to define an extensional equality on types: $A =_e B$ *iff* A and B have the same membership and equality relations. However, in NuPRL type theory the main equality relation on types is intensional. For example, if A and B are two non-empty types, then $(A \rightarrow \bot) = (B \rightarrow \bot)$ only when $A = B$, even though we have $(A \rightarrow \bot) =_e (B \rightarrow \bot)$ since they are both empty types.[2]

Example 1.6. When introducing certain type constructors, such as a set (cf. Section 4) or a quotient (cf. Section 7) one, into NuPRL type theory, there are often two choices for an equality definition:
Completely Intensional. The predicates have to be equal, for example
$\{x : A \mid B[x]\} = \{x : A' \mid B'[x]\}$ iff $A = A'$ and for all $a = a' \in A$, $B[a]{=}B'[a']$.
Somewhat Extensional. The predicates have to imply one another, for example
$\{x : A \mid B[x]\} = \{x : A' \mid B'[x]\}$ iff $A = A'$ and for all $a = a' \in A$, $B[a]{\Leftrightarrow}B'[a']$.

Essentially, in the intensional case the map $x \rightsquigarrow B[x]$ has to respect A's equality relation in order for $\{x : A \mid B[x]\}$ to be well–formed and in the extensional case $B[x]$ only needs to respect it up to $\Leftrightarrow$ (logical *iff*).

We will continue the discussion of the differences between these two choices in Sections 2.3 and 7.1.

2 Squash Operator

2.1 Squash Operator: Introduction

The first concept that is needed for our formalization of quotient types is that of hiding the witnessing information of a certain true proposition thus only retaining the information that the proposition is known to be true while hiding the information on *why* it is true.

To formalize such notion, for each type A we define a type $[A]$ ("squashed A") which is empty *if and only if* A is empty and contains a single canonical element $\bullet$ [3] when A is inhabited. Informally one can think of $[A]$ as a proposition that says that A is a *non-empty type*, but "squashes down to a point" all the information on *why* A is non-empty. The **squash** operator is *intensional*, e.g. $[A] = [B]$ iff $A = B$ (see also Remark 2.2).

Remark 2.1. We could define **squash** operator as $[A] := \{x : Unit \mid A\}$. Note that it does not make sense for us to actually add such a definition to the system since we want to formalize the set type using the **squash** operator and not the other way around.

---

[2] Strictly speaking, NuPRL type theory does not contain Martin-Löf's "$A = B$" judgment form. Instead, NuPRL uses proposition of the form $A = B \in \mathbb{U}_i$ where $\mathbb{U}_i$ is the i-th universe of types. However in this paper we will often omit "$\in \mathbb{U}_i$" for simplicity.

[3] MetaPRL system uses the unit element () or "it" as a $\bullet$, NuPRL uses `Ax` and [27] uses *Triv*.

The squash operator (sometimes also called hide) was introduced in [9]. It is also used in MetaPRL [11,12,14] [4].

In the next section we will present the axiomatization we chose for the squash operator and we will explain our choices in Section 3.

2.2 Squash Operator: Axioms

First, whenever A is non-empty, $[A]$ must be non-empty as well:

$$\frac{\Gamma \vdash A}{\Gamma \vdash [A]} \qquad (SquashIntro)$$

Second, if we know $[A]$ and we are trying to prove an equality (or a membership) statement, we can allow "unhiding" contents of A and continue with the proof:

$$\frac{\Gamma;\, x : A;\, \Delta \vdash t_1 = t_2 \in C}{\Gamma;\, x : [A];\, \Delta \vdash t_1 = t_2 \in C} \qquad (SquashElim)$$

(assuming x does not occur free in Δ, t_i and C [5]). This rule is valid because in Martin-Löf type theory equality has no *computational context* and is always witnessed by $\bullet$, so knowing the witness of A does not add any "additional power".

Finally, the only possible element of a squash type is $\bullet$:

$$\frac{\Gamma;\, x : [A];\, \Delta[\bullet] \vdash C[\bullet]}{\Gamma;\, x : [A];\, \Delta[x] \vdash C[x]} \qquad (SquashMemElim)$$

As mentioned in the introduction, all these new axioms can be proved sound in Allen's semantics of type theory [1,2]. All soundness proofs are very straightforward, and we omit them in this paper. We also omit some purely technical axioms (such as well–formedness ones) that are unnecessary for understanding this work.

2.3 Squash Operator: Derived Rules

Here are the rules that can be derived from the axioms we have introduced above. First, whenever $[A]$ is non-empty, $\bullet$ must be in it:

$$\frac{\Gamma \vdash [A]}{\Gamma \vdash \bullet \in [A]} \qquad (SquashMemIntro)$$

[4] In MetaPRL squash was first introduced by J.Hickey as a replacement for NuPRL's hidden hypotheses mechanism, but eventually it became clear that it gives a mechanism substantially widely useful than NuPRL's hidden hypotheses.

[5] We use the sequent schema syntax of [23] for specifying rules. Essentially, variables that are explicitly mentioned may occur free only where they are explicitly mentioned.

Second, using (*SquashMemElim*) we can prove a stronger version of (*SquashElim*):

$$\frac{\Gamma;\, x : A;\, \Delta[\bullet] \;\vdash\; t_1[\bullet] = t_2[\bullet] \in B[\bullet]}{\Gamma;\, x : [A];\, \Delta[x] \;\vdash\; t_1[x] = t_2[x] \in B[x]} \qquad (SquashElim2)$$

Third, we can prove that squashed equality implies equality:

$$\frac{\Gamma \;\vdash\; [t_1 = t_2 \in A]}{\Gamma \;\vdash\; t_1 = t_2 \in A} \qquad (SquashEqual)$$

Remark 2.2. Note that if we would have tried to make the **squash** operator extensional, we would have needed an extra well–typedness assumption in the (*SquashElim*) rule (as we had to do in (*EsquashElim*) rule in Section 7.2) which would have made it useless for proving well–typedness and membership statements. In particular, the (*SquashEqual*) rule (as well as any reasonable modification of it) would not have been valid.

Next, we can prove that if we can deduce a witness of a type A just by knowing that some unknown x is in A (we call such A a *squash–stable* type), then $[A]$ implies A:

$$\frac{\Gamma \;\vdash\; [A] \qquad \Gamma;\, x : A \;\vdash\; t \in A}{\Gamma \;\vdash\; A} \qquad (SquashStable)$$

Remark 2.3. The notion of *squash stability* is also discussed in [18] and is very similar to the notion of *computational redundancy* [5, Section 3.4].

Finally, we can prove that we can always eliminate the squashes in hypotheses not only when the conclusion is an equality (as in (*SquashElim*) and (*SquashElim2*)), but also when it is a squash [6]:

$$\frac{\Gamma;\, x : A;\, \Delta[\bullet] \;\vdash\; [C[\bullet]]}{\Gamma;\, x : [A];\, \Delta[x] \;\vdash\; [C[x]]} \qquad (Unsquash)$$

3 Choosing the Rules

For each of the concepts and type operators we discuss in this paper there might be numerous different ways of axiomatizing it. When choosing a particular set of axioms we were using several general guidelines.

First, in a context of an interactive tactic-based theorem prover it is very important to ensure that each rule is formulated in a *reversible* way whenever possible. By reversible rule we mean a rule where conclusion is valid *if and only if* the premises are valid. This means that it is always "safe" to apply such

[6] In general, it is true whenever the conclusion is squash–stable.

a rule when (backward) searching for a proof of some sequent — there is no "danger" that back–chaining through the rule would turn a provable statement into a statement that is no longer true. This property allows us to add such rules to proof tactics more freely without having to worry about a possibility that applying such tactic can bring the proof into a "dead end" [7]. For example, among the `squash` axioms of Section 2.2 only (*SquashIntro*) is irreversible and the other axioms are reversible.

Second, we wanted to make sure that each rule makes the smallest "step" possible. For example, the (*SquashElim*) rule only eliminates the `squash` operator, but does not attempt to eliminate the witness of the `squash` type while the (*SquashMemElim*) only eliminates the witness of the `squash` type and does not attempt to eliminates the `squash` operator. This gives users a flexibility to "steer" proofs exactly where they want them to go. Of course, we often do want to make several connected steps at once, but that can be accomplished by providing derived rules [8] while still retaining the flexibility of the basic axioms. For example, the (*SquashElim2*) allows one to both eliminate the `squash` operator and its witness in a single step, while still using (*SquashElim*) or (*SquashMemElim*) when only one and not the other is needed. As we will see in Section 4 this "small step" requirement is especially important for the irreversible rules.

Finally, it is important for elimination rules to match corresponding introduction rules in their "power"[9]. Such balance helps insure that most rules are reversible not only with respect to validity, but also with respect to provability (which is obviously needed to make applying such rules truly "safe" in a theorem prover).

4 Intensional Set Type

4.1 Set Type: Introduction

The decomposition of the axiomatization of quotient types into smaller pieces has an additional advantage (besides making quotient types easier to reason about) of making the theory more modular. The type operators that we use for formalizing quotient types can be now reused when formalizing other types as well. To illustrate this, we will show how the traditional formalization of the set types can be greatly improved and simplifies using the `squash` operator.

[7] Of course if a tactic is designed to fall back when it fails to find a complete derivation for the statement being proved, it would not become dangerous when we allow it to use an irreversible rule (although it might become more likely to fail). But if a tactic is only meant to propel the proof further without necessarily completing it (such as for example NuPRL's `Auto` and MetaPRL's `autoT`), then allowing such tactic to use irreversible rules can make things substantially less pleasant to the user.

[8] See [23] for a description of MetaPRL's derived rules mechanism.

[9] More specifically, elimination rules should be *locally complete* and *locally sound* with respect to the introduction rules, as described in [26]. But since we believe that this third guideline is not as crucial as the first two, we chose not provide a detailed discussion of it.

Informally, $\{x : A \mid B[x]\}$ is a type containing all elements $x \in A$ such that $B[x]$ is a true proposition. The key property of set type is that when we have a witness $w \in \{x : A \mid B[x]\}$, we know that $w \in A$ and we know that $B[w]$ is non-empty; but in general we have no way of reconstructing a witness for $B[w]$.

4.2 Set Type: Traditional Approach

Set types were first introduced in [7] and were also formalized in [3,9,12,24]. In those traditional implementations of type theory the rules for set types are somewhat asymmetric. When proving something like

$$\frac{\Gamma;\, y : A \vdash y \in A' \qquad \Gamma;\, y : A;\, z : B[y] \vdash B'[y]}{\Gamma;\, y : \{x : A \mid B[x]\} \vdash y \in \{x : A' \mid B'[x]\}}$$

one was forced to apply the set elimination rule before the set introduction rule. As we will see in a moment, the problem was that the traditional set introduction rule is irreversible and would go "too far" if one applies it right away. It would yield a subgoal $\Gamma;\, y : \{x : A \mid B[x]\} \vdash B'[y]$ that would only be valid if one could reconstruct a proof witness of $B'[y]$ without having access to the witness of $B[y]$.

4.3 Set Type: A New Approach

Using the `squash` operator we only need[10] the following two simple axioms to formalize the set type:

$$\frac{\Gamma;\, y : A;\, z : [B[y]];\, \Delta[y] \vdash C[y]}{\Gamma;\, y : \{x : A \mid B[x]\};\, \Delta[y] \vdash C[y]} \qquad (\mathit{SetElim})$$

$$\frac{\Gamma \vdash t \in A \qquad \Gamma \vdash [B[t]] \qquad \Gamma \vdash \{x : A \mid B[x]\}\,\mathrm{Type}}{\Gamma \vdash t \in \{x : A \mid B[x]\}} \qquad (\mathit{SetIntro})$$

Now we can explain the problem with the traditional approach [8,9,24,27] — there the set introduction rule is somewhat analogous to applying (*SetIntro*) and then as much (*Unsquash*) as possible and then (*SquashIntro*). Such rule does too many things at once and one of those things (*SquashIntro*) is irreversible. With such rule we can only deconstruct the set operator in the conclusion when the irreversible part of this rule would not render the resulting subgoals unprovable.

The reason this traditional formalization required a rule that does so much at once was the lack of a way to express the intermediate results. In a sense, in that implementation, set (and quotient) types had at least two separate "jobs" — one was to change the type membership (equality) and another — to hide the proof of the membership (equality) predicate. And there was only a single collection of rules for both of the "jobs", which made the rules hard to use.

[10] As in Section 2.2 we omit some unessential axioms.

The **squash** operator now takes over the second "job" which allows us to express the properties of each of the two jobs in a separate set of rules. Our rules (*SetElim*) and (*SetIntro*) are now both reversible, both perform only a singly small step of the set type and they exactly match each other. The set introduction rule now does only that — introduces the set type into the conclusion of the sequent and leaves it to the **squash** rules (such as (*Unsquash*) and (*SquashIntro*)) to manage the "hiding/unhiding the proof predicate" aspect. We believe this makes the theory more modular and easier to use.

5 Extensional Squash Operator (Esquash)

5.1 Esquash Operator: Introduction

The second concept that is needed for our formalization of quotient types is that of "hiding" the intensional structure of a certain proposition; essentially we need the concept of "being extensional" — as we will see in Section 7, even the intensional quotient type has some extensionality in it. In order to make the theory modular, we want to express the concept of the extensionality directly, not through some complex operator for which the extensionality is just a "side-effect". As we mentioned in Remark 2.2, the **squash** operator needs to be intensional, so we will need to define a new operation.

The operation we will use, called **esquash**, acts very similar to **squash** except that two "esquashed" types are equal whenever they are simultaneously non-empty or simultaneously empty. This way **esquash** completely "hides" both the witnesses of a type and its intensional structure, leaving only the information on whether a type is non-empty or not.

5.2 Esquash Operator: Axioms

First, equality — two **esquash** types are equal *iff* they are simultaneously true or simultaneously false:

$$\frac{\Gamma \vdash [\![A]\!] \Leftrightarrow [\![B]\!]}{\Gamma \vdash [\![A]\!] = [\![B]\!]} \qquad (\textit{EsquashEquality})$$

Second, **esquash** of an intensional type is equivalent to **squash** [11]:

$$\frac{\Gamma \vdash [\![A]\!] \qquad \Gamma \vdash A\,\text{Type}}{\Gamma \vdash [A]} \qquad (\textit{EsquashElim})$$

$$\frac{\Gamma \vdash [A]}{\Gamma \vdash [\![A]\!]} \qquad (\textit{EsquashIntro})$$

[11] $x : T \vdash [\![A[x]]\!]$ only requires $A[x]$ to be non-empty when $x \in T$. However since **squash** is intensional, $x : T \vdash \big[A[x]\big]$ also requires $A[x] = A[x']$ when $x = x' \in T$. Because of this we need the well–typedness condition in (*EsquashElim*).

Finally, the only member of a non-empty **esquash** type is •:

$$\frac{\Gamma;\, x : [\![A]\!];\, \Delta[\bullet] \vdash C[\bullet]}{\Gamma;\, x : [\![A]\!];\, \Delta[x] \vdash C[x]} \qquad (\mathit{EsquashMemElim})$$

Remark 5.1. We could define the **esquash** operator as

$$[\![A]\!]_i \;:=\; A = \mathrm{True} \in \left(x, y : \mathbb{U}_i /\!/ (x \Leftrightarrow y)\right) .$$

Unfortunately, this definition increases the universe level. With this definition if $A \in \mathbb{U}_i$, then $[\![A]\!]_i$ is in $\mathbb{U}_{i+1}$. This can create many difficulties, especially when we want to be able to iterate the **esquash** operator. And in any case we want to formalize quotient types using the **esquash** operator, not the other way around.

Remark 5.2. In MetaPRL J. Hickey had initially defined an **esquash** operator using the extensional quotient [12]: $[\![A]\!] \;:=\; \mathtt{tt} = \mathtt{ff} \in (x, y : \mathbb{B} \underset{e}{/\!/} (x = y \in \mathbb{B} \vee A))$ [13]. This definition does not increase the universe level like the previous one, but on the other hand it requires an extensional quotient type while the previous one works with both intensional and extensional quotients. Another problem with this definition is that almost all NuPRL-4 rules on quotient types require one to prove that the equality predicate is actually intensional, so it would be impossible to prove the properties of **esquash** from this definition using NuPRL-4 rules.

5.3 Esquash Operator: Derived Rules

First, using (*EsquashMemElim*) we can prove that any non-empty **esquash** type has an • in it:

$$\frac{\Gamma \vdash [\![A]\!]}{\Gamma \vdash \bullet \in [\![A]\!]} \qquad (\mathit{EsquashMemIntro})$$

Second, we can derive a more general and complex version of (*EsquashElim*):

$$\frac{\Gamma;\, x : [A];\, \Delta[x] \vdash B[x] \qquad \Gamma;\, x : [\![A]\!];\, \Delta[x] \vdash A\,\mathrm{Type}}{\Gamma;\, x : [\![A]\!];\, \Delta[x] \vdash B[x]} \qquad (\mathit{EsquashElim2})$$

6 Explicit Nondeterminicity

6.1 Explicit Nondeterminicity: Introduction

The idea of introducing explicit nondeterminicity first came up as a way to be able to express the elimination rules for quotient types in a more natural way, but it seems that it is also a useful tool to have on its own.

[12] See Section 7.1 for more on extensional and intensional quotient types.

[13] Where $\mathbb{B}$ is the type of booleans and tt ("true") and ff ("false") are its two members.

At first, we considered adding the nd operation similar to *amb* in [21] and to the approach used in [17]. The idea was to have $\mathsf{nd}\{t_1; t_2\}$ which can be either t_1 or t_2 nondeterministically. Then we were going to say that the expression that contains nd operators is well–formed iff its meaning does not depend on choosing which of nd's arguments to use. The problem with such an approach is that we need some way of specifying that several occurrences of the same $\mathsf{nd}\{t_1; t_2\}$ have to be considered together — either all of them would go to t_1 or all of them would go to t_2. For example, we can say that $\mathsf{nd}\{1; -1\}^2 = 1 \in \mathbb{Z}$ (which is true), but if we expand the ^{2} operator, we will get $\mathsf{nd}\{1; -1\} * \mathsf{nd}\{1; -1\} = 1 \in \mathbb{Z}$ which is only true if we require both nd's in it to expand to the same thing.

The example above suggests using some index on nd operator, which would keep track of what occurrences of nd should go together. In such a case it is natural for that index to be of the type $(\mathbb{B}//\mathrm{True})$ and as it turns out, this type represents a key idea that is worth formalizing on its own. As "usual", since we want to express the properties of the quotient types using the $\mathsf{ND} == (\mathbb{B}//\mathrm{True})$ type, it can not be defined using the quotient operator and needs to be introduced as a primitive.

The basic idea behind this ND type is that it contains two elements, say `tt` and `ff` and $\mathtt{tt} = \mathtt{ff} \in \mathsf{ND}$. In addition to these two constants we also need the `if ... then ... else ... fi` operator such that `if tt then` t_1 `else` t_2 `fi` is computationally equivalent to t_1 and `if ff then` t_1 `else` t_2 `fi` is computationally equivalent to t_2. A natural approach would be to "borrow" these constants and this operator from $\mathbb{B}$ (the type of booleans) [14], but we can create new ones, it does not matter. We will write "$\mathsf{nd}_x\{t_1; t_2\}$" as an abbreviation for "`if` x `then` t_1 `else` t_2 `fi`".

6.2 Explicit Nondeterminicity: Axioms

$$\frac{\Gamma;\ u : A[\mathtt{tt}] = A[\mathtt{ff}];\ y : A[\mathtt{tt}];\ x : \mathsf{ND};\ \Delta[x; y]\ \vdash\ C[x; y]}{\Gamma;\ x : \mathsf{ND};\ y : A[x];\ \Delta[x; y]\ \vdash\ C[x; y]} \quad (\textit{ND-elim})$$

$$\frac{\Gamma \vdash C[\mathtt{tt}] = C[\mathtt{ff}] \qquad \Gamma \vdash C[\mathtt{tt}]}{\Gamma;\ x : \mathsf{ND} \vdash C[x]} \quad (\textit{ND-elim2})$$

Notice that (*ND-elim*) does not completely eliminate the ND hypothesis, but only "moves" it one hypothesis to the right, so to completely eliminate the ND hypothesis, we will need to apply (*ND-elim*) repeatedly and then apply (*ND-elim2*) in the end.

For the purpose of formalizing the quotient operators we only need the two rules above. A complete formalization of ND would also include the axiom

$$\frac{}{\vdash \mathtt{tt} = \mathtt{ff} \in \mathsf{ND}} \quad (\textit{ND-intro})$$

[14] This would mean that ND is just $\mathbb{B}//\mathrm{True}$.

6.3 Explicit Nondeterminicity: Derived Rule

$$\frac{\Gamma \vdash t[\mathtt{tt}] = t[\mathtt{ff}] \in A}{\Gamma;\, x : \mathsf{ND} \vdash t[x] \in A} \qquad \textit{(ND-memb)}$$

7 Intensional Quotient Type

7.1 Quotient Type: Introduction

The quotient types were originally introduced in [9]. They are also presented in [27].

While extensional quotient type can be useful sometimes, usually the intensional quotient type is sufficient and the extensionality just unnecessary complicates proofs by requiring us to prove extra well–typedness statements. In addition to that NuPRL formalization of quotient types (see [9] and [22, Appendix A]) does not allow one to take full advantage of extensionality since most of the rules for the quotient type have an assumption that the equality predicate is in fact intensional. While early versions of NuPRL type theory considered extensional set and quotient types, these problems forced the change of set constructor (which is used substantially more often than the quotient) into an intensional one.

In order to avoid the problems outlined above, in this paper we introduce the intensional quotient type as primitive, and we concentrate our discussion of quotient types on intensional quotient types. But since we have the **esquash** operator in our theory, an extensional quotient type can be naturally defined if needed, using $A \underset{e}{/\!/} E \;:=\; A \underset{i}{/\!/} [\![E]\!]$ and an extensional set type can be defined the same way: $\{x : A \mid_e P[x]\} \;:=\; \{x : A \mid_i [\![P[x]]\!]\}$.

7.2 Intensional Quotient Type: Axioms

Two intensional quotient types are equal when both the quotiented types are equal, and the equality relations are equal:

$$\frac{\Gamma \vdash A = A' \qquad \Gamma;\, x : A;\, y : A \vdash E[x;y] = E'[x;y] \qquad \text{“}E \text{ is an ER over } A\text{”}}{\Gamma \vdash (A/\!/E) = (A'/\!/E')}$$

(IquotEqualIntro)

where “E is an ER over A” is just an abbreviation for conditions that force E to be an equivalence relation over A.

Next, when two elements are equal in a quotient type, the equality predicate must be true on those elements. However, we know neither the witnesses of this predicate nor its intensional structure, therefore the equality in a quotient type only implies the **esquash** of the equality predicate:

$$\frac{\Gamma;\, u : x = y \in (A/\!/E);\, v : [\![E[x;y]]\!];\, \Delta[u] \vdash C[u]}{\Gamma;\, u : x = y \in (A/\!/E);\, \Delta[u] \vdash C[u]} \qquad \textit{(IquotEqualElim)}$$

The opposite is also true — we only need to prove the **esquash** of the equality predicate to be able to conclude that corresponding elements are equal in the quotient type:

$$\frac{\Gamma \vdash [\![E[x;y]]\!] \qquad \Gamma \vdash x \in (A//E) \qquad \Gamma \vdash y \in (A//E)}{\Gamma \vdash x = y \in (A//E)} \qquad (\textit{IquotMemEqual})$$

Note that this rule has equality[15] in the quotient type in both the conclusion and the assumptions, so we still need a "base case" — an element of a type base type will also be an element of any well-typed quotient of that type:

$$\frac{\Gamma \vdash x \in A \qquad \Gamma \vdash (A//E)\,\mathrm{Type}}{\Gamma \vdash x \in (A//E)} \qquad (\textit{IquotMemIntro})$$

Finally, we need to provide an elimination rule for quotient types. It turns out that being functional over some equivalence class of a quotient type is the same as being functional over an ND of any two elements of such class, so we can formulate the elimination rule as follows:

$$\frac{\Gamma;\, u_1 : A;\, u_2 : A;\, v : E[u_1;u_2];\, x : \mathsf{ND};\, \Delta[\mathsf{nd}_x\{u_1;u_2\}] \vdash [C[\mathsf{nd}_x\{u_1;u_2\}]]}{\Gamma;\, u : A//E;\, \Delta[u] \vdash [C[u]]} \qquad (\textit{IquotElim})$$

7.3 Intensional Quotient Type: Derived Rules

From (*IquotElim*) and (*SquashEqual*) we can derive

$$\frac{\Gamma;\, u_1 : A;\, u_2 : A;\, v : E[u_1;u_2];\, x : \mathsf{ND};\, \Delta[\mathsf{nd}_x\{u_1;u_2\}] \vdash t[u_1] = t[u_2] \in C}{\Gamma;\, u : A//E;\, \Delta[u] \vdash t[u] \in C} \qquad (\textit{IquotElim2})$$

From (*IquotEqualElim*) and (*EsquashElim2*), we can derive

$$\frac{\begin{array}{l}\Gamma;\, u : x = y \in (A//E);\, v : [E[x;y]];\, \Delta[u] \vdash C[u] \\ \Gamma;\, u : x = y \in (A//E);\, \Delta[u] \vdash E[x;y]\,\mathrm{Type}\end{array}}{\Gamma;\, u : x = y \in (A//E);\, \Delta[u] \vdash C[u]} \qquad (\textit{IquotEqualElim2})$$

which is equivalent to NuPRL's (*quotient_equalityElimination*). However, (*IquotEqualElim*) is more general than (*quotient_equalityElimination*).

Example 7.1. We now can prove things like $x : \mathsf{ND};\, y : \mathsf{ND} \vdash \mathsf{nd}_x\{2;4\} = \mathsf{nd}_y\{4;6\} \in \mathbb{Z}_2$ where $\mathbb{Z}_2$ is $\mathbb{Z}$ quotiented over a "`mod 2`" equivalence relation.

[15] As we explained in Remark 1.1, in Martin-Löf type theory membership is just a particular case of the equality.

8 Indexed Collections

8.1 Indexed and Predicated Collections

Consider an arbitrary type T in universe $\mathbb{U}$. We want to define the type of collections of elements of T. Such a type turned out to be very useful for various verification tasks as a natural way of representing sets of objects of a certain type (states, transitions, *etc*). We also want to formalize collections in the most general way possible, without assuming anything about T. In particular, we do not want to assume that T is enumerable or that equality on T is decidable. And in fact, the constant problems we were facing when trying to formalize collections properly were the main reason for the research that lead to this paper.

There are at least two different approaches we can take to start formalizing such collections.

1. We can start formalizing collections as pairs consisting of an index set $I : \mathbb{U}$ and an index function $f : (I \to T)$. In other words, we can start with the type $I : \mathbb{U} \times (I \to T)$.
2. We can start formalizing collections by concentrating on membership predicates of collections. In other words, we can start with the type $T \to \mathbf{Prop}$.

It is easy to see that these two approaches are equivalent. Indeed, if we have a pair $\langle I, f \rangle$, we can get a predicate $\lambda t.\, \exists i \in I.\, f(i) = t \in T$ and if we have a predicate P, we can take a pair $\langle \{t : T \mid P(t)\}, \lambda t.\, t \rangle$. Because of this isomorphism, everywhere below we will allow ourselves to use $T \to \mathbf{Prop}$ as a base for the collection type even though $I : \mathbb{U} \times (I \to T)$ is a little closer to our intuition about collections.

Clearly, the type $T \to \mathbf{Prop}$ is not quite what we want yet since two different predicates from that type can represent the same collection. An obvious way of addressing this problem is to use a quotient type. In other words, we want to define the type of collections as

$$\mathsf{Col}(T) := c_1, c_2 : (T \to \mathbb{U}) /\!/ (\forall t \in T.\, c_1(t) \Leftrightarrow c_2(t)). \tag{1}$$

8.2 Collections: The Problem

Once we have defined the type of collections, the next natural step is to start defining various basic operations on that type. In particular, we want to have the following operations:

- A predicate telling us whether some element is a member of some collection: $\forall c \in \mathsf{Col}(T).\, \forall t \in T.\, \mathtt{mem}(c; t) \in \mathbf{Prop}$
- An operator that would produce a union of a family of collections: $\forall I \in \mathbb{U}.\, \forall C \in (I \to \mathsf{Col}(T)).\, \bigcup_{i \in I} C(i) \in \mathsf{Col}(T)$

And we want our operators to have the following natural properties:

- $\forall c \in T \to \mathbb{U}. \forall t \in T. c(t) \Rightarrow \mathtt{mem}(c; t)$
- $\forall c \in T \to \mathbb{U}. \forall t \in T. \neg c(t) \Rightarrow \neg\mathtt{mem}(c; t)$
- $\forall c_1, c_2 \in \mathsf{Col}(T). \Big(\big(\forall t : T. \left(\mathtt{mem}(c_1; t) \Leftrightarrow \mathtt{mem}(c_2; t)\right)\big) \Leftrightarrow c_1 = c_2 \in \mathsf{Col}(T) \Big)$
  (note that the $\Leftarrow$ direction follows from the typing requirement for `mem`).
- $\forall I \in \mathbb{U}. \forall C : I \to \mathsf{Col}(T). \forall t \in T. \left(\exists i : I. \mathtt{mem}(C(i); t) \Rightarrow \mathtt{mem}(\bigcup_{i \in I} C(i); t)\right)$
  Note that we do not require an implication in the opposite direction since that would mean that we will have to be able to reconstruct i constructively just from some indirect knowledge that it exists. Instead we only require
- $\forall I \in \mathbb{U}, C \in I \to \mathsf{Col}(T), t \in T. \Big(\neg\big(\exists i : I. \mathtt{mem}(C(i); t)\big) \Rightarrow \neg\big(\mathtt{mem}(\bigcup_{i \in I} C(i); t)\big)\Big)$

It turned out that formulating these operations with these properties is very difficult [16] in NuPRL-4 type theory with its monolithic approach to quotient types. The problems we were constantly experiencing when trying to come up with a solution included `mem` erroneously returning an element of $\mathbf{Prop}// \Leftrightarrow$ instead of $\mathbf{Prop}$, `union` being able to accept only arguments of type

$$C_1, C_2 : (I \to (T \to \mathbf{Prop}))//(\forall i \in I. \forall t \in T\, C_1(i; t) \Leftrightarrow C_2(i; t))$$

(which is a subtype of the $I \to \mathsf{Col}(T)$ type that we are interested in), *etc.*

8.3 Collections: A Possible Solution

Now that we have $[\,]$ and $[\![\,]\!]$ operators, it is relatively easy to give the proper definitions. If we take $\mathtt{mem}(c; t) := [\![c(t)]\!]$ and $\bigcup_{i \in I} C(i) := \lambda t. \exists i \in I. \mathtt{mem}(C(i); t)$, we can prove all the properties listed in Section 8.2. These proofs were successfully carried out in the MetaPRL proof development system [12,14,13].

9 Related Work

Semantically speaking, in this paper we formalize exactly the same quotient types as NuPRL does. However the formalization presented here strictly subsumes the NuPRL's one. All the NuPRL rules can be derived from the rules presented in this paper. Consequently, in a system that supports a derived rules mechanism, any proof that uses the original NuPRL axiomatization for quotient types would still be valid under our modular axiomatization. For a more detailed comparison of the two axiomatizations see [22, Section 7.4].

In [10] Pierre Courtieu attempts to add to Coq's Calculus of Constructions a notion very similar to quotient type. Instead of aiming at "general" quotient

[16] Several members of NuPRL community made numerous attempts to come up with a satisfactory formalization. The formalization presented in this paper was the only one that worked.

type, [10] considers types that have a "normalization" function that, essentially, maps all the members of each equivalence class to a canonical member of the class. Courtieu shows how by equipping a quotient type with such normalization function, one can substantially simplify handling of such a quotient type. In a sense, **esquash** works the same way — it acts as a normalization function for the **Prop**//$\Leftrightarrow$. The main difference here is that instead of considering a normalization function that returns *an existing* element of each equivalence class, with **esquash** we utilize the open–ended nature of the type theory to equip each equivalence class with *a new* normal element.

In [15,16] Martin Hofmann have studied the intensional models for quotient types in great detail. This paper is in a way complimentary to such studies — here we assume that we already have some appropriate semantical foundation that allows quotient types and try to come up with an axiomatization that would be most useful in an automated proof assistant.

Acknowledgments

This work would not be possible without Mark Bickford, who came up with the problem of formalizing collections. This work was also a result of very productive discussions with Robert Constable, Alexei Kopylov, Mark Bickford, and Jason Hickey. The author is also very grateful to Vladimir Krupski who was the first to notice the modularity consequences of this work. The text of this paper would not be nearly as good as it currently is without the very useful critique of Sergei Artemov, Robert Constable, Stuart Allen, Alexei Kopylov, and anonymous reviewers.

References

1. Stuart F. Allen. *A Non-Type-Theoretic Semantics for Type-Theoretic Language.* PhD thesis, Cornell University, 1987.
2. Stuart F. Allen. A Non-type-theoretic Definition of Martin-Löf's Types. In *Proceedings of the Second Symposium on Logic in Computer Science*, pages 215–224. IEEE, June 1987.
3. Roland Backhouse. A note of subtypes in Martin-Löf's theory of types. Technical Report CSM-90, University of Essex, November 1984.
4. Roland Backhouse. On the meaning and construction of the rules in Martin-Löf's theory of types. In A. Avron, editor, *Workshop on General Logic, Edinburgh, February 1987*, number ECS-LFCS-88-52. Department of Computer Science, University of Edinburgh, May 1988.
5. Roland C. Backhouse, Paul Chisholm, Grant Malcolm, and Erik Saaman. Do-it-yourself type theory. *Formal Aspects of Computing*, 1:19–84, 1989.
6. Ken Birman, Robert Constable, Mark Hayden, Jason J. Hickey, Christoph Kreitz, Robbert van Renesse, Ohad Rodeh, and Werner Vogels. The Horus and Ensemble projects: Accomplishments and limitations. In *DARPA Information Survivability Conference and Exposition (DISCEX 2000)*, pages 149–161. IEEE, 2000.

7. Robert L. Constable. Mathematics as programming. In *Proceedings of the Workshop on Programming and Logics, Lectures Notes in Computer Science 164*, pages 116–128. Springer-Verlag, 1983.
8. Robert L. Constable. Types in logic, mathematics, and programming. In S. R. Buss, editor, *Handbook of Proof Theory*, chapter X, pages 683–786. Elsevier Science B.V., 1998.
9. Robert L. Constable, Stuart F. Allen, H.M. Bromley, W.R. Cleaveland, J.F. Cremer, R.W. Harper, Douglas J. Howe, T.B. Knoblock, N.P. Mendler, P. Panangaden, James T. Sasaki, and Scott F. Smith. *Implementing Mathematics with the NuPRL Development System*. Prentice-Hall, NJ, 1986.
10. Pierre Courtieu. Normalized types. In L. Fribourg, editor, *Computer Science Logic, Proceedings of the 10th Annual Conference of the EACSL*, volume 2142 of *Lecture Notes in Computer Science*, pages 554–569. Springer-Verlag, 2001. `http://link.springer-ny.com/link/service/series/0558/tocs/t2142.htm`.
11. Jason J. Hickey. NuPRL-Light: An implementation framework for higer-order logics. In William McCune, editor, *Proceedings of the 14th International Conference on Automated Deduction*, volume 1249 of Lecture Notes on Artificial Intelligence, pages 395–399, Berlin, July 13–17 1997. Springer. CADE'97. An extended version of the paper can be found at `http://www.cs.caltech.edu/~jyh/papers/cade14_nl/default.html`.
12. Jason J. Hickey. *The MetaPRL Logical Programming Environment*. PhD thesis, Cornell University, Ithaca, NY, January 2001.
13. Jason J. Hickey, Brian Aydemir, Yegor Bryukhov, Alexei Kopylov, Aleksey Nogin, and Xin Yu. A listing of MetaPRL theories. `http://metaprl.org/theories.pdf`.
14. Jason J. Hickey, Aleksey Nogin, Alexei Kopylov, et al. MetaPRL home page. `http://metaprl.org/`.
15. Martin Hofmann. *Extensional concepts in intensional Type theory*. PhD thesis, University of Edinburgh, Laboratory for Foundations of Computer Science, July 1995.
16. Martin Hofmann. A simple model for quotient types. In *Typed Lambda Calculus and Applications*, volume 902 of *Lecture Notes in Computer Science*, pages 216–234, 1995.
17. Douglas J. Howe. Semantic foundations for embedding HOL in NuPRL. In Martin Wirsing and Maurice Nivat, editors, *Algebraic Methodology and Software Technology*, volume 1101 of Lecture Notes in Computer Science, pages 85–101. Springer-Verlag, Berlin, 1996.
18. Alexei Kopylov and Aleksey Nogin. Markov's principle for propositional type theory. In L. Fribourg, editor, *Computer Science Logic, Proceedings of the 10th Annual Conference of the EACSL*, volume 2142 of *Lecture Notes in Computer Science*, pages 570–584. Springer-Verlag, 2001. `http://link.springer-ny.com/link/service/series/0558/tocs/t2142.htm`.
19. Xiaoming Liu, Christoph Kreitz, Robbert van Renesse, Jason J. Hickey, Mark Hayden, Kenneth Birman, and Robert Constable. Building reliable, high-performance communication systems from components. In *17th ACM Symposium on Operating Systems Principles*, December 1999.
20. Per Martin-Löf. Constructive mathematics and computer programming. In *Proceedings of the Sixth International Congress for Logic, Methodology, and Philosophy of Science*, pages 153–175, Amsterdam, 1982. North Holland.
21. J. McCarthy. A basis for a mathematical theory of computation. In P. Braffort and D. Hirschberg, editors, *Computer Programming and Formal Systems*, pages 33–70. Amsterdam:North-Holland, 1963.

22. Aleksey Nogin. Quotient types — A modular approach. Department of Computer Science `http://cs-tr.cs.cornell.edu/Dienst/UI/1.0/Display/ncstrl.cornell/TR2002-1869` TR2002-1869, Cornell University, April 2002. See also `http://nogin.org/papers/quotients.html`.
23. Aleksey Nogin and Jason Hickey. Sequent schema for derived rules. Accepted to TPHOLs 2002, 2002.
24. Bengt Nordström and Kent Petersson. Types and specifications. In *IFIP'93*. Elsvier, 1983.
25. Bengt Nordström, Kent Petersson, and Jan M. Smith. *Programming in Martin-Löf's Type Theory*. Oxford Sciences Publication, Oxford, 1990.
26. Frank Pfenning and Rowan Davies. Judgmental reconstruction of modal logic. *Mathematical Structures in Computer Science*, 11(4), August 2001.
27. Simon Thompson. *Type Theory and Functional Programming*. Addison-Wesley, 1991.
28. Anne Sjerp Troelstra. *Metamathematical Investigation of Intuitionistic Mathematics*, volume 344 of *Lecture Notes in Mathematics*. Springer-Verlag, 1973.

Sequent Schema for Derived Rules

Aleksey Nogin[1]* and Jason Hickey[2]

[1] Department of Computer Science
Cornell University, Ithaca, NY 14853
nogin@cs.cornell.edu
http://nogin.org/
[2] Department of Computer Science
California Institute of Technology 256-80
Pasadena, CA 91125
jyh@cs.caltech.edu
http://www.cs.caltech.edu/~jyh

Abstract. This paper presents a general *sequent schema* language that can be used for specifying sequent–style rules for a logical theory. We show how by adding the *sequent schema* language to a theory we gain an ability to prove new inference rules within the theory itself. We show that the extension of any such theory with our *sequent schema* language and with any new rules found using this mechanism is *conservative*.
By using the *sequent schema* language in a theorem prover, one gets an ability to allow users to derive new rules and then use such derived rules as if they were primitive axioms. The conservativity result guarantees the validity of this approach. This property makes it a convenient tool for implementing a derived rules mechanism in theorem provers, especially considering that the application of the rules expressed in the *sequent schema* language can be efficiently implemented using MetaPRL's fast rewriting engine.

1 Introduction

In an interactive theorem prover it is very useful to have a mechanism to allow users to prove some statement in advance and then reuse the derivation in further proofs. Often it is especially useful to be able to *abstract away* the particular derivation. For example, suppose we wish to formalize a data structure for labeled binary trees. If binary trees are not primitive to the system, we might implement them in several ways, but the details are irrelevant. The more important feature is the inference rule for induction. In a sequent logic, the induction principle would be similar to the following.

$$\frac{\Gamma \vdash P(leaf) \qquad \Gamma, a\colon btree, P(a), b\colon btree, P(b) \vdash P(node(a,b))}{\Gamma, x\colon btree \vdash P(x)}$$

* This work was partially supported by ONR grant N00014-01-1-0765 and AFRL grants F49620-00-1-0209 and F30602-98-2-0198.

V.A. Carreño, C. Muñoz, S. Tahar (Eds.): TPHOLs 2002, LNCS 2410, pp. 281–297, 2002.

If this rule can be established, further proofs may use it to reason about binary trees *abstractly* without having to unfold the btree definition. This leaves the user free to replace or augment the implementation of binary trees as long as she can still prove the same induction principle for the new implementation. Furthermore, in predicative logics, or in cases where well-formedness is defined logically, the inference rule is strictly more powerful than its propositional form.

If a mechanism for establishing a derived rule is not available, one alternative is to construct a proof "script" or tactic that can be reapplied whenever a derivation is needed. There are several problems with this. First, it is inefficient — instead of applying the derived rule in a single step, the system has to run through the whole proof each time. Second, the proof script would have to unfold the btree definition, exposing implementation detail. Third, proof scripts tend to be fragile, and must be reconstructed frequently as a system evolves. Finally, by looking at a proof script or a tactic code, it may be hard to see what exactly it does, while a derived rule is essentially self-documenting.

This suggests a need for a *derived rules mechanism* that would allow users to derive new inference rules in a system and then use them as if they were *primitive rules* (i.e. axioms) of the system. Ideally, such mechanism would be general enough not to depend on a particular logical theory being used. Besides being a great abstraction mechanism, derived rules facilitate the proof development by making proofs and partial proofs easily reusable. Also, a derived rule contains some information on how it is supposed to be used [1] and such information can be made available to the system itself. This can be used to substantially reduce the amount of information a user has to provide in order for the system to know how to use such a new rule in the proof automation procedures.

In this paper we propose a purely syntactical way of dealing with derived rules. The key idea of our approach is in using a special higher–order language for specifying rules; we call it a *sequent schema* language. From a theoretical point of view, we take some logical theory and express its rules using sequent schema. Next we add the same language of sequent schema to the theory itself. After that we allow extending our theory with a new rule $\frac{S_1 \quad \cdots \quad S_n}{S}$ whenever we can prove S from S_i in the expanded theory (we will call such a rule a *derived rule*). We show that no matter what theory we started with, such a double–extended theory is always a conservative extension of the original one, so our maneuver is always valid.

In case of a theorem prover (such as MetaPRL [4,6], which implements this approach), the user only has to provide the axioms of the base theory in a sequent schema language and the rest happens automatically. The system immediately allows the user to mix the object language of a theory with the sequent schema meta-language. Whenever a derived rule is proven in a system, it allows using that rule in further proofs as if it were a basic axiom of the theory[2]. This paper

[1] For example, an implication $A \Rightarrow B$ can be stated and proved as an A elimination rule or as a B introduction rule, depending on how we expect it to be used.

[2] A system might also allow use in the reverse order — first state a derived rule, use it, and later "come back" and prove the derived rule. Of course in such system a

shows that such theorem prover would not allow one to derive anything that is not derivable in a conservative extension of the original theory.

For the sake of brevity and clarity this paper focuses on a simple version of sequent schema (which is still sufficient for formalizing Martin–Löf style type theories, including the NuPRL one [1]). In Section 7 we present several ways of extending the sequent schema language; in particular we explain how to represent constants in the sequent schema framework.

This paper is arranged as follows — first we will describe the syntax of sequent schema (Section 3). Then in Section 4 we explain what a sequent scheme stands for, that is when a *plain* sequent matches some sequent scheme. In Section 5 we show how rules are specified using sequent schema and what it means for the rule to be admissible for sequent schema rather than just being admissible for plain sequents. In Section 6 we argue what properties of the sequent schema can guarantee admissibility of the rules derived using sequent schema. Finally, in Section 8 we discuss related work and compare our derived rules approach to approaches taken by several other theorem provers.

A variant of the sequent schema language is implemented in MetaPRL proof assistant [4,6] using its fast rewriting engine [5]. In Appendices A through C we discuss some aspects of that implementation. In Appendix A we show how we can simplify the syntax by omitting some redundant information. In Appendix B we describe the way MetaPRL sequent schema are used to specify not only the set of valid instances of a rule, but also a way to point at a particular instance of that rule. Finally, in Appendix C we present the MetaPRL concrete syntax for the sequent schema language.

Throughout the paper (except for Appendix C) we will ignore the issues of concrete syntax and will assume we are only dealing with abstract syntax.

2 Introduction to Sequent Schema

The sequent schema language resembles higher–order abstract syntax presented in Pfenning and Elliott [12]. The idea of sequent schema is to use higher–order context variables to describe sequent and other contexts as well as to use second–order variables in the style of Huet and Lang [7] to describe sequent and rule schema. In all rule specifications, these meta-variables will be implicitly universally quantified.

As we mention in the introduction, in this paper we present a simple version of our sequent schema language. It assumes that all syntactically well–formed sentences of the theory we are extending have the form

$$x_1 : A_1; x_2 : A_2; \cdots; x_n : A_n \vdash C \tag{2.I}$$

where

proof would not be considered complete until all the derived rules used in it are also proven. Such an approach allows one to "test–drive" a derived rule before investing time into establishing its admissibility.

(A) Each A_i and C is a *term*. A term is constructed from variables and a set of operators (including 0-arity operators — constants). Each operator may potentially introduce binding occurrences (for example λ would be a unary operator that introduces one binding occurrence).
(B) Each hypothesis A_i introduces a binding occurrence for variable x_i. In other words, x_i is bound in hypotheses A_j $(j > i)$ and the conclusion C.
(C) All sequents are closed. In other words, each free variable of C must be one of x_i $(i = 1, \ldots, n)$ and each free variable of A_j must be one of x_i $(i < j)$.

Example 2.1. $x : \mathbb{Z} \vdash x \geq 0$ and $x : \mathbb{Z}; y : \mathbb{Z} \vdash x > y$ and $\vdash 5 > 4$ are syntactically well-formed (but not necessarily true) sequents in a theory that includes 0-ary operator (constant) $\mathbb{Z}$ and binary operators $>$ and $\geq$.

Example 2.2. $x : \mathbb{Z}; y : \mathbb{Z} \vdash x = y \in \mathbb{Z}$ and $x : \mathbb{Z} \vdash \exists y : \mathbb{Z}.(x = y \in \mathbb{Z})$ are well-formed (but not necessarily true) sequents that include a 0-ary operator $\mathbb{Z}$, a ternary operator $\cdot = \cdot \in \cdot$ and a binary operator $\exists v \in \cdot . \cdot$ that introduces a binding occurrence in its second argument.

Remark 2.3. We assume that the language of the theory under consideration contains at least one closed term [3]; we will use $\bullet$ when we need to refer to an arbitrary closed term.

As in LF [3] we assume that object level variables are mapped to meta–theory variables (denoted by x, y, z). We will call these meta–theory variables *first–order*, but of course they may have a higher–order meaning in the object theory. Similarly we assume that the object–level binding structure is mapped to the meta–level binding structure and we will introduce a separate context binding structure in the sequent schema language. We also consider alpha–equal terms (including sequents and sequent schema) to be identical and we assume that substitution avoids capture by renaming.

3 Language of Sequent Schema

We assume there exist countable sets of context variables $\mathcal{C}$ and second order variables $\mathcal{V}$, disjoint from each other and from ordinary object–level variables. We will assume an arity function $\alpha : (\mathcal{C} \cup \mathcal{V}) \to \mathbb{N}$ and a function $\beta : (\mathcal{C} \cup \mathcal{V}) \to$ {finite subsets of $\mathcal{C}$} that determines the contexts that a variable may depend on. We will assume that for each value of α and β there are infinitely many variables in each of the two classes.

Remark 3.1. The language of sequent schema is essentially untyped, although we could describe all the terms free of higher–order variables as having the type `term`, the second–order variables as having the type $\mathtt{term}^n \to \mathtt{term}$ and the context variables as having the type $(\mathtt{term}^m \to \mathtt{term}) \to \mathtt{term}^n \to \mathtt{term}$.

[3] This is not normally a limitation because the λ operator or an arbitrary constant can normally be used to construct a closed term.

In this paper we will use abstract syntax to present the language of sequent schema[4]. We will denote context variables by $\mathbf{C}$, $\mathbf{H}$, $\mathbf{J}$ and second–order variables by A, B, C and V. For clarity we will write the value of β as a subscript for all the variables (although it can usually be deduced from context and in the MetaPRL system we rarely need to specify it explicitly — see Appendix A).

The language of sequent schema is outlined in the following table:

Syntax	**Intended Meaning**
$S ::= \vdash T \mid Hs;\ S.$ In other words $S ::= Hs_1; \cdots; Hs_n \vdash T\ (n \geq 0)$	**Sequent Scheme.** *Hs*'s specify the *hypotheses* of a sequent and T specifies its *conclusion*.
$Hs ::= \mathbf{C}_{\{\mathbf{C}_1;\cdots;\mathbf{C}_k\}}[T_1; \cdots; T_n] \mid x : T$ where $\mathbf{C}, \mathbf{C}_i \in \mathcal{C}$, $\alpha(\mathbf{C}) = n$, T and T_i are terms and x is an object–level variable.	**Hypotheses Specification.** The first variant is used to specify a sequent context — a sequence (possibly empty) of hypotheses. In general, a context may depend on some arguments and T_i specify the values of those arguments. $\mathbf{C}_i$ are the contexts that introduce the variables that are allowed to occur free in $\mathbf{C}$ itself (not in its arguments). The second variant specifies a single hypothesis that introduces a variable x. As in (2.I) this is a binding occurrence and x becomes bound in all the hypotheses specifications following this one, as well as in the conclusion.
$SOV ::= V_{\{\mathbf{C}_1;\cdots;\mathbf{C}_k\}}[T_1; \cdots; T_n]$ where $V \in \mathcal{V}$, $\alpha(V) = n$, $\mathbf{C}_i \in \mathcal{C}$, T and T_i are terms.	**Second-Order Variable Occurrences.** Second–order variables in sequent schema language are the ordinary second–order variables as in [7] except that we need $\beta(V)$ to be able to specify the names of contexts which introduced the variables that can occur free in V.
T — a term built using operators from variables and second–order variables.	**Term Specification.** Term specifications are ordinary terms except that they may contain second–order variables.

Remark 3.2. It is worth mentioning that a *plain* sequent (as described in (2.I)) is just a particular case of a sequent scheme. Namely, a plain sequent is a sequent scheme that contains neither sequent contexts, nor second–order variables.

Example 3.3.

$$\mathbf{H}_{\{\}}[];\ x : A_{\{\mathbf{H}\}}[];\ \mathbf{J}[x]_{\{\mathbf{H}\}} \vdash C_{\{\mathbf{H};\mathbf{J}\}}[x^2]$$

is a sequent scheme in a language that contains the $\cdot^2$ unary operator. In this scheme $\mathbf{H}$ and $\mathbf{J}$ are context variable, x is an ordinary (first–order) variable, A

[4] For concrete syntax used in MetaPRL system, see Appendix C.

and C are second–order variables. This scheme matches many plain sequents, including the following ones:

$$x : \mathbb{Z} \vdash (x^2) \in \mathbb{Z}$$
$$y : \mathbb{Z};\ x : \mathbb{Z};\ z : \mathbb{Z} \vdash (z + y) \in \mathbb{Z}$$

Here is how the same scheme would look in a simplified syntax of Appendix A:

$$\mathbf{H};\ x : A;\ \mathbf{J}[x] \vdash C[x^2]$$

4 Semantics — Sequent Schema

Informally speaking, the main idea of sequent schema is that whenever a binding occurrence of some variable is explicitly mentioned (either *per se* or as a context), it can only occur freely in places where it is explicitly mentioned, but can not occur freely where it is omitted. Second–order variable are meant to stand for contexts that can not introduce new bindings and context variable are meant to stand for contexts that may introduce an arbitrary number of new bindings (with β being used to restrict in which subterms those new bound variables may potentially occur freely).

To make the above more formal, we first need to define what we mean by free and bound variables in sequent schema.

Definition 4.1 (Free Variables). *The notions of free variables (or, more specifically, free first–order variables)* **FOV** *and free context variables* **CV** *are defined as follows:*

- *Free variables of plain terms (free of higher–order variables) are defined as usual. Plain terms do not have free context variables.*
- *For* $\Upsilon \in (\mathcal{C} \cup \mathcal{V})$, $\mathbf{FOV}(\Upsilon_{\{\cdots\}}[T_1; \cdots ; T_n]) = \bigcup\limits_{1 \le i \le n} \mathbf{FOV}(T_i)$.
- $\mathbf{CV}(\Upsilon_{\{\mathbf{C}_1;\cdots;\mathbf{C}_k\}}[T_1; \cdots ; T_n]) = \{\mathbf{C}_1; \cdots ; \mathbf{C}_n\} \cup \bigcup\limits_{1 \le i \le n} \mathbf{CV}(T_i)$.
- $\mathbf{FOV}(\vdash T) = \mathbf{FOV}(T)$ *and* $\mathbf{CV}(\vdash T) = \mathbf{CV}(T)$
- *In a sequent scheme* $x : T;\ S$*, the hypothesis specification* $x : T$ *binds all free occurrences of* x *in* S*. Hence,*

$$\begin{aligned}\mathbf{FOV}(x : T;\ S) &= (\mathbf{FOV}(S) - \{x\}) \cup \mathbf{FOV}(T)\\ \mathbf{CV}(x : T;\ S) &= \mathbf{CV}(S) \cup \mathbf{CV}(T).\end{aligned}$$

- *In a sequent scheme* $\mathbf{C}_{\{\mathbf{C}_1;\cdots;\mathbf{C}_k\}}[T_1; \cdots ; T_n];\ S$ *the hypothesis specification* $\mathbf{C}_{\{\cdots\}}[\cdots]$ *binds the free occurrences of* $\mathbf{C}$ *in* S*. Hence,*

$$\mathbf{CV}(\mathbf{C}_{\{\mathbf{C}_1;\cdots;\mathbf{C}_k\}}[T_1; \cdots ; T_n];\ S) = \{\mathbf{C}_1; \cdots ; \mathbf{C}_k\} \cup (\mathbf{CV}(S) - \{\mathbf{C}\}) \cup \bigcup_{1 \le i \le n} \mathbf{CV}(T)$$

$$\mathbf{FOV}(\mathbf{C}_{\{\mathbf{C}_1;\cdots;\mathbf{C}_k\}}[T_1; \cdots ; T_n];\ S) = \mathbf{FOV}(C) \cup \mathbf{FOV}(S) \cup \bigcup_{1 \le i \le n} \mathbf{CV}(T).$$

– *For all objects* $\star$ *such that* $\mathbf{FOV}(\star)$ *and* $\mathbf{CV}(\star)$ *are defined (see also Definitions 4.3 and 4.4 below) we will denote* $\mathbf{FOV}(\star) \cup \mathbf{CV}(\star)$ *as* $\mathbf{Vars}(\star)$.

Definition 4.2. *We will call a sequent scheme* S closed *when* $\mathbf{Vars}(S) = \emptyset$.

Definition 4.3. *A substitution function is a pair of a term and a list of first–order variables. For a substitution function* $\sigma = \langle T; x_1, \cdots, x_n \rangle$ $(n \geq 0)$ *we will say that the arity of* σ *is* n*; the free occurrences of* x_i *in* T *are bound in* σ*, so* $\mathbf{FOV}(\sigma) = \mathbf{FOV}(T) - \{x_1; \cdots; x_n\}$ *and* $\mathbf{CV}(\sigma) = \mathbf{CV}(T)$*. By* $\sigma(T_1, \cdots, T_n)$ *we will denote* $[T_1/x_1, \cdots, T_n/x_n]T$ *— the result of simultaneous substitution of* T_i *for* x_i $(1 \leq i \leq n)$ *in* T*. As usual, we will consider two alpha–equal substitutions to be identical.*

We will say that the substitution function is trivial, when T *is a closed term.*

Definition 4.4. *Similarly, a context substitution function is a pair of a list of hypothesis specification and a list of first–order variables. For a context substitution function* $\Sigma = \langle Hs_1, \cdots, Hs_k; x_1, \cdots, x_n \rangle$ $(k, n \geq 0)$ *we will say that the arity of* Σ *is* n, $\mathbf{FOV}(\Sigma) = \mathbf{FOV}(Hs_1; \cdots; Hs_k \vdash \bullet) - \{x_1; \cdots; x_n\}$ *and* $\mathbf{CV}(\Sigma) = \mathbf{CV}(Hs_1; \cdots; Hs_k \vdash \bullet)$*, where* $\bullet$ *is an arbitrary closed term (see Remark 2.3). We will say that* Σ *introduces a set of bindings*

$$\mathbf{BV}(\Sigma) = \bigcup_{1 \leq i \leq k} \begin{cases} \{x\}, & \text{when } Hs_i \text{ is } x : T \\ \{\mathbf{C}\}, & \text{when } Hs_i \text{ is } \mathbf{C}_{\{\cdots\}}[\cdots] \end{cases}$$

We will say that a context substitution function is trivial when $k = 0$.

Definition 4.5. *A function* R *on* $\mathcal{C} \cup \mathcal{V}$ *is a scheme* refinement function *if it "respects"* α *and* β*. Namely, the following conditions must hold:*

(A) *For all* $V \in \mathcal{V}$, $R(V)$ *is a substitution function of arity* $\alpha(V)$.
(B) *For all* $\mathbf{C} \in \mathcal{C}$, $R(\mathbf{C})$ *is context substitution function of arity* $\alpha(\mathbf{C})$.
(C) *For all* $\Upsilon \in (\mathcal{C} \cup \mathcal{V})$, $\mathbf{Vars}(R(\Upsilon)) \subseteq \mathbf{BV}(R(\beta(\Upsilon)))$ *where* $\mathbf{BV}(R(\beta(\Upsilon)))$ *is a notation for* $\bigcup_{\mathbf{C} \in \beta(\Upsilon)} \mathbf{BV}(R(\mathbf{C}))$.

Definition 4.6. *If* R *is a refinement function, we will define* R *on terms, hypothesis specifications, sequent schema and substitution functions as follows:*

(A) $R(T) = T$*, when* T *is a plain term.*
(B) $R(V_{\{\cdots\}}[T_1; \cdots; T_n]) = R(V)(R(T_1), \cdots, R(T_n))$*. From Definition 4.5 we know that* $R(V)$ *is a substitution function of an appropriate arity and from Definition 4.3 we know how to apply it to a list of terms, so this definition is valid.*
(C) $R(x : T) = x : R(T)$ *and*
$R(\mathbf{C}_{\{\cdots\}}[T_1; \cdots; T_n]) = R(\mathbf{C})(R(T_1), \cdots, R(T_n))$.

(D) $R(\vdash T) = \vdash R(T)$ *and* $R(Hs;\, S) = R(Hs);\, R(S)$.
(E) $R(\langle T; x_1, \cdots, x_n\rangle) = \langle R(T); x_1, \cdots, x_n\rangle$ *and* $R(\langle Hs_1, \cdots, Hs_k; x_1, \cdots, x_n\rangle) = \langle R(Hs_1), \cdots, R(Hs_k); x_1, \cdots, x_n\rangle$ *(as usual, we assume that x_i are automatically alpha–renamed to avoid capture).*

Lemma 4.7. *If S is a closed sequent scheme and R is a refinement function, then $R(S)$ is also a closed sequent scheme.*

Proof. This property follows immediately from condition (C) in Definition 4.5 of refinement functions.

Definition 4.8. *We say that a sequent scheme (possibly a plain sequent) S* matches *a sequent scheme S'* iff *S is $R(S')$ for some refinement function R.*

Example 4.9.

(A) Scheme $\vdash \lambda x.A_{\{\}}[x]$ is matched by every well–formed $\vdash \lambda x.(\cdots)$ no matter what $\cdots$ is.
(B) Scheme $\vdash \lambda x.A_{\{\}}$ is matched by every well–formed $\vdash \lambda x.(\cdots)$ as long as $\cdots$ has no free occurrences of x. But in case $\cdots$ does have free occurrences of x, we would not be able to come up with a refinement function without violating Lemma 4.7.
(C) Scheme $\mathbf{H};\, x : A_{\{\mathbf{H}\}} \vdash C_{\{\mathbf{H}\}}$ essentially specifies that any matching sequent must have at least one hypothesis and that the variable introduced by the last hypothesis can not occur freely in the conclusion of the sequent. In particular, it is matched by $x : \mathbb{Z} \vdash 5 \in \mathbb{Z}$ using refinement function

$$\mathbf{H} \rightsquigarrow \langle;\rangle \qquad A \rightsquigarrow \mathbb{Z} \qquad C \rightsquigarrow 5 \in \mathbb{Z}$$

It is also matched by $\mathbf{H};\, x : \mathbb{Z};\, J_{\{\mathbf{H}\}}[x];\, y : \mathbb{Z} \vdash x \in \mathbb{Z}$ using refinement function

$$\mathbf{H} \rightsquigarrow \langle \mathbf{H}\,,\, x : \mathbb{Z}\,,\, J_{\{\mathbf{H}\}}[x]\ ;\ \rangle \qquad A \rightsquigarrow \mathbb{Z} \qquad C \rightsquigarrow x \in \mathbb{Z}$$

However it is not matched by neither $x : \mathbb{Z} \vdash x \in \mathbb{Z}$ nor $\vdash 5 \in \mathbb{Z}$.
(D) Every sequent scheme (including every plain sequent) matches the scheme $\mathbf{H} \vdash A_{\{\mathbf{H}\}}$.
(E) See Example 3.3.

5 Rule Specifications

Definition 5.1. *If S_i $(0 \le i \le n,\ n \ge 0)$ are closed sequent schema, then* $\dfrac{S_1 \ \cdots\ S_n}{S_0}$ *is* a rule scheme *(or just* a rule*).*

Definition 5.2. *We will say that*

a rule $\dfrac{S_1 \cdots S_n}{S_0}$ *is* an instance *of* $\dfrac{S'_1 \; \cdots \; S'_n}{S'_0}$

when for some refinement function R we have $S_i = R(S'_i)$ $(0 \le i \le n)$. We will call such an instance plain *if all S_i are plain sequents.*

Example 5.3. The following is the rule specification for the weakening (thinning) rule:

$$\frac{\mathbf{H}_{\{\}}[];\ \mathbf{J}_{\{\mathbf{H}\}}[] \vdash B_{\{\mathbf{H};\mathbf{J}\}}[]}{\mathbf{H}_{\{\}}[];\ x : A_{\{\mathbf{H}\}}[];\ \mathbf{J}_{\{\mathbf{H}\}}[] \vdash B_{\{\mathbf{H};\mathbf{J}\}}[]}$$

Note that condition (C) of Definition 4.5 guarantees that x will not occur free in whatever would correspond to $\mathbf{J}$ and B in instances of this rule.

Here is how this rule would look if we take all the syntax shortcuts described in the Appendix A:

$$\frac{\mathbf{H};\ \mathbf{J} \vdash B}{\mathbf{H};\ x : A;\ \mathbf{J} \vdash B}$$

This is exactly how this rule is written in the MetaPRL system.

Example 5.4. A `Cut` rule might be specified as

$$\frac{\mathbf{H}_{\{\}}[];\ \mathbf{J}_{\{\mathbf{H}\}}[] \vdash C_{\{\mathbf{H}\}}[] \qquad \mathbf{H}_{\{\}}[];\ x : C_{\{\mathbf{H}\}}[];\ \mathbf{J}_{\{\mathbf{H}\}}[] \vdash B_{\{\mathbf{H};\mathbf{J}\}}[]}{\mathbf{H}_{\{\}}[];\ \mathbf{J}_{\{\mathbf{H}\}}[] \vdash B_{\{\mathbf{H};\mathbf{J}\}}[]}$$

Note that in the first assumption of this rule C is inside the context $\mathbf{J}$, however $\beta(C)$ can not contain $\mathbf{J}$ since otherwise the second assumption would not be closed. In fact, this particular example shows why we need to keep track of β and can not just always deduce it from context — see also Remark 6.7.

With Appendix A syntax shortcuts `Cut` would look as

$$\frac{\mathbf{H};\ \mathbf{J} \vdash C_{\{\mathbf{H}\}} \qquad \mathbf{H};\ x : C;\ \mathbf{J} \vdash B}{\mathbf{H};\ \mathbf{J} \vdash B}$$

We assume that the object theory has some notion of derivability of closed plain sequents. We do not assume anything special about that notion, so everything we say here should be applicable to arbitrary derivability notions.

Definition 5.5. *We say that a rule* $\dfrac{S_1 \; \cdots \; S_n}{S_0}$ $(n \ge 0)$ *is* admissible *when for every plain instance* $\dfrac{S'_1 \; \cdots \; S'_n}{S'_0}$*, whenever S'_i $(1 \le i \le n)$ are all derivable, S'_0 is also derivable.*

Remark 5.6. Informally speaking, Definitions 5.2 and 5.5 say that in rule schema all second–order and context variables are implicitly universally quantified.

Now we can give a complete description of the *mental* procedure (in case of an automated system, the system will do most of the work) for adding derived rules support to a logical theory. First, we formalize the theory using admissible rule schema. Second, we expand the language of the theory from plain sequents to plain schema and we allow the use of arbitrary instances of the rules in the proofs, not just the plain instances. And whenever we can prove S_0 from S_i $(1 \le i \le n)$, we allow adding the rule $\frac{S_1 \cdots S_n}{S_0}$ to the list of rules and allow using such *derived rules* in addition to the rules present in the initial formalization of the theory.

More formally, whenever a rule $\frac{S_1 \cdots S_n}{S_0}$ is present in a theory, and R is a refinement function, we allow a new derived rule $\frac{R(S_1) \cdots R(S_n)}{R(S_0)}$ to be added to that theory. Additionally, when two rules $\frac{S_1 \cdots S_m}{S}$ and $\frac{S\ S'_1 \cdots S'_n}{S'}$ are present in a theory, we allow a new derived rule $\frac{S_1 \cdots S_m\ S'_1 \cdots S'_n}{S'}$ to be added to the theory. Note that in a practical system only "interesting" rules have to be explicitly added to the system while the intermediate rules that are needed to derive such "interesting" ones may be added implicitly.

In the next section we prove that this approach leads to a conservative extension of the original theory.

From the point of view of a user of an automated system that implements this approach, in order to start formal reasoning in some logical theory the user only needs to describe the syntax of the theory (for example, give a list of the operators and their corresponding arities) and then to provide the system with the list of the primitive rules (e.g. axioms) of the theory in the sequent schema syntax[5]. After that the system would immediately allow the user to start using the combined language of sequent schema and of base theory in derivations and whenever user would prove schema S from assumptions S_i $(1 \le i \le n)$, the system would allow using the rule $\frac{S_1 \cdots S_n}{S_0}$ in further derivations on an equal footing with the primitive rules of the base theory [6]. The main theorem of the next sections guarantees that the derived rules would not allow proving anything that is not derivable directly from the primitive rules.

6 Conservativity

Theorem 6.1. *The matching relation is transitive. That is, if R_1 and R_2 are refinement functions, then $R = R_2 \circ R_1$ is also a refinement function.*

[5] This does not imply that the set of rules have to be finite. MetaPRL, for example, provides the facility to specify an infinite family of rules by providing the code capable of recognizing valid instances.

[6] As we mentioned earlier, some systems might chose to allow user to "test–drive" a derived rule before forcing the user to prove the rule.

Proof. This property follows from the definitions of Section 4. We will show how to establish that.

Since application of a refinement function does not change the arity of a substitution function (see Definitions 4.3, 4.4 and 4.6 (E)) and since R_1 is a refinement function, the arity conditions of Definition 4.5 will hold for R. The remaining part is to prove that the free variables condition (C) of Definition 4.5 holds for R.

Consider $\Upsilon \in (\mathcal{C} \cup \mathcal{V})$. The proofs of condition (C) for the cases $\Upsilon \in \mathcal{C}$ and $\Upsilon \in \mathcal{V}$ are very similar, so we will only present the latter. We know that $R_1(\Upsilon)$ is some substitution function $\langle T; x_1, \cdots, x_n\rangle$. We also know that $R(\Upsilon) = R_2(R_1(\Upsilon)) = \langle R_2(T); x_1, \cdots, x_n\rangle$.

From Definitions 4.3 and 4.4 we know that $\mathbf{Vars}(R(\Upsilon)) = \mathbf{Vars}(R_2(T)) - \{x_1; \cdots; x_n\}$. From Definitions 4.1 and 4.6(A,B) we can deduce that

$$\mathbf{Vars}(R_2(T)) = \mathbf{FOV}(T) \cup \left(\bigcup_{\{V \in \mathcal{V} | V \text{ occurs in } T\}} \mathbf{Vars}(R_2(V)) \right). \tag{6.I}$$

Let us consider $\mathbf{FOV}(T)$ first. Since R_1 is a refinement function, from the condition (4.5) (C) for it we know that $\mathbf{FOV}(T)$ (except for x_i, but we do not need to consider x_i since they are not in $\mathbf{Vars}(R(\Upsilon))$) are all in $\mathbf{BV}(R_1(\beta(\Upsilon)))$ and from the definition of $\mathbf{BV}$ and the Definition 4.6 (C) we know that an application of a refinement function preserves all the first–order variables of $\mathbf{BV}$, so $\mathbf{FOV}(T) \subseteq \mathbf{BV}(R(\beta(\Upsilon)))$.

Now to cover the rest of (6.I) consider some $V \in \mathcal{V}$ such that V occurs in T. From Definition 4.1 we know that $\beta(V) \subseteq \mathbf{CV}(T) \subseteq \mathbf{Vars}(R_1(\Upsilon))$. And since R_1 is a refinement function, $\mathbf{Vars}(R_1(\Upsilon)) \subseteq \mathbf{BV}(R_1(\beta(\Upsilon)))$, so

$$\beta(V) \subseteq \mathbf{BV}(R_1(\beta(\Upsilon))) \tag{6.II}$$

Now consider some $\mathbf{C} \in \beta(V)$. We know from (6.II) that $\mathbf{C} \in \mathbf{BV}(R_1(\beta(\Upsilon)))$ and from the Definitions 4.4 and 4.6 (C) that if $\mathbf{C} \in \mathbf{BV}(R_1(\beta(\Upsilon)))$, then $\mathbf{BV}(R_2(\mathbf{C})) \subseteq \mathbf{BV}(R_2(R_1(\beta(\Upsilon))))$. This means that $\forall \mathbf{C} \in \beta(V). \mathbf{BV}(R_2(\mathbf{C})) \subseteq \mathbf{BV}(R_2(R_1(\beta(\Upsilon))))$ and if we take the union over all $\mathbf{C} \in \beta(\Upsilon)$ and recall that $R = R_2 \circ R_1$, we will get $\mathbf{BV}(R_2(\beta(\Upsilon))) \subseteq \mathbf{BV}(R(\beta(\Upsilon)))$. Since R_2 is a refinement function, condition (C) dictates that $\mathbf{Vars}(R_2(V)) \subseteq \mathbf{BV}(R_2(\beta(V)))$, so $\mathbf{Vars}(R_2(V)) \subseteq \mathbf{BV}(R(\beta(\Upsilon)))$ which takes care of the remaining part of (6.I). □

Note that there are two alternative ways of defining the value of $R_2 \circ R_1$ on terms, hypothesis specifications, sequent schema and substitution functions. The first is to define $(R_2 \circ R_1)(\Upsilon) = R_2(R_1(\Upsilon))$ and the second is to define $(R_2 \circ R_1)$ based on its value on $\mathcal{C} \cup \mathcal{V}$ using Definition 4.6.

Lemma 6.2. *The two definitions above are equivalent.*

Proof. This follows trivially from Definition 4.6 and the fact that

$$\sigma([T_1/x_1; \cdots; T_n/x_n]T) = [\sigma(T_1)/x_1; \cdots; \sigma(T_n)/x_n](\sigma(T))$$

(again, we rely on the assumption that x_i's are automatically alpha–renamed as needed to avoid capture).

Theorem 6.3. *Every instance of an admissible rule is also an admissible rule.*

Proof. If $\mathcal{R}$ is an admissible rule scheme and $\mathcal{R}'$ is its instance, then from Theorem 6.1 we know that every instance of $\mathcal{R}'$ will also be an instance of $\mathcal{R}$. Because of that and by Definition 5.5, if $\mathcal{R}$ is admissible, $\mathcal{R}'$ must be admissible too. □

Lemma 6.4. *Let us define $R_{\mathtt{triv}}$ to be a function that always returns a trivial substitution (see Definitions 4.3 and 4.4), namely*

$$R_{\mathtt{triv}}(\Upsilon) = \begin{cases} \langle \bullet; x_1, \cdots, x_{\alpha(\Upsilon)} \rangle, & \textit{when } \Upsilon \in \mathcal{V} \\ \langle ; x_1, \cdots, x_{\alpha(\Upsilon)} \rangle, & \textit{when } \Upsilon \in \mathcal{C} \end{cases}$$

Such $R_{\mathtt{triv}}$ would be a refinement function.

Proof. The conditions of Definition 4.5 are obviously true for $R_{\mathtt{triv}}$ — by construction it returns substitutions of the right arity and condition (C) is satisfied since $\mathbf{Vars}(R_{\mathtt{triv}}(\Upsilon))$ will always be an empty set.

Theorem 6.5. *If* $\mathcal{R} = \dfrac{S_1 \;\cdots\; S_m}{S}$ *and* $\mathcal{R}' = \dfrac{S\; S'_1 \;\cdots\; S'_n}{S'}$ $(m, n \geq 0)$ *are admissible rules, then the rule* $\dfrac{S_1 \;\cdots\; S_m\; S'_1 \;\cdots\; S'_n}{S'}$ *is also admissible.*

Proof. Suppose that R is a refinement function,

$$\frac{R(S_1) \;\cdots\; R(S_m) \;\; R(S'_1) \;\cdots\; R(S'_n)}{R(S')}$$

is a plain instance of the rule in question and all $R(S_i)$ $(1 \leq i \leq m)$ and $R(S'_j)$ $(1 \leq j \leq n)$ are derivable. We need to establish that $R(S')$ is also derivable.

Let R' be $R_{\mathtt{triv}} \circ R$. We know that whenever R returns a plain sequent, R' will return the same plain sequent (since an application of a refinement function does not change plain terms and plain sequents). Also, $R_{\mathtt{triv}}$ and hence R' will always turn every scheme into a plain sequent.

We know that $R(S_i) = R'(S_i)$ are derivable plain sequents and we know that $\mathcal{R}$ is an admissible rule, so by applying R' to $\mathcal{R}$ we get that $R'(S)$ is also derivable. Similarly, we know that $R(S'_j) = R'(S'_j)$ are derivable and by applying R' to admissible rule $\mathcal{R}'$ we get that $R'(S) = R(S)$ is derivable. □

Corollary 6.6. *Our procedure of extending a logical theory with derived rules (as described at the end of Section 5) will produce a conservative extension of the original theory.*

Proof. Theorems 6.3 and 6.5 tell us that every new derived rule that we add to the system will be admissible. This means that any plain sequent we could derive with the help of these new rules has already been derivable in the original theory. □

Remark 6.7. In Example 5.4 we saw that β can not always be deduced from context. In fact it was the `Cut` example that illuminated the need for an explicit β — without it we could construct a counterexample for Theorem 6.5 by taking $\mathcal{R}'$ to be the `Cut`.

7 Extending the Language of Sequent Schema

For simplicity in this paper we presented a minimal calculus necessary for adding a derived rule mechanism to a type theory. However, by appropriately extending the language of sequent schema we can apply our technique to other logical theories (including more complicated versions of type theories). Of course, for each of these extensions we need to make sure that the matching relation is still transitive and that refinements can not turn a closed scheme into a non-closed one. However these proofs are usually very similar to the ones presented in this paper.

One obvious extension is an addition of meta-variables that would range over some specific kind of constants. In particular, in MetaPRL formalization of NuPRL type theory [4] we use meta-variables that range over integer constants, meta-variables that range over string constants, *etc.* As usual, we consider all these meta-variables to be universally quantified in every rule scheme.

Before we can present other extensions of the language of sequent schema, we first need to better understand the way we are using sequent contexts to describe sequents. If we introduce special "sequent" operators **hyp** and **concl**, a sequent described in (2.I) can be rewritten as

$$\mathbf{hyp}\Bigg(A_1;\, x_1.\,\mathbf{hyp}\Bigg(A_2;\, x_2.\,\cdots\mathbf{hyp}\Big(A_n;\, x_n.\,\mathbf{concl}(C)\Big)\cdots\Bigg)\Bigg) \qquad (7.\mathrm{I})$$

Notice that **hyp** is a binary operator that introduces a binding occurrence in its second argument, so (7.I) has the same binding structure as (2.I). Now if we rewrite $\mathbf{C}_{\{\ldots\}}[T_1;\cdots;T_n];\, S$ in the same style, we get $\mathbf{C}_{\{\ldots\}}[S';T_1;\cdots;T_n]$ where S' is a rewrite of S. In this notation, we can say that $\mathbf{C}$ acts like an ordinary second–order variable on its 2-nd through n-th arguments, but on its first argument it acts in a following way:

(A) $\mathbf{C}$ is bound in its first argument, but not in others. In other words, while a second–order variable can not bind free variables of its arguments, a context instance can bind free variables of its first argument.
(B) The substitution function $R(C)$ has to use its first argument exactly once. In other words, each instance of $\mathbf{C}_{\{\ldots\}}[S';T_1;\cdots;T_n]$ would have exactly one occurrence of S' (while it may have arbitrary number of occurrences of each T_i including 0 occurrences of it).

(C) **C** stands for a chain of **hyp** operators with the S' at the end of the chain. In some instances of $\mathbf{C}_{\{\cdots\}}[S'; T_1; \cdots ; T_n]$ parts of that chain may be represented by other context variables, but in a plain instance it will always be a chain.

It can be easily shown that the language of sequent schema would still have all necessary properties if we add some (or all) of the following:

- "General" contexts that do not have restriction (C) or even restriction (B).
- Contexts for operators that have the same binding structure as **hyp** (note that we do not have to add a new kind of contexts for each operator — we can just make the operator name be a parameter). In particular, in type theory it might be useful to have contexts tied to the dependent product operator and to the dependent intersection operator [8].

8 Related Work

Most modern proof assistants allow their users to prove some statement in advance and then reuse it in further proofs. However, most of those mechanisms have substantial limitations.

NuPRL [1] allows its users to prove and reuse theorems (which must be closed sentences). Unfortunately, many elimination and induction principles can not be stated as theorems.

In addition to a NuPRL-like theorems mechanism, HOL [2] also has a derived rules mechanism, but in reality HOL's "derived rules" are essentially tactics that apply an appropriate sequence of primitive axioms. While this approach guarantees safety, it is extremely inefficient. According to [9], in a recent version of HOL many of the basic derived rules were replaced with the primitive versions (presumably to boost the efficiency of the system), and currently there is nothing in the system that attempts to check whether these rules are still truly derivable. In fact, the commented out tactic code that would "properly" implement those derived rules is no longer compatible with the current version of the HOL system.

Logical frameworks (such as Isabelle [10] and LF [3] based systems, including Elf [11]) use a rather complex meta-theory that can be used to allow its users to prove and reuse derived rules. Still, we believe that by directly supporting sequents–based reasoning and directly supporting derived rules, the sequent schema language can provide a greater ease of use (by being able to make more things implicit) and can be implemented more efficiently (by optimizing the logical engine for the refinement functions as defined in 4.5 and 4.6 - see [5]). Additionally, sequent schema allow us to prove the conservativity result once and for all the logical theories, while in a more general approach we might have to make these kinds of arguments every time as a part of proving an *adequacy* theorem for each logical theory being formalized.

Acknowledgments

Authors would like to thank Robert Constable and Alexei Kopylov for many productive discussions we had with them and Sergei Artemov and anonymous reviewers for their valuable comments on preliminary versions of this document.

References

1. Robert L. Constable, Stuart F. Allen, H.M. Bromley, W.R. Cleaveland, J.F. Cremer, R.W. Harper, Douglas J. Howe, T.B. Knoblock, N.P. Mendler, P. Panangaden, James T. Sasaki, and Scott F. Smith. *Implementing Mathematics with the NuPRL Development System.* Prentice-Hall, NJ, 1986.
2. Michael Gordon and T. Melham. *Introduction to HOL: A Theorem Proving Environment for Higher-Oder Logic.* Cambridge University Press, Cambridge, 1993.
3. Robert Harper, Furio Honsell, and Gordon Plotkin. A framework for defining logics. *Journal of the Association for Computing Machinery*, 40(1):143–184, January 1993. A revised and expanded verion of '87 paper.
4. Jason J. Hickey. *The MetaPRL Logical Programming Environment.* PhD thesis, Cornell University, Ithaca, NY, January 2001.
5. Jason J. Hickey and Aleksey Nogin. Fast tactic-based theorem proving. In J. Harrison and M. Aagaard, editors, *Theorem Proving in Higher Order Logics: 13th International Conference, TPHOLs 2000*, volume 1869 of *Lecture Notes in Computer Science*, pages 252–266. Springer-Verlag, 2000.
6. Jason J. Hickey, Aleksey Nogin, Alexei Kopylov, et al. MetaPRL home page. `http://metaprl.org/`.
7. Gérard P. Huet and Bernard Lang. Proving and applying program transformations expressed with second-order patterns. *Acta Informatica*, 11:31–55, 1978.
8. Alexei Kopylov. Dependent intersection: A new way of defining records in type theory. Department of Computer Science TR2000-1809, Cornell University, 2000.
9. Pavel Naumov, Mark-Olivar Stehr, and José Meseguer. The HOL/NuPRL proof translator: A practical approach to formal interoperability. In *The 14th International Conference on Theorem Proving in Higher Order Logics, Edinburgh, Scotland*, Lecture Notes in Computer Science, pages 329–345. Springer-Verlag, September 2001.
10. Lawrence C. Paulson. *Isabelle: A Generic Theorem Prover*, volume 828. Springer-Verlag, New York, 1994.
11. Frank Pfenning. Elf: A language for logic definition and verified metaprogramming. In *Proceedings of Fourth Annual Symposium on Logic in Computer Science*, pages 313–322, Pacific Grove, California, June 1989. IEEE Computer Society Press.
12. Frank Pfenning and Conal Elliott. Higher-order abstract syntax. In *Proceedings of the ACM SIGPLAN'88 Conference on Programming Language Design and Implementation (PLDI)*, pages 199–208, Atlanta, Georgia, June 1988. ACM Press.

A Syntax Simplifications

Since a sequent scheme must be closed, for every instance of a context or a second–order variable $\Upsilon_{\beta(\Upsilon)}[\cdots]$ we know that all the variables of $\beta(\Upsilon)$ must be bound in that scheme. This gives us an upper limit on $\beta(\Upsilon)$. In cases that the actual $\beta(\Upsilon)$ is equal to that upper limit, we may allow ourselves to omit the $\beta(\Upsilon)$ subscript and deduce the value of β from the rest of the scheme. Out experience shows that we can take this shortcut in almost all of the case, but not in all of them — see Example 5.4, for instance.

When a context or a second–order variable has an arity of zero, then according to Section 3 we have to write $\Upsilon_{\beta(\Upsilon)}[]$, but in this case it is natural to omit the argument part $[]$.

Remark A.1. When both of the omissions mentioned above are performed for a second–order variable V of arity 0, it becomes possible to confuse it with a first–order variable V. However, we can still distinguish them in a context of a rule scheme — if a particular occurrence of V is bound, it must be a first–order variable (since second–order variables can not be bound) and if it is free, it must be a second–order variable (since sequent schema in a rule can not have free first–order variables — see Definition 5.1).

Example A.2. See Example 5.3.

Remark A.3. Note that while these simplifications describe a way of reducing redundancies of the language we use, our experience suggests that the simplified language still contains enough redundancy to be able to successfully detect most common typographical errors and other common mistakes. In particular, by using the second–order variables to specify the substitutions indirectly and by requiring sequent schema to be closed, we make it much harder to forget to specify that a certain substitution needs to be performed or a certain variable needs to be renamed — a mistake that may be easy to make in a system where rule specification language contains an explicit substitution operator.

B MetaPRL Rule Specifications

In the MetaPRL system rule specifications have a dual purpose. First, they provide a rule scheme as described in Section 5 which specifies valid instances of that rule (see Definitions 5.1 and 5.2). Second, they specify a way to point at a particular instance of that rule (using the current goal the rule is being applied to as part of the input). It is important to realize that the core of the MetaPRL system expects the rule instance to be uniquely specified. Of course, users can always write advanced tactics that would use, say, a higher–order unification (or type inference, or heuristics, or something else) to find out the right rule instance and than instruct the system to apply the appropriate instance. However such a tactic would be outside of the kernel of the system. This approach gives users the freedom of choosing whatever algorithms they want to come up with the best rule instance. Additionally, this means that the core of the system does not need to be able to find the right instance itself; it only needs to be able to check whether a particular instance is applicable.

In MetaPRL a rule specification contains a rule scheme $\frac{S_1 \cdots S_n}{S}$ together with a sequence of argument specifications $As_1, \cdots, As_m$ $(m \geq 0)$ where each As_i is either a closed term scheme or a context variable name[7]. MetaPRL requires each context variable occurring in one of the S_i or S to be one of the arguments. It also restricts the arguments of the context variable occurrences to bound first–order variables. MetaPRL also requires second–order and context variable occurring in one of the S_i, S or As_i has to occur in one of the As_i or S in the following way:

(A) This occurrence should not be inside the argument of another context or second–order variable occurrence.
(B) All arguments of this occurrence of our variable must be distinct bound first–order variables.

When a user (or a higher–level tactic) needs to apply a particular instance of a rule, it will pass arguments $A_1, \ldots, A_m$ to the rule (and the system itself will provide a goal sequent scheme G that the rule is being applied to). For each $1 \leq i \leq m$, if As_i is a context variable, A_i would be *an address* that would specify which hypothesis specifications of G would be matched to that context variable and is As_i is a term scheme, A_i would be a close term that is supposed to match As_i. Because of the conditions (A)–(B) above and by knowing that G must match S and every A_i must match As_i, the system can determine the correct instance (if one exits) in a deterministic and straightforward manner (see [5] for more information).

C Concrete Syntax of Sequent Schema

The following table describes the concrete syntax used by the MetaPRL system:

Description	Abstract syntax	Concrete syntax
Sequent turnstyle	$\vdash$	`>-`
First–order variable	a	`'a`
Second–order variable	$A_{\{\mathbf{H};\mathbf{J}\}}[x;y]$	`'A<H;J>['x;'y]`
Sequent context	$\mathbf{J}_{\{\mathbf{H}\}}[x]$	`'J<H>['x]`
Sequent example	$\mathbf{H};\ x : A;\ \mathbf{J}[x] \vdash C[x]$	`'H; x:'A; 'J['x] >- 'C['x]`
Rule	$\frac{S_1 \cdots S_n}{S}$	S_1 `-->` $\cdots$ `-->` S_n `-->` S

[7] This is a simplification. There are other kinds of arguments that are used to specify the refinement of the parameter meta–variables described in Section 7.

Algebraic Structures and Dependent Records

Virgile Prevosto[1,2], Damien Doligez[1], and Thérèse Hardin[1,2]

[1] I.N.R.I.A – Projet Moscova
B.P. 105 – F-78153 Le Chesnay, France
Damien.Doligez@inria.fr
[2] L.I.P. 6 – Equipe SPI
8 rue du Cap. Scott – 75015 PARIS, France
{therese.hardin,virgile.prevosto}@lip6.fr

Abstract. In mathematics, algebraic structures are defined according to a rather strict hierarchy: rings come up after groups, which rely themselves on monoids, and so on. In the FOC project, we represent these structures by *species*. A species is made up of algorithms as well as proofs that these algorithms meet their specifications, and it can be built from existing species through inheritance and refinement mechanisms.
To avoid inconsistencies, these mechanisms must be used carefully. In this paper, we recall the conditions that must be fulfilled when going from a species to another, as formalized by S. Boulmé in his PhD [3]. We then show how these conditions can be checked through a static analysis of the FOC code. Finally, we describe how to translate FOC declarations into COQ.

1 Introduction

1.1 The FOC Project

Although computer algebra is based upon strong mathematical grounds, errors are not so rare in current computer algebra systems. Indeed, algorithms may be very complex and there is an abundance of corner cases. Moreover, preconditions may be needed to apply a given algorithm and errors can occur if these preconditions are not checked.

In the FOC language[1], any implementation must come with a proof of its correctness. This includes of course pre- and post- condition statements, but also proofs of purely mathematical theorems. In a computer algebra library, a single proof is of little use by itself. Indeed numerous algorithms, and thus their proofs, can be reused in slightly different contexts. For example a tool written for groups can be used in rings, provided that the system knows every ring is a group. Thus, we need a completely formalized representation of the relations between the mathematical structures, which will serve as a common framework for both proofs and implementations.

In his PhD thesis [3], S. Boulmé gives a formal specification of both the hierarchy of the library and the tools used to extend it. This formalization of the

[1] http:www-spi.lip6.fr/~foc.

V.A. Carreño, C. Muñoz, S. Tahar (Eds.): TPHOLs 2002, LNCS 2410, pp. 298–313, 2002.

specification, briefly presented below (Sec. 2), points out that some invariants must be preserved when extending an existing structure. In particular, the *dependencies* between the functions and the properties of a given structure must be analyzed carefully, as well as *dependencies* between structures.

We have elaborated a syntax that allows the user to write programs, statements and proofs. This syntax is restrictive enough to prevent some inconsistencies, but not all. In this paper we describe the core features of this syntax (Sec. 3), and present code analyses to detect remaining inconsistencies (Sec. 4). Then, we show how to use the results of this analysis to translate FOC sources into COQ, in order to have FOC proofs verified by the COQ system (Sec 5).

1.2 FOC's Ground Concepts

Species. Species are the nodes of the hierarchy of structures that makes up the library. They correspond to the algebraic structures in mathematics. A species can be seen as a set of *methods*, which are identified by their names. In particular, there is a special method, called the *carrier*, which is the type of the representation of the underlying set of the algebraic structure.

Every method can be either *declared* or *defined*. *Declared* methods introduce the constants and primitive operations. Moreover, axioms are also represented by declared methods, as would be expected in view of the Curry-Howard isomorphism. *Defined* methods represent implementations of operations and proofs of theorems. The declaration of a method can use the carrier.

As an example, a monoid is built upon a set represented by its carrier. It has some declared operations, (specified by their *signature*), namely $=$, $+$, and *zero*. These operations must satisfy the axioms of monoids, which are expressed in FOC by *properties*. We can then define a *function*, *double*, such that $double(x) = x+x$, and prove some *theorems* about it, for instance that $double(zero) = zero$.

Interface. An *interface* is attached to each species: it is simply the list of all the methods of the species considered as only declared. As S. Boulmé pointed out, erasing the definitions of the methods may lead to inconsistencies. Indeed, some properties may depend on previous definitions, and become ill-typed if these definitions are erased. This is explained in more detail in section 2.2. Interfaces correspond to the point of view of the end-user, who wants to know which functions he can use, and which properties these functions have, but doesn't care about the details of the implementation.

Collection. A *collection* is a completely defined species. This means that every field must be defined, and every parameter instantiated. In addition, a collection is "frozen". Namely, it cannot be used as a parent of a species in the inheritance graph, and its carrier is considered an abstract data type. A collection represents an implementation of a particular mathematical structure, such as $(\mathbb{Z}, +, *)$ implemented upon the GMP library.

Parameters. We also distinguish between "atomic" species and "parameterized" species. There are two kinds of parameters: entities and collections. For instance, a species of matrices will take two integers (representing its dimensions) as parameters. These integers are entities of some collection. For its coefficients, the species of matrices will also take a collection as argument, which must have at least the features specified by the interface of `ring`. Of course, it can be a richer structure, a `field` for instance.

A species s_1 parameterized by an interface s_2 can call any method declared in s_2. Thus, the parameter must be instantiated by a completely defined species, *i.e.* a collection.

2 Constraints on Species Definition

S. Boulmé, in [3], specified different conditions that must be fulfilled when building the species hierarchy. These conditions are required to define a model of the hierarchy in the calculus of inductive constructions. By building a categorical model of the hierarchy, S. Boulmé also showed that they were necessary conditions. One of the objectives of this paper is to show how the implementation of FOC fulfills these conditions.

2.1 Decl- and Def- Dependencies

We will now present these conditions through an example. We can take for instance the species of `setoid`, a set with an equality relation. More precisely, the species has the following methods: a carrier `rep`, an abstract equality `eq`, and a property `eq_refl` stating that `eq` is reflexive. From `eq`, we define its negation `neq`, and prove by the theorem `neq_nrefl` that it is irreflexive. Using a COQ-like syntax, we can represent `setoid` like this:

$$\left\{\begin{array}{l}
\texttt{rep} : \textbf{Set} \\
\texttt{eq} : \texttt{rep} \texttt{->} \texttt{rep} \texttt{->} \textbf{Prop} \\
\texttt{neq} : \texttt{rep} \texttt{->} \texttt{rep} \texttt{->} \textbf{Prop} := [\texttt{x},\texttt{y} : \texttt{rep}](\texttt{not}\ (\texttt{eq x y})) \\
\texttt{eq_refl} : (\texttt{x} : \texttt{rep})(\texttt{eq x x}) \\
\texttt{neq_nrefl} : (\texttt{x} : \texttt{rep})(\texttt{not}\ (\texttt{neq x x})) := \\
\qquad\qquad [\texttt{x} : \texttt{rep};\ \texttt{H} : (\texttt{not}\ (\texttt{eq x x}))](\texttt{H}\ (\texttt{eq_refl x}))
\end{array}\right\}$$

Thanks to the Curry-Howard isomorphism, functions and specifications are treated the same way. We must first verify that the methods have well-formed types. In addition, the body of every defined method must have the type given in its declaration. We can remark that the order in which we introduce the methods of the species is important: in order to write the type of `eq`, we must know that there exists a `Set` called `rep`. Similarly, the body of `neq` refers to `eq`, as does the property `eq_refl`. These three cases are very similar: a method $m2$ uses $m1$, and in order to typecheck $m2$, we need $m1$ in the typing environment. In this case, we speak of a *decl-dependency* of $m2$ upon $m1$.

On the other hand, in order to typecheck `neq_nrefl`, it is not enough to have the type of `neq` in the environment. Indeed, we must know that it is defined as `(not (eq x y))`, because hypothesis `H` in the body of `neq_nrefl` must match the definition of `neq`. Thus, `neq` must be unfolded during the typechecking of `neq_nrefl`. We identify this case as a *def-dependency*. When dealing with inheritance, this new kind of dependency has a major drawback: if we want to redefine `neq` in a species that inherits from `setoid`, then we will have to provide a new proof for `neq_nrefl`. There is no such problem with *decl-dependencies*: the definition of `neq` remains valid for any definition of `eq`, provided it has the right type.

2.2 Purely Abstract Interface

Def-dependencies do not occur only in proofs. They can also appear at the level of types. For instance, take the following species definition (again in a Coq-like syntax, `0` being a constant of type `nat`).

$$\{\texttt{rep}: \texttt{Set} := \texttt{nat};\ \texttt{p}: (\exists \texttt{x} : \texttt{rep} \mid \texttt{x} = \texttt{0})\}$$

Here, in order to accept the property `p` as well-typed, we have to know that `rep` is an alias of `nat`. If we remove the definition of `rep`, then the resulting interface is clearly inconsistent. Thus we cannot accept such a species definition, because any species must receive a valid interface. In a correctly written species, the type of a method cannot def-depend upon another method. This restriction was identified by S. Boulmé when representing species by records with dependent fields.

3 Syntax

In this section, we present the core syntax of Foc and an intuitive explanation of its semantics. The complete syntax is built upon the core syntax by adding syntactic sugar without changing its expressive power, so the properties of the core language are easily extended to the full language. In the rest of the paper, we will use the following conventions concerning variable names. Lambda-bound variables, function names and method names are usually denoted by x or y. Species names are denoted by s, and collection names by c. There is also a keyword, **self**, which can be used only inside a species s. It represents the "current" collection (thus **self** is a collection name), allowing to call its methods by **self**!**x** (see 3.5).

3.1 Types

type ::= $c \mid \alpha \mid$ *type* −> *type* | *type* * *type*

A type can be a collection name c (representing the carrier of that collection), a type variable α, or a function or a product type.

3.2 Expressions and Properties

identifier ::= x, y
declaration ::= x [**in** *type*]
expression ::= x | $c!x$ | **fun** *declaration* −> *expression*
| *expression*(*expression* { ,*expression* }*)
| **let** [**rec**] *declaration* = *expression* **in** *expression*

An expression can be a variable (x, y), a method x of some collection c, a functional abstraction, a function application, or a local definition with an expression in its scope.

Properties are boolean expressions with first-order quantifiers:

prop ::= *expr* | *prop* **and** *prop* | *prop* **or** *prop* | *prop* → *prop* | **not** *prop*
| **all** x **in** *type*, *prop* | **ex** x **in** *type*, *prop*

3.3 Fields of a Species

def_field ::= **let** *declaration* = *expression*
| **let rec** { *declaration* = *expression*; }+
| **rep** = *type* | **theorem** x : *prop* **proof:** [deps] *proof*
decl_field ::= **sig** x **in** *type* | **rep** | **property** x : *prop*
field ::= *def_field* | *decl_field*
deps ::= { (**decl:** | **def:**) { x }* }*

A *field* ϕ of a species is usually a declaration or a definition of a method name. In the case of mutually recursive functions, a single field defines several methods at once (using the **let rec** keyword).

The carrier is also a method, introduced by the **rep** keyword. Each species must have exactly one **rep** field.

The proof language (the *proof* entry of the grammar) is currently under development. For the time being, proofs can be done directly in Coq, although the properties themselves are translated automatically. The dependencies (Sec. 2) of a proof must be stated explicitely in the *deps* clause of a theorem definition.

3.4 Species and Collection Definitions

species_def ::= **species** *s* [(*parameter* { , *parameter* }*)]
[**inherits** *species_expr* { , *species_expr* }*]
= { *field*; }* **end**
collection_def ::= **collection** *c* **implements** *species_expr*
parameter ::= x **in** *type*
| *c* **is** *species_expr*
species_expr ::= *s* | *s* (*expr_or_coll* { , *expr_or_coll* }*)
expr_or_coll ::= *c* | *expression*

A *species_expr* is a species identifier (for an atomic species), or a species identifier applied to some arguments (for a parameterized species). The arguments can be collections or expressions. Accordingly, in the declaration of a parameterized species, a formal parameter can be a variable (with its type) or a collection name (with its interface, which is a *species_ expr*). A collection definition is simply giving a name to a *species_ expr*.

3.5 Method and Variable Names

As pointed out in [10], method names can not be α-converted, so that they must be distinguished from variables. The notation **self!x** syntactically enforces this distinction, as we can remark in the following example.

```
let y = 3;;
species a (x in int) = let y = 4; let z = x; let my_y = self!y; end
collection a_imp implements a(y)
```

Here, a_imp!z returns 3, while a_imp!my_y returns 4.

3.6 A Complete Example

We will illustrate the main features of FOC with an example of species definition. Assume that the species setoid and monoid have already been defined, and that we have a collection integ that implements $\mathbb{Z}$. We now define the cartesian products of two setoids and of two monoids.

```
species cartesian_setoid(a is setoid, b is setoid)
  inherits setoid =
   rep = a * b;
   let eq = fun x -> fun y -> and(a!eq(fst(x),fst(y)),b!eq(snd(x),snd(y)));
   theorem refl : all x in self, self!eq(x,x)
     proof:
       def: eq;
       {* (* A Coq script that can use the definition of self!eq *) *} ;
end

species cartesian_monoid(a1 is monoid, b1 is monoid)
  inherits monoid, cartesian_setoid(a1,b1) =
    let bin_op = fun x -> fun y ->
       let x1 = fst(x) in let x2 = snd(x) in
       let y1 = fst(y) in let y2 = snd(y) in
         create_pair(a!bin_op(x1,y1),b!bin_op(x2,y2));
    let neutral = create_pair(a!neutral,b!neutral);
end

collection z_square implements cartesian_monoid(integ,integ)
```

4 Finding and Analyzing Dependencies

As said above, the syntax of FOC prevents some kinds of inconsistencies, but not all. To eliminate the remaining ones, we perform a static analysis on the species definitions.

4.1 Informal Description of Static Analysis

Given a species definition, we must verify that it respects the following constraints.

- All expressions must be well-typed in an ML-like type system. Redefinitions of methods must not change their type.
- When creating a collection from a species, all the fields of the species must be defined (as opposed to simply declared).
- The **rep** field must be present or inherited in every species.
- Recursion between methods is forbidden, except within a **let rec** field.

4.2 Classifying Methods

As said in section 2, when defining a species s, it is important to find the dependencies of a method x upon the other methods of s, in order to check the correctness of s. It is syntactically impossible for some dependencies to occur in FOC source. For instance, we can not write a type that depends upon a function or a property, so that the carrier of s never depends upon another method. Thus, while in the work of S. Boulmé there is only one sort of method, we distinguish here three kinds of methods: the carrier, the functions, and the specifications. Each of these can be declared or defined.

All the dependencies that can be found in a FOC definition are summed up in Fig. 1. In particular, note that a def-dependency can occur between a statement

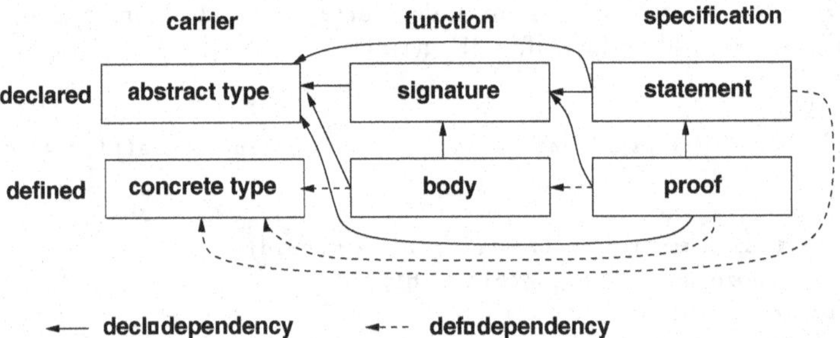

Fig. 1. Possible Dependencies between Methods

and the carrier. Indeed, the example of section 2.2 can be written in FOC:

```
species a =
  rep = nat;
  property p : ex x in self, base_eq(x,0);
end
```

where base_eq is the built-in equality primitive.

Since we want to have a fully abstract interface for each species written in FOC, such a species definition will be rejected by the dependency analysis.

4.3 Identifying the Dependencies

The dependencies between the various kinds of methods cannot be computed in a uniform way. For instance, the decl-dependencies of a function body b are found by simply listing all sub-expressions of the form **self**!m in b. On the other hand, to identify a def-dependency of b upon `rep`, we need to typecheck b. We will now describe the computation of the dependencies.

Syntactic Criterion. We mention here all the dependencies that are found by simply looking at the Abstract Syntax Tree (AST) of the method.

- m_1 is a function body and m_2 is a function (either declared or defined): m_1 decl-depends upon m_2 if **self**!m_2 is a sub-expression of m_1.
- m_1 is a statement and m_2 a function: m_1 decl-depends upon m_2 if **self**!m_2 is a sub-expression of m_1.
- m_1 is a proof and m_2 a function or a statement: m_1 decl-depends upon m_2 if m_2 appears in the **decl** clause of m_1. It def-depends upon m_2 if m_2 appears in the **def** clause of m_1.

Typing Criterion. Some dependencies require a finer analysis to be caught. This is done in the typing phase (Sec. 4.7) and concerns the dependencies upon **rep**.

- m_1 decl-depends upon **rep** if the type of a subexpression of m_1 contains **self**.
- m_1 def-depends upon **rep** if **rep** is defined to τ, and when typing m_1, a unification step uses the equality **self** $= \tau$. In this case, the unification returns **self**.

Notations. $\lceil m_1 \rfloor_s$ is the set of names upon which m_1, considered as a method of the species s, decl-depends. Similarly, $\lceil\lceil m_1 \rfloor\rfloor_s$ is the set of names upon which m_1 def-depends. **rep** is considered as a name. Note that $\lceil\lceil m_1 \rfloor\rfloor_s \subseteq \lceil m_1 \rfloor_s$.

4.4 Name Unicity

A name can not belong to two distinct fields of a species body. We take this condition as an invariant, which is easy to check syntactically. From a programming point of view, such situation would be an error, since one of the two declarations (or definitions) would be ignored.

Notations. Let $\mathcal{N}(\phi)$ be the names of methods introduced in a field ϕ (only one name when no mutual recursion), and $\mathcal{D}(\phi)$, the names that are introduced in a field definition. In the following, we will consider this general form of a species definition (*defspec*), which respects the invariant:

$$\textbf{species } \mathrm{s} \textbf{ inherits } s_1, \ldots s_n = \phi_1 \ldots \phi_m, \\ \text{such that } \forall i, j \leq m, \mathcal{N}(\phi_i) \cap \mathcal{N}(\phi_j) = \emptyset.$$

Then we define the set of names of the species s by

$$\mathcal{N}(s) = \left(\bigcup_{i=1}^{n} \mathcal{N}(s_i)\right) \cup \left(\bigcup_{j=1}^{m} \mathcal{N}(\phi_j)\right)$$

4.5 Binding of a Method

Let $x \in \mathcal{N}(s)$. The binding $\mathcal{B}_s(x)$ of x is, roughly speaking, the body of the definition of x, if any. But, in case of multiple inheritance, x may be associated to several inherited definitions. Then $\mathcal{B}_s(x)$ is the last such definition in the order specified by the **inherits** clause.

Definition 1 (Binding of a Method). *Let s be a species defined by defspec, and $x \in \mathcal{N}(s)$. $\mathcal{B}_s(x)$, $\mathbb{I}_s(x)$ and $\mathcal{D}(s)$ are recursively defined as follows.*

- *if $\forall i \leq n,\ x \notin \mathcal{D}(s_i) \wedge \forall j \leq m,\ x \notin \mathcal{D}(\phi_j)$ then $\mathcal{B}_s(x) = \bot$.*
- *if $\exists i \leq m,\ \phi_i$ is **let** $x = expr$ then $\mathcal{B}_s(x) = expr$, and $\mathbb{I}_s(x) = n + 1$.*
- *if $\exists i \leq m,\ \phi_i$ is **let rec** $\{x_1 = expr_1 \ldots x_l = expr_l\}$, and $x_j = x$ then $\mathcal{B}_s(x) = expr_j$ and $\mathbb{I}_s(x) = n + 1$*
- *if $\exists i \leq m,\ \phi_i$ is **theorem** x : ...proof then $\mathcal{B}_s(x) = proof$, and $\mathbb{I}_s(x) = n + 1$*
- *else let i_0 be the greatest index such that $x \in \mathcal{D}(s_{i_0})$ then $\mathcal{B}_s(x) = \mathcal{B}_{s_{i_0}}(x)$, and $\mathbb{I}_s(x) = i_0$*

$\mathcal{D}(s) = \{x \in \mathcal{N}(s), \mathcal{B}_s(x) \neq \bot\}$

4.6 Normal Form of a Species

To ensure that a species s meets all the constraints, we compute its *normal form* (def. 3), in which inheritance resolution, dependency analysis and typing are performed. A species in normal form has no **inherits** clause, and all its fields are ordered in such a way that a field depends only upon the preceding ones.

Since **rep** has no dependencies, we choose **rep** as the first field of the normal form. Then, any other field may depend upon **rep**. To study dependencies between functions, we distinguish between **let** and **let rec** definitions. If m_1 and m_2 are defined inside the same **let rec** field, they are allowed to mutually depend upon each other – provided that a termination proof is given[2]. Thus, for a **let rec** definition ϕ, the mutual dependencies between the methods m_i of ϕ are not recorded in $\wr m_i \int_s$.

Definition 2 (Well-Formedness).
A species s defined by defspecis said to be well-formed *if:*

- *the s_i are well-formed.*
- *All the definitions are well-typed.*

[2] Note that this termination proof def-depends upon m_1 and m_2.

– *The different fields introduce different names:*

$$\forall i, j, i \neq j \Rightarrow \mathcal{N}(\phi_i) \cap \mathcal{N}(\phi_j) = \emptyset$$

– *A given definition decl-depends only upon previous fields:*

$$\forall i \leq n, \forall x \in \mathcal{N}(\phi_i), \, \lceil x \rfloor_s \subset \bigcup_{j=1}^{i-1} \mathcal{N}(\phi_j)$$

Requiring that definitions are well-typed implies that def-dependencies are correctly handled. Indeed, typechecking will fail if the definition is missing.

Definition 3 (Normal Form). *A species s is said to be in normal form if it is well-formed and it has no* ***inherits*** *clause.*

Definition 4. *changed(y, x) is a relation over $\mathcal{N}(s)$, s being defined by*

$$\textbf{species } s \textbf{ inherits } s_1 \ldots s_m = \phi_1 \ldots \phi_n \textbf{ end}$$

$$\begin{aligned} changed(y,x) \iff & (\exists j > \mathbb{I}_s(x),\, y \in \mathcal{D}(s_j) \wedge \mathcal{B}_{s_j}(y) \neq \mathcal{B}_{\mathbb{I}_s(x)}(y)) \\ \vee & \; (\exists k,\, y \in \mathcal{D}(\phi_k) \wedge \mathbb{I}_s(x) \neq n+1) \end{aligned}$$

Theorem 1 (Normal Form of Well-Formed Species). *For each well-formed species s, there exists a species nfs, which is in* normal form *and enjoys the following properties:*

- *names:* $\mathcal{N}(nfs) = \mathcal{N}(s)$
- *defined names:* $\mathcal{D}(nfs) \subseteq \mathcal{D}(s)$
- *definitions:* $\forall x \in \mathcal{D}(nfs), \mathcal{B}_s(x) = \mathcal{B}_{nfs}(x)$
- $\forall x \in \mathcal{D}(s) \setminus \mathcal{D}(nfs), \exists y \in \lceil\!\lceil x \rfloor\!\rfloor_s$ s.t.
  - $y \notin \mathcal{D}(nfs)$ *or*
  - $y \in \mathcal{D}(nfs)$ *and changed(y, x).*

The last clause ensures that we erase as few method bindings as possible, namely only the ones that def-depend upon methods that have *changed* during inheritance lookup, or upon a method that must itself be erased.

The proof gives all the steps of the static analysis performed on species (inheritance lookup, dependency analysis and typing). In the proof, we assimilate a species in normal form and the *ordered* sequence of all its methods. $s_1 @ s_2$ denotes the concatenation of two sequences.

Let $norm(s_i)$ be a normal form of s_i. We first build the following sequence: $\mathbb{W}_1 = norm(s_1)@...@norm(s_n)@[\phi_1, ..., \phi_m]$. $\mathbb{W}_1$ may contain several occurrences of the same name, due to multiple inheritance or redefinition. To solve such conflicts, we introduce a function $\ominus$, which merges two fields sharing some names.

- If the two fields ϕ_1 and ϕ_2 are declarations, $\phi_1 \oslash \phi_2$ is a declaration too. If only one of the field is defined, $\oslash$ takes this definition. If both ϕ_1 and ϕ_2 are definitions, then $\oslash$ selects ϕ_2.
- Two **let rec** fields ϕ_1 and ϕ_2 can be merged even if they do not introduce exactly the same sets of names, because you can inherit a **let rec** field and then redefine only some of its methods (keeping the inherited definition for the others), or even add some new methods to this recursion. Merging two **let rec** fields is not given for free, though. Indeed, it implies that the user provides a new termination proof, that involves *all* the methods defined in $\phi_1 \oslash \phi_2$.

Our analysis builds a sequence $\mathbb{W}_2$ of definitions from $\mathbb{W}_1 = \phi_1 \dots \phi_n$, starting with $\mathbb{W}_2 = \emptyset$. This is done with a loop; each iteration examines the first field remaining in $\mathbb{W}_1$ and updates $\mathbb{W}_1$ and $\mathbb{W}_2$. The loop ends when $\mathbb{W}_1$ is empty. The loop body is the following:
Let $\mathbb{W}_1 = \phi_1, \mathbb{X}$ and $\mathbb{W}_2 = \psi_1 \dots \psi_m$

- if $\mathcal{N}(\phi_1) \cap (\cup_{i=1}^{m} \mathcal{N}(\psi_i)) = \emptyset$ then $\mathbb{W}_1 \leftarrow \mathbb{X}$ and $\mathbb{W}_2 \leftarrow (\psi_1 \dots \psi_n, \phi_1)$: if the analyzed field does not have any name in common with the ones already processed, we can safely add it at the end of $\mathbb{W}_2$.
- else let i_0 be the smallest index such that $\mathcal{N}(\phi_1) \cap \mathcal{N}(\psi_{i_0}) \neq \emptyset$, then we do $\mathbb{W}_1 \leftarrow ((\phi_1 \oslash \psi_{i_0}), \mathbb{X})$ and $\mathbb{W}_2 \leftarrow (\psi_1 \dots \psi_{i_0-1}, \psi_{i_0+1} \dots \psi_m)$. In the case of mutually recursive definitions, ϕ_1 can have some names in common with more than one ψ_i, so that $\phi_1 \oslash \psi_{i_0}$ is kept in $\mathbb{W}_1$. In addition, we abstract all the fields $\{\psi_i\}_{i>i_0}$ such that $\exists x \in \mathcal{N}(\phi_1)$, $y \in \mathcal{N}(\psi_i), y <_s^{def} x$, where $<_s^{def}$ is the transitive closure of $\llbracket \cdot \rrbracket_s$.

The complete proof that this algorithm computes effectively a well-formed normal form that satisfies the conditions of theorem 1 can be found in [14]. In fact, the algorithm can be applied to any species, provided that fields can be reordered according to the last clause of def. 3. If it succeeds, then s is indeed well-formed. If it fails, the definition of s is inconsistent, and thus rejected.

4.7 Typing a Normal Form

Once inheritance resolution and dependency analyses have been done, we have to type the definitions of a species in normal form. The typing algorithm for functions is basically the same as the Hindley-Milner type inference algorithm used in the ML family of languages. We also check that specifications are well-typed, but the verification of proofs is left to Coq (Sec. 5).

The only trick here is that types must be preserved through inheritance so that, if a method is redefined, we have to check that the inferred type for the new definition is compatible with the old one. Moreover, we may detect a def-dependency upon **rep**, as said in 4.3, while typing the statement of a property or a theorem. In this case, we must reject the species definition, as explained in Sec. 2.2, since such a species can not have a fully abstract interface.

The typing inference rules are given in [14].

4.8 Parameterized Species

Let s be a parameterized species, written **species** $s(c \textbf{ is } a) \ldots$ where c is a fresh name. The typing environment of the body of s contains a binding $(c, \mathcal{A}(a, c))$, where $\mathcal{A}(a, c)$ is an interface defined as follows.

If $a = \{x_i : \tau_i = e_i\}_{i=1..n}$, then

$$\mathcal{A}(a, c) = \langle x_i : \tau_i[\textbf{self} \leftarrow c]\rangle_{i=1..n}$$

A collection parameter may be instantiated by a richer structure than expected. For instance, polynomials must be defined over a ring, but may perfectly be given a field instead. So we define a *sub-species* relation $\preccurlyeq$ in order to instantiate a collection parameter with arguments of the right interface.

Definition 5 (Sub-species). *Let $\mathcal{T}_s(x)$ be the type of x in s. Let s_1, s_2 be two species.*

$$s_1 \preccurlyeq s_2 \iff \mathcal{N}(s_2) \subset \mathcal{N}(s_1) \wedge \forall x \in \mathcal{N}(s_2), \mathcal{T}_{s_1}(x) = \mathcal{T}_{s_2}(x)$$

Thanks to the type constraints during inheritance lookup, if a inherits from b, then $a \preccurlyeq b$. Since only the types of the methods are concerned, the relation is easily extended to interfaces.

5 Certification: The Translation into Coq

5.1 Interfaces

As in [3] interfaces are represented by Coq's `Records`, and collections by instances of the corresponding `Records`. In Coq, a `Record` is a n-uple in which every component is explicitly named:

Record `my_record` := { `label_1` : `type_1`; `label_2` : `type_2`; ... }.

The main issue here is that we are dealing with *d*ependent `Records`: `type_2` can use `label_1`, as in the following example:

Record `comparable` :=
{ `my_type` : **Set**; `less_than` : `my_type` -> `my_type` -> **Prop** }.

So the order in which the different labels appear is important.

We define a `Record` type in Coq, which denotes the interface of the species. If the species is $\{x_i : \tau_i = e_i\}$, then the `Record` is defined as

Record `name_spec` : Type := `mk_spec`$\{x_i : \tau_i\}$

We explicitly give *all* the fields of the `Record`, including the inherited ones. They have to be given in the order of the normal form because decl-dependencies can be present even at the level of types.

We also provide `coercions` between the `Record` we have just built and the `Record(s)` corresponding to the interface(s) of the father species. Such `coercions` reflect the inheritance relations of Foc.

5.2 Species

Unlike [3], a species s is not represented by a *MixDRec*, that is a `Record` that mix concrete and abstract fields. For any method m defined in s, we introduce a *method generator*, gen_m. If a method is inherited, the corresponding generator has been defined in a preceding species, and does not need to be recompiled. This offers a kind of modularity.

For instance, in the following species

```
species a =
 sig eq in self -> self -> bool;
 let neq = fun x -> fun y -> notb(self!eq(x,y));
end
```

The *method generator* for `neq` is

$$\lambda abst\_T : Set. \lambda abst\_eq : abst\_T -> abst\_T -> bool.$$

$$\lambda x, y : abst\_T.notb(abst\_eq\ x\ y)$$

Then, each species that inherits from `setoid` can use this definition of `neq`, instantiating `abst_eq` with its *own* definition of `eq`. This way, we can handle *late-binding*.

More formally, Let Σ be a normal form of a species s, (sequence $\Sigma = \{x_i : \tau_i = e_i\}$ of methods). Let e be an expression occuring in a field ϕ (e being a declaration, a statement, a binding, or a proof). We define below $\Sigma \Cap e$, which is the minimal environment needed to typecheck e (or ϕ). Due to def-dependencies, $\Cap$ can not simply select the methods ϕ depends upon. Each time ϕ def-depends upon ψ, we must also keep the methods upon which ψ itself decl-depends.

Definition 6 (Minimal Environment). *Let $\Sigma = \{x_i : \tau_i = e_i\}$ and e be an expression. $\Sigma \Cap e$ is the environment needed to typecheck e, and is defined as follows.*

$$\Sigma \sqcap e = \{x_j : \tau_j = new\_e_j | x_j \in \langle e \rangle \wedge (x_j : \tau_j = e_i) \in \Sigma\}$$

$$\text{where } new\_e_j = \begin{cases} e_j & \text{if } x_j \in [\![e]\!] \\ \bot & \text{otherwise} \end{cases}$$

$$\begin{aligned} U_1 &= \Sigma \sqcap e \\ U_{k+1} &= U_k \cup \bigcup_{(x_j : \tau_j = e_j) \in U_k} \Sigma \sqcap e_j \\ \Sigma \Cap e &= \bigcup_{k>0} U_k \end{aligned}$$

$$\text{where } \{x : \tau = \bot\} \cup \{x : \tau = e\} = \{x : \tau = e\}$$

We turn now to the translation of the definition d of a method y in Coq, according to the environment $\Sigma \Cap d$. This is done by recursion on the structure of the environment. $[d]_{coq}$ is the straightforward translation of d in Coq, each call **self**!x being replaced by the introduced variable $abst\_s$.

Definition 7 (Method Generator).

$$\begin{aligned} [\![\emptyset, d]\!] &= [d]_{coq} \\ [\![\{x:\tau = e; l\}, d]\!] &= \textbf{\textit{Let}}\ abst\_x : \tau := (gen_x\ abst\_x_i)\ \textbf{\textit{in}}\ [\![l, d]\!] \\ [\![\{x:\tau = \bot; l\}, d]\!] &= \lambda abst\_x : \tau.\ [\![l, d]\!] \end{aligned}$$

where $gen_x = [\![\Sigma \Cap \mathcal{B}_s(x), \mathcal{B}_s(x)]\!]$ *and* $abst\_x$ *is a fresh name.*

The second case treats def-dependencies. The method x being defined in Σ as $\{x : \tau = e\}$ has already been compiled to Coq. Thus its method generator gen_x has been obtained by abstracting the names x_i of $\Sigma \Cap \mathcal{B}_s(x)$ (note that $\Sigma \Cap \mathcal{B}_s(x) \subseteq \Sigma \Cap d$). Here, gen_x is applied to the corresponding $abst\_x_i$.

The third case concerns –simple– decl-dependencies. We only abstract x.

5.3 Collections

Collections are defined using the method generators. Namely, if c implements $s = \{x_i : \tau_i = e_i\}$, the Coq translation is the following:

$$\textbf{Definition}\ c\_x_1 := gen\_x_1.$$

$$\dots$$

$$\textbf{Definition}\ c\_x_n := (gen\_x_n\ ([\![x_n]\!]_s).$$

$$\textbf{Definition}\ c := (mk\_s\ c\_x_1 \dots c\_x_n).$$

where $[\![x]\!]_s = \{x_i \in \mathcal{N}(s) | \, (x_i, \tau_i, \bot) \in \Sigma \Cap \mathcal{B}_s(x)\}$. $[\![x]\!]_s$ represents the definitions that must be provided to the method generator in order to define x. mk_s is the function that generates the record corresponding to the interface of s.

5.4 Parameters

A natural way to handle parameters in Coq would be to create functions that take `Records` as arguments and return `Records`. For instance, (the interface of) a cartesian product can be defined like this:

```
Record cartesian [ A, B : basic_object] : Type :=
        { T : Set; fst : T -> A ...}
```

Another solution is to take the parameters as the first fields of the `Record`:

```
Record cartesian : Type :=
        { A : basic_object; B: basic_object; ...}
```

These two translations are quite similar for Coq. In the first one, `cartesian` will be a parameterized type, while it is not the case in the second one: `A` and `B` are only the first two arguments of its unique constructor. The second solution seems to have some practical advantages over the first one:

- Parameters can be accessed directly as fields of the record
- Fields accesses (the equivalent of methods call) do not need extra arguments, as it would be the case in the first solution.
- Coercions between parameterized records are easier to define too.
- More important, it reflects the fact that collections can not have parameters: in an implementation of `cartesian`, the fields `A` and `B` must be defined as well as `T` and `fst`.

6 Related Work

Other projects use Coq's `Records` to represent algebraic structure. In particular, L. Pottier [13] has developed quite a large mathematical library, up to fields. H. Geuvers and the FTA project [9] have defined abstract and concrete representations of reals and complex numbers. In addition, R. Pollack [12] and G. Betarte [1] have given their own embedding of dependent records in Type Theory. We can also mention Imps [8], a proof system which aims at providing a computational support for mathematical proofs. However, none of these works include a computational counterpart, similar to the OCaml translation of Foc. P. Jackson [11] implemented a specification of multivariate polynomials in Nuprl. His approach is quite different from Foc, as in his formalism, a group can not be directly considered as a monoid, for instance.

7 Conclusion and Future Work

To sum up, we can say that Foc has now achieved a quite good expressive power. The static analyses that are discussed in Sec. 4 have been implemented [14] in a compiler that generates Coq and OCaml code. An important number of mathematical structures have been implemented, and performances are good.

It seems to us that we provide a well-adapted framework to prove the properties needed for each species' implementation. It it is now necessary to define a proof language for Foc, dedicated to users of computer algebra systems. This is currently under development.

Building mathematical structures requires the whole power of the Calculus of Inductive Constructions, but higher-order features are mostly needed only to handle dependencies. Once we have succeeded to build an appropriate environment, the proofs themselves stay in first order logic most of the time. This may lead to a quite high level of automatization in the proof part of the project, leading to proofs in deduction modulo [6,7]. We could then try to delegate some part of the proofs to rewriting tools. Similarly, it would be interesting to offer powerful tools that allow the user of Foc to define his own Foc proof *tactics*.

D. Delahaye's PhD [5] presents very promising developments in this area and may be of great help here. From the Curry-Howard point of view this future work is the counterpart in the proof universe of the basic expressions of the FOC language.

References

1. G. Betarte. *Dependent Record Types and Formal Abstract Reasoning: Theory and Practice.* PhD thesis, University of Göteborg, 1998.
2. S. Boulmé, T. Hardin, and R. Rioboo. Polymorphic data types, objects, modules and functors: is it too much ? Research Report 14, LIP6, 2000. available at `http://www.lip6.fr/reports/lip6.2000.014.html`.
3. S. Boulmé. *Spécification d'un environnement dédié à la programmation certifiée de bibliothèques de Calcul Formel.* PhD thesis, Université Paris 6, december 2000.
4. B. Buchberger and all. A survey on the theorema project. In W. Kuechlin, editor, *Proceedings of ISSAC'97.* ACM Press, 1997.
5. D. Delahaye. *Conception de langages pour décrire les preuves et les automatisations dans les outils d'aide à la preuve.* PhD thesis, Université Paris 6, 2001.
6. G. Dowek, T. Hardin, and C. Kirchner. Theorem proving modulo. Research Report 3400, INRIA, 1998.
7. G. Dowek, T. Hardin, and C. Kirchner. Hol-$\lambda\sigma$: an intentional first-order expression of higher-order logic. *Mathematical Structures in Computer Science*, 11(1):21–45, 2001.
8. W. M. Farmer, J. D. Guttman, and F. J. Thayer. The IMPS user's manual. Technical Report M-93B138, The MITRE Corporation, 202 Burlington Road, Bedford, MA 01730-1420, USA, November 1995. Available at `ftp://math.harvard.edu/imps/doc/`.
9. H. Geuvers, R. Pollack, F. Wiedijk, and J. Zwanenburg. The algebraic hierarchy of the fta project. In *Proceedings of the Calculemus Workshop*, 2001.
10. R. Harper and M. Lillibridge. A type-theoretic approach to higher-order modules with sharing. In *21st Symposium on Principle of Programming Languages*, 1994.
11. P. Jackson. Exploring abstract algebra in constructive type theory. In *Proceedings of 12th International Conference on Automated Deduction*, July 1994.
12. R. Pollack. Dependently typed records for representing mathematical structures. In *TPHOLs 2000.* Springer-Verlag, 2000.
13. L. Pottier. contrib algebra pour coq, mars 1999. `http://pauillac.inria.fr/coq/contribs-eng.html`.
14. V. Prevosto, D. Doligez, and T. Hardin. Overview of the FOC compiler. to appear as a research report, LIP6, 2002. available at `http://www-spi.lip6.fr/~prevosto/papiers/foc2002.ps.gz`>.
15. The Coq Development Team. *The Coq Proof Assistant Reference Manual.* Projet LogiCal, INRIA-Rocquencourt – LRI Paris 11, Nov. 1996.

Proving the Equivalence of Microstep and Macrostep Semantics

Klaus Schneider

University of Kaiserslautern, Department of Computer Science
P.O. Box 3049, 67653 Kaiserslautern, Germany
Klaus.Schneider@informatik.uni-kl.de

Abstract. Recently, an embedding of the synchronous programming language Quartz (an Esterel variant) in the theorem prover HOL has been presented. This embedding is based on control flow predicates that refer to macrosteps of the programs. The original semantics of synchronous languages like Esterel is however normally given at the more detailed microstep level. This paper describes how a variant of the Esterel microstep semantics has been defined in HOL and how its equivalence to the control flow predicate semantics has been proved. Beneath proving the equivalence of the micro- and macrostep semantics, the work presented here is also an important extension of the existing embedding: While reasoning at the microstep level is not necessary for code generation, it is sometimes advantageous for understanding programs, as some effects like schizophrenia or causality problems become only visible at the microstep level.

1 Introduction

Reasoning about programming languages is one of the main applications of higher order logic theorem provers [9, 10, 33, 13, 26]. This requires to formalize the programming language in the theorem prover's logic in advance. In general, one distinguishes thereby between deep and shallow embeddings [3]: Using a deep embedding requires to define constants for the statements of the programming language; hence programs become higher order logic terms. In contrast, a shallow embedding simply translates programs (outside the theorem prover) into formulas of its logic. In general, a deep embedding is more complicated, but offers additional advantages like reasoning about the programming language at a meta level. In particular, such a meta level reasoning is of interest when functions on programs like program transformations or code generation are to be verified. This is mandatory for safety-critical applications. Synchronous languages like Esterel [5, 7] or variants thereof [19, 18, 29] are more and more used for designing safety-critical reactive real time systems [14]. As synchronous languages have furthermore a formally well-grounded semantics, we have chosen Esterel-like synchronous languages to reason about reactive systems in HOL.

The basic paradigm of synchronous languages is the *perfect synchrony*, which follows from the fact that most of the statements are executed as 'microsteps', i.e. without taking time. Consumption of time must be explicitly programmed with special statements like Esterel's **pause** statement: The execution of a **pause** statement consumes one logical unit of time, and therefore separates different macrosteps from each other.

V.A. Carreño, C. Muñoz, S. Tahar (Eds.): TPHOLs 2002, LNCS 2410, pp. 314–331, 2002.

As the **pause** statement is the only (basic) statement that consumes time, all threads run in lockstep: they execute the microsteps between two **pause** statements in zero time, and automatically synchronize at their next **pause** statements. The distinction between micro- and macrosteps is therefore the reason for the synchronous execution of threads. As a result, the synchronous languages are the only programming languages that allow both multi-threaded and deterministic programs.

The abstraction to macrosteps makes synchronous programming so attractive. It is not only an ideal programmer's model, it additionally allows the direct translation of programs into synchronous hardware circuits, where macrosteps directly correspond to clock cycles[1]. As the same translation is also used for software generation, many optimizations known for hardware circuits can be used to optimize software as well [6, 15]. For this reason, synchronous programs are a good basis for HW/SW codesign [25].

However, the abstraction to macrosteps is not for free: Schizophrenia problems and causality problems are the two major problems that must be solved by a compiler. *Causality cycles* arise due to conflicts between enabling conditions and immediate data manipulations of a statement (in a hardware circuit, this corresponds to combinatorial loops). Algorithms for causality analysis, that check if such cycles yield stable values, are already available [21, 17, 12, 31, 11, 6]. *Schizophrenia problems* are due to multiple execution of the same statement at the same point of time (i.e. in the same macrostep). In particular, this becomes problematic if a local declaration is multiply entered with different values for the local variable. Schizophrenia problems are also well-known and efficient solutions exist as well[2] [24, 6, 30].

A lot of different ways have already been studied to define the semantics of imperative synchronous languages (see Section 2 for more details). At least, there are (1) semantics based on process algebras [8, 6, 32], (2) semantics based on hardware circuits or equations [1, 24, 6, 28], and (3) semantics based on control flow predicates [29]. While the SOS semantics reflects microsteps of the program, the other two mentioned semantics do only consider macrosteps. Hence, the SOS semantics has a finer granularity and is therefore able to explain the behavior of a program in a more detailed manner. While this is neither necessary for code generation nor for verification, it is sometimes helpful for understanding programs, in particular those that have causality or schizophrenia problems. For this reason, both the microstep and the macrostep semantics are important for different reasons. Using different semantics raises however the question whether both were equivalent, which is the topic of this paper.

Before the equivalence of two semantics can be proved with a theorem prover, we have to first embed the language into the theorem prover. A crucial problem for embedding languages (or more general theories) in already existing logics is to avoid inconsistencies: Simply postulating a set of axioms may lead to inconsistent theories so that everything could be derived. State-of-the-art theorem provers like HOL [16] therefore use certain definition principles to preserve the consistency. One main definition principle that guarantees such a conservative extension is primitive recursion [22]. Primitive

---

[1] As the hardware circuit is a synchronous one, all macrosteps require then the same amount of physical time. This is normally not the case for a software implementation.

[2] We must however admit that the transformation given in [30] is not correct in any case.

recursive definitions have moreover the advantage that the theorem prover can automatically derive suitable induction rules.

However, it is not straightforward to define the semantics of a language by means of primitive recursion only. In particular, the SOS semantics of Esterel can not be directly defined by primitive recursion, since the rule for loops recursively calls itself without 'decreasing' the program (see Section 2.4). Therefore, the control flow predicate semantics has been prefered in [29] to embed the Esterel variant Quartz in HOL. This embedding has been used so far for various purposes like formal synthesis [28], reasoning about schizophrenia problems [30], and of course, for the formal verification of program properties. The difficulty to define a SOS semantics is due to the fact that the SOS rules follow the potentially nonterminating execution of the program. Therefore, primitive recursive definitions are not adequate (since they always terminate). There is however a simple trick to circumvent this problem: Instead of using directly the SOS rules, one may use a haltset encoding of them [8] (see Section 3 for our version).

In this paper, we define a haltset encoding of the SOS rules of our Esterel variant Quartz in HOL. Some intrinsic technical problems have to be solved for such a definition (see Section 4). Having solved them, we have then proved the equivalence of (1) the SOS semantics defined in this paper, (2) the previous semantics based on control flow predicates [29], and (3) the semantics based on circuits/equations [28]. As a result, we can now also reason about microsteps of programs which is sometimes necessary for understanding programs with causality or schizophrenia problems. We do however not consider these issues in this paper (the presented semantics does not yet consider causality and is therefore equivalent to the logical semantics defined in [6]). Moreover, we do neither consider local declarations nor schizophrenia problems in this paper.

The paper is organized as follows: in the next section, we define the syntax and semantics of Quartz. We will also briefly consider the different kinds of semantics that have been developed so far. In Section 3, we then define our haltset encoding of the SOS semantics. The results of the paper are given in Section 4: After discussing some technical problems for the embedding of the haltset encoding in HOL, we will list there the proved theorems that show the equivalence to the other semantics.

2 Syntax and Semantics

Quartz [28, 29] is a variant of Esterel [5, 7, 14] that extends Esterel by delayed assignments and emissions, asynchronous concurrency, nondeterministic choice, and inline assertions. Asynchronous concurrency is important to model distributed systems, or to allow the compiler to schedule the threads in an optimal way. The same holds for nondeterministic choice. Delayed assignments and emissions are often convenient, since they follow the traditional sequential programming style and therefore allow simpler translations from conventional programming languages like C. In the following, we briefly describe the syntax and semantics of Quartz. For more details, the reader is referred to [29, 28, 30, 20] or to the Esterel primer, which is an excellent introduction to synchronous programming [7].

2.1 Syntax and Informal Semantics

Logical time is modeled by the natural numbers $\mathbb{N}$, so that the semantics of a data type expression is a function of type $\mathbb{N} \to \alpha$ for some type α. Quartz distinguishes between two kinds of variables, namely *event variables* and *state variables*. The semantics of an event variable is a function of type $\mathbb{N} \to \mathbb{B}$, while the semantics of a state variable may have the more general type $\mathbb{N} \to \alpha$. The main difference between event and state variables is however the data flow: the value of a state variable y is 'sticky', i.e. if no data operation has been applied to y, then its value does not change. On the other hand, the value of an event variable x is not sticky: its value is reset to false (we denote Boolean values as true and false) in the next macrostep, if it is not explicitly made true there. Hence, the value of an event variable is true at a point of time if and only if there is a thread that emits this variable at this point of time. Event variables are made present with the **emit** statement, while state variables are manipulated with assignments (:=). Of course, any event or state variable may also be an input, so that the values are determined by the environment only. Emissions and assignments are all data manipulating statements. The remaining basic statements of Quartz are given below:

Definition 1. (Basic Statements of Quartz**)** *The set of basic statements of* Quartz *is the smallest set that satisfies the following rules, provided that S, S_1, and S_2 are also basic statements of* Quartz*, ℓ is a location variable, x is an event variable, y is a state variable of type α, σ is a Boolean expression, and τ an expression of type α:*

- **nothing** *(empty statement)*
- **emit** x *and* **emit next**(x) *(emissions)*
- $y := \tau$ *and* $y :=$ **next**(τ) *(assignments)*
- ℓ : **pause** *(consumption of time)*
- **if** σ **then** S_1 **else** S_2 **end** *(conditional)*
- $S_1; S_2$ *(sequential composition)*
- $S_1 \parallel S_2$ *(synchronous parallel composition)*
- $S_1 \parallel\mid S_2$ *(asynchronous parallel composition)*
- **choose** S_1 $[\!]$ S_2 **end** *(nondeterministic choice)*
- **do** S **while** σ *(iteration)*
- **suspend** S **when** σ *(suspension)*
- **weak suspend** S **when** σ *(weak suspension)*
- **abort** S **when** σ *(abortion)*
- **weak abort** S **when** σ *(weak abortion)*
- **local** x **in** S **end** *(local event variable)*
- **local** $y : \alpha$ **in** S **end** *(local state variable)*
- **now** σ *(instantaneous assertion)*
- **during** S **holds** σ *(invariant assertion)*

In general, a statement S may be started at a certain point of time t_1, and may terminate at time $t_2 \geq t_1$, but it may also never terminate. If S immediately terminates when it is started ($t_2 = t_1$), it is called *instantaneous*, otherwise we say that the execution of S takes time, or simply that S *consumes time*.

Let us now discuss the above basic statements: **nothing** simply does nothing, i.e., it neither consumes time, nor does it affect any data values. Executing **emit** x makes the event variable x present for the current macrostep, i.e., the value of x at that point of time is true. Executing an assignment $y := \tau$ means that y and τ have the same values in the current macrostep. The variants **emit next**(x) and $y :=$ **next**(τ) are similarly defined as **emit** x and $y := \tau$, respectively, but with a delay of one macrostep. In the latter statement, τ is evaluated at the current point of time, and its value is passed to y at the next point of time. We emphasize that none of these statements consumes time, although the delayed versions affect values of variables at the next point of time.

There is only one basic statement that consumes time, namely the **pause** statement. It does not affect any data values. We endow **pause** statements with unique location variables ℓ that we will use as state variables to encode the control flow automaton. The location variables are HOL variables of type $\mathbb{N} \to \mathbb{B}$, which yields some problems for the definition of the haltset encoded SOS rules. The uniqueness is not guaranteed by the embedding and must therefore be checked separately.

Depending on the value of σ in the current macrostep, the conditional statement **if** σ **then** S_1 **else** S_2 **end** either S_1 or S_2 is immediately executed. $S_1; S_2$ is the sequential execution of S_1 and S_2, i.e., we first enter S_1 and execute it. If S_1 never terminates, then S_2 is never executed at all. If, on the other hand, S_1 terminates, we immediately start S_2 and proceed with its execution.

$S_1 \parallel S_2$ denotes the synchronous parallel execution of S_1 and S_2: If $S_1 \parallel S_2$ is entered, we enter both S_1 and S_2, and proceed with the execution of both statements. As long as both S_1 and S_2 are active, both threads are executed in lockstep. If S_1 terminates, but S_2 does not, then $S_1 \parallel S_2$ behaves further as S_2 does (and vice versa). If finally S_2 terminates, then $S_1 \parallel S_2$ terminates.

Beneath the synchronous concurrency, Quartz additionally has asynchronous concurrency $S_1 \mid\mid\mid S_2$. The difference is that one of the threads may execute more than one macrostep while the other one executes a single one or even none. One may argue that the presence of asynchronous parallel execution contradicts the definition of a synchronous language. However, it is not too difficult to replace $S_1 \mid\mid\mid S_2$ by standard Esterel statements using additional inputs [29] (that are called control variables). Another Quartz statement that does not belong to Esterel is the nondeterministic choice: **choose** $S_1 \; [\!] \; S_2$ **end** will nondeterministically execute either S_1 or S_2. Again, using additional input (control) variables, nondeterministic choice can be reduced to other statements [29], so that we neither consider nondeterministic choice nor asynchronous concurrency in the following.

do S **while** σ implements iteration: if this statement is entered, S is executed until it terminates. If then σ holds, S is once more executed, otherwise the loop terminates. It is required that for any input, the loop body S must not be instantaneous.

suspend S **when** σ implements process suspension: S is entered when the execution of this statement starts (regardless of the current value of σ). For the following points of time, however, the execution of S only proceeds if σ evaluates to false, otherwise its execution is 'frozen' until σ releases the further execution. Beneath suspension, abortion of processes is an important means for the process management. This is realized with the **abort** statements: **abort** S **when** σ immediately enters S at starting time (regardless of the current value of σ). Then, S is executed as long as σ is false. If σ becomes true during the execution of S, then S is immediately aborted. The 'weak' variants of suspension and abortion differ on the data manipulations at suspension or abortion time: While the strong variants ignore *all* data manipulations at abortion or suspension time, *all of them* are performed by the weak variants. There are also **immediate** variants of suspension and abortion that consider the condition σ additionally at starting time. These can be defined in terms of the other variants [29].

The statements **local** x **in** S **end** and **local** $y : \alpha$ **in** S **end** are used to define local event and local state variables, respectively. Their meaning is that they behave like S, but the scope of the variable x or y is limited to S. This means that the local variable is not seen outside the **local** statement. We do not consider local declarations in this paper to avoid the difficulties with schizophrenia problems.

Quartz allows us to demand assertions that must hold when the control flow reaches certain locations: **now** σ demands that σ must hold in the current macrostep. **during** S **holds** σ behaves like S, but additionally demands that whenever the control flow is inside S, then σ must hold. There is no further execution if the condition σ does not hold; the behavior is not defined in this case.

2.2 Semantics Based on Control Flow Predicates

The separation between control and data flow is a well-known technique for hardware designers. The semantics of Quartz has been defined using such a separation, where the definition of the control flow is based on the control flow predicates $\mathsf{enter}(S)$, $\mathsf{move}(S)$, and $\mathsf{term}(S)$, that describe entering conditions, conditions for internal moves, and termination conditions of a statement S, respectively. The data flow of a statement is defined by its *guarded commands*. Some more details are given below:

$\mathsf{in}(S)$ is the disjunction of the **pause** labels occurring in S. Therefore, $\mathsf{in}(S)$ holds at some point of time iff at this point of time, the control flow is at some location inside S.

$\mathsf{inst}(S)$ holds iff the control flow can not stay in S when S would now be started. This means that the execution of S would be instantaneous at this point of time.

$\mathsf{enter}(S)$ describes where the control flow will be at the next point of time, when S would now be started.

$\mathsf{term}(S)$ describes all conditions where the control flow is currently somewhere inside S and wants to leave S. Note however, that the control flow might still be in S at the next point of time since S may be entered at the same time, for example, by a surrounding loop statement.

$\mathsf{move}(S)$ describes all internal moves, i.e., all possible transitions from somewhere inside S to another location inside S.

$\mathsf{guardcmd}(\varphi, S)$ is a set of pairs of the form $(\gamma, \mathcal{C})$, where $\mathcal{C}$ is a data manipulating statement, i.e., either an emission or an assignment. The meaning of $(\gamma, \mathcal{C})$ is that $\mathcal{C}$ is immediately executed whenever the guard γ holds.

Note that the above control flow predicates depend on time, i.e. are of type $\mathbb{N} \to \mathbb{B}$. Detailed definitions of the above predicates and the set of guarded commands are given in [29], the important thing to notice here is that all of them can be defined by simple primitive recursive definitions of the set of Quartz statements.

2.3 Semantics Based on Hardware Circuits

The translation of synchronous programs to equation systems, which may directly be interpreted as hardware circuits, is the essential means of the currently dominating compilation techniques (both for software and hardware). The idea is quite old [4, 27] and many other variants have been defined since then [24, 1, 15, 6, 28, 30]. Similar to the control flow predicates, only macrosteps are considered in this kind of semantics. The generated hardware circuit is a synchronous one that executes in each clock cycle a macrostep of the program. Clearly, **pause** statements correspond to flipflops, and microsteps are implemented with combinatorial logic gates. If software is to be generated, then the software is more or less a simulator of the particular hardware circuit: the software simply holds local variables for the flipflops, and computes the current outputs as well as the next values of the flipflops from the current inputs and the current values of the flipflops. Clearly, arbitrary data types including pointer structures can be manipulated in software, which is however not possible in a pure hardware design.

2.4 Semantics Based on Process Algebraic Rules

The original semantics of Esterel [8, 5, 6, 32] is given by a structural operational semantics (SOS) [23, 2] which can be written as rules of the form $S \xrightarrow[\mathcal{E}]{\mathcal{D}, b} S'$, where S and S' are statements, $\mathcal{D}$ are actions (emissions or assignments), b is a Boolean value[3] (the completion flag), and $\mathcal{E}$ is the current environment. The rules describe executions of microsteps, and indicate with the completion flag b the beginning/end of a macrostep. The values of the inputs and outputs during that macrostep are thereby known by the environment $\mathcal{E}$ and can be asked for the definition of the rules. For example, the rules for starting a conditional are as follows, which means that depending on whether σ holds in the current environment $\mathcal{E}$, we either execute the 'then' or the 'else' branch:

$$\frac{S_1 \xrightarrow[\mathcal{E}]{\mathcal{D}_1, b_1} S_1' \qquad \mathcal{E} \models \sigma}{\textbf{if } \sigma \textbf{ then } S_1 \textbf{ else } S_2 \textbf{ end} \xrightarrow[\mathcal{E}]{\mathcal{D}_1, b_1} S_1'} \qquad \frac{S_2 \xrightarrow[\mathcal{E}]{\mathcal{D}_2, b_2} S_2' \qquad \mathcal{E} \not\models \sigma}{\textbf{if } \sigma \textbf{ then } S_1 \textbf{ else } S_2 \textbf{ end} \xrightarrow[\mathcal{E}]{\mathcal{D}_2, b_2} S_2'}$$

The flag b indicates whether the execution completes the macrostep. If this is not the case, further microsteps can be executed in that macrostep. For example, this is used in the following rules to capture the semantics of sequences:

$$\frac{S_1 \xrightarrow[\mathcal{E}]{\mathcal{D}_1, \mathsf{false}} S_1' \qquad S_2 \xrightarrow[\mathcal{E}]{\mathcal{D}_2, b_2} S_2'}{S_1; S_2 \xrightarrow[\mathcal{E}]{\mathcal{D}_1 \cup \mathcal{D}_2, b_2} S_2'}$$

$$\frac{S_1 \xrightarrow[\mathcal{E}]{\mathcal{D}_1, \mathsf{true}} \textbf{nothing}}{S_1; S_2 \xrightarrow[\mathcal{E}]{\mathcal{D}_1, \mathsf{true}} S_2} \qquad \frac{S_1 \xrightarrow[\mathcal{E}]{\mathcal{D}_1, \mathsf{true}} S_1' \qquad S_1' \neq \textbf{nothing}}{S_1; S_2 \xrightarrow[\mathcal{E}]{\mathcal{D}_1, \mathsf{true}} S_1'; S_2}$$

[3] As we do not consider **trap** statements, Boolean values are sufficient. Although not trivial, **trap** statements can always be replaced with corresponding immediate weak abortion statements whose abortion conditions respect the priority rules of the **trap** statements.

The first rule means that we first execute a part of S_1 so that the further execution has to proceed with statement S_1'. If this partial execution does not complete a macrostep ($b =$ false, which implies $S_1' =$ **nothing**), then we immediately start S_2 to further execute microsteps of that macrostep. On the other hand, if the execution of S_1 completes a macrostep ($b =$ true), then S_2 is not started in that macrostep. Instead, we either start S_2 in the next macrostep or resume the execution of S_1 by starting S_1'.

The actions $\mathcal{D}$ of a rule $S \xrightarrow[\mathcal{E}]{\mathcal{D}, b} S'$ are the emissions and assignments that are executed during the execution from S to S'. Clearly, this must be consistent with the environment $\mathcal{E}$: output event variables that are emitted must be present, and all present output event variables must be emitted somewhere in the macrostep. Similar conditions have to hold for state variables: Changes are only allowed with corresponding assignments, and write conflicts are forbidden.

The problem with the SOS rules is that they describe the potentially infinite execution of the program by recursive definitions. This leads to the following problem with the rules for loops (we consider here while-do loops instead of the do-while loops to keep the paper more concise):

$$\frac{\mathcal{E} \not\models \sigma}{\textbf{while}\ \sigma\ \textbf{do}\ S\ \textbf{end} \xrightarrow[\mathcal{E}]{\{\}, \mathsf{false}} \textbf{nothing}}$$

$$\frac{\mathcal{E} \models \sigma \quad S \xrightarrow[\mathcal{E}]{\mathcal{D}, \mathsf{true}} S' \quad S' \neq \textbf{nothing}}{\textbf{while}\ \sigma\ \textbf{do}\ S\ \textbf{end} \xrightarrow[\mathcal{E}]{\mathcal{D}, \mathsf{true}} S'; \textbf{while}\ \sigma\ \textbf{do}\ S\ \textbf{end}}$$

Note that bodies of loop statements must not be instantaneous, i.e. for all possible inputs they must consume time. Otherwise, the program would have to execute infinitely many microsteps in one macrostep, which is never the case for synchronous programs. This property makes synchronous programs well-suited for worst-case execution time analysis.

As can be seen, the second rule for loops refers to itself, since in the last rule the entire while-statement still appears on the right hand side. For this reason, the SOS rules do not allow a simple definition by primitive recursion and can therefore not be directly used for an embedding of the language. We will however describe in the next section a simple trick to circumvent this problem that goes back to [8].

3 The Haltset Encoding of SOS Rules

The SOS rules discussed in the previous section were of the form $S \xrightarrow[\mathcal{E}]{\mathcal{D}, b} S'$, i.e., the transitions connect statements S and S'. These statements are often called derivatives of an original statement and are used to formalize where the control flow currently rests, or more precisely from which locations it will proceed with the execution.

Instead, one could also encode the current control flow position simply by collecting the labels of those **pause** statements that currently hold the control flow. Then, rules of the form $(S,\mathcal{E}) \rhd H \xrightarrow{\mathcal{D}} H'$ can be defined in the same manner as before. The meaning is as follows: if the control flow is currently located at the **pause** statements whose labels are contained in H, then the execution in the macrostep with environment $\mathcal{E}$ will invoke the data manipulating actions collected in $\mathcal{D}$, and will then stop at the **pause** statements whose labels are contained in H'. Note that the execution completes a macrostep iff H' will be empty. For this reason, there is no longer the need for using a completion flag in the haltset encoded SOS rules.

As a result, the SOS rules that are encoded with haltsets are now defined by recursion over the haltsets which turns them now into primitive recursive rules. For example, the rules for loops now look as follows:

$$\frac{\mathcal{E} \models \sigma \qquad (S,\mathcal{E}) \rhd \{\} \xrightarrow{\mathcal{D}} H}{(\textbf{while}\ \sigma\ \textbf{do}\ S\ \textbf{end}, \mathcal{E}) \rhd \{\} \xrightarrow{\mathcal{D}} H}\ \textit{(enter loop)}$$

$$\frac{\mathcal{E} \not\models \sigma}{(\textbf{while}\ \sigma\ \textbf{do}\ S\ \textbf{end}, \mathcal{E}) \rhd \{\} \xrightarrow{\{\}} \{\}}\ \textit{(bypas loop)}$$

$$\frac{(S,\mathcal{E}) \rhd H \xrightarrow{\mathcal{D}} H' \qquad H \neq \{\} \qquad H' \neq \{\}}{(\textbf{while}\ \sigma\ \textbf{do}\ S\ \textbf{end}, \mathcal{E}) \rhd H \xrightarrow{\mathcal{D}} H'}\ \textit{(start loop)}$$

$$\frac{H \neq \{\} \qquad (S,\mathcal{E}) \rhd H \xrightarrow{\mathcal{D}} \{\} \qquad \mathcal{E} \models \sigma \qquad (S,\mathcal{E}) \rhd \{\} \xrightarrow{\mathcal{D}'} H'}{(\textbf{while}\ \sigma\ \textbf{do}\ S\ \textbf{end}, \mathcal{E}) \rhd H \xrightarrow{\mathcal{D} \cup \mathcal{D}'} H'}\ \textit{(reenter loop)}$$

$$\frac{H \neq \{\} \qquad (S,\mathcal{E}) \rhd H \xrightarrow{\mathcal{D}} \{\} \qquad \mathcal{E} \not\models \sigma}{(\textbf{while}\ \sigma\ \textbf{do}\ S\ \textbf{end}, \mathcal{E}) \rhd H \xrightarrow{\mathcal{D}} \{\}}\ \textit{(exit loop)}$$

As a result, a transition system is obtained whose nodes are labeled with haltsets, i.e., subsets of the set of labels of the program. Clearly, a program with n **pause** statements will have at most 2^n reachable states in the transition system. The important matter of fact is that this will always be a finite number of states and therefore the recursion of the SOS rules will always terminate.

4 Technical Problems Concerning the Implementation

In principle, SOS rules with the haltset encoding could be used to define the semantics of Quartz in HOL. However, the already available embedding does not use metavariables, i.e., event and state variables as well as location variables and all other expressions that appear in a Quartz program are directly taken from HOL, i.e. these are already available HOL variables and expressions. This is very convenient for defining the control flow predicate semantics and also very convenient for the definition of the hardware circuit semantics. However, it makes it impossible to use haltsets as given in the previous section, because the labels of **pause** statements are just HOL variables of type $\mathbb{N} \to \mathbb{B}$ (i.e. anonymous functions). If two labels of different **pause** statements are active at the same points of time, then we can not distinguish them.

To solve this problem, we assign to each **pause** statement a unique number (its index) by a simple left-to-right traversal over the program. The haltsets consist then of the corresponding indices instead of the anonymous labels. The price to pay is however that when we proceed from one substatement to another one, we often have to add or subtract an offset from the corresponding index set. We furthermore found it simpler to use lists of indices instead of sets of indices. This added no further problems, but allowed to use simple induction over the lengths of the lists. To be concise in the following, we use the following abbreviations[4]:

- $H \uparrow_p := \mathsf{MAP}\ (\lambda x.x + p)\ H$
- $H \downarrow_p := \mathsf{MAP}\ (\lambda x.x - p)\ H$
- $H \mid_a^b := \mathsf{FILTER}\ (\lambda x.a \leq x \wedge x \leq b)\ H$

The variables and data expressions that appear in the program are still taken from HOL, i.e. data expressions in the program of type α are HOL terms of type $\mathbb{N} \to \alpha$ (the natural numbers are used to model time). Therefore, there is no need to consider the environment $\mathcal{E}$ in an explicit manner. Instead, the environment $\mathcal{E}$ of a point of time t is simply replaced with t. Assumptions like $\mathcal{E} \models \sigma$ are then simply replaced with $\sigma(t)$.

Finally, we need the number $\mathsf{NP}(P)$ of **pause** statements of a program P, and define furthermore a predicate $\mathsf{InHS}(H, P)$ so that $\mathsf{InHS}(H, P)$ holds[5] iff the haltset H contains an index that corresponds to a **pause** statement of program P. Given a point of time t, a haltset H of a program P, we now define a haltset $\mathsf{NextHS}(H, t, P)$ so that $(P, t) \rhd H \xrightarrow{\mathcal{D}} \mathsf{NextHS}(H, t, P)$ holds for some set $\mathcal{D}$ (recall that $\mathcal{E}$ has been replaced with the point of time t). $\mathsf{NextHS}(H, t, P)$ represents the control flow part of the SOS rules and is defined as follows by primitive recursion over the set of statements:

- $\mathsf{NextHS}(H, t, \textbf{nothing}) = []$
- $\mathsf{NextHS}(H, t, \textbf{emit}\ x) = []$
- $\mathsf{NextHS}(H, t, \textbf{emit next}(x)) = []$
- $\mathsf{NextHS}(H, t, y := \tau) = []$
- $\mathsf{NextHS}(H, t, y := \textbf{next}(\tau)) = []$
- $\mathsf{NextHS}(H, t, \ell : \textbf{pause}) = \text{if } \mathsf{MEM}(1, H) \text{ then } [] \text{ else } [1]$

[4] We use the constants MAP, MEM, APPEND, and FILTER that are defined in the HOL theory for lists with the intended meaning.

[5] The definition of $\mathsf{InHS}(H, P)$ has been given by primitive recursion over P. Then, we have proved that $\mathsf{InHS}(H, S) \Leftrightarrow \left(H \mid_1^{\mathsf{NP}(S)} \neq []\right)$ holds.

- $\mathsf{NextHS}(H, t, \textbf{if}\ \sigma\ \textbf{then}\ P\ \textbf{else}\ Q\ \textbf{end}) =$
$$\left(\begin{array}{l}\mathsf{let}\ H_P = H\ |_1^{\mathsf{NP}(P)}\ \mathsf{in}\\ \mathsf{let}\ H_Q = H\ \downarrow_{\mathsf{NP}(P)}|_1^{\mathsf{NP}(Q)}\ \mathsf{in}\\ \mathsf{let}\ H'_P = \mathsf{NextHS}(H_P, t, P)\ \mathsf{in}\\ \mathsf{let}\ H'_Q = \mathsf{NextHS}(H_Q, t, Q) \uparrow_{\mathsf{NP}(P)}\ \mathsf{in}\\ \quad \mathsf{if}\ \mathsf{InHS}(H_P, P)\ \mathsf{then}\ H'_P\\ \quad \mathsf{elseif}\ \mathsf{InHS}(H_Q, Q)\ \mathsf{then}\ H'_Q\\ \quad \mathsf{elseif}\ \sigma(t)\ \mathsf{then}\ H'_P\\ \quad \mathsf{else}\ H'_Q\end{array}\right)$$
- $\mathsf{NextHS}(H, t, P; Q) =$
$$\left(\begin{array}{l}\mathsf{let}\ H_P = H\ |_1^{\mathsf{NP}(P)}\ \mathsf{in}\\ \mathsf{let}\ H_Q = H\ \downarrow_{\mathsf{NP}(P)}|_1^{\mathsf{NP}(Q)}\ \mathsf{in}\\ \mathsf{let}\ H'_P = \mathsf{NextHS}(H_P, t, P)\ \mathsf{in}\\ \mathsf{let}\ H'_Q = \mathsf{NextHS}(H_Q, t, Q) \uparrow_{\mathsf{NP}(P)}\ \mathsf{in}\\ \quad \mathsf{if}\ \mathsf{InHS}(H_Q, Q)\ \mathsf{then}\ H'_Q\\ \quad \mathsf{elseif}\ H'_P = []\ \mathsf{then}\ H'_Q\\ \quad \mathsf{else}\ H'_P\end{array}\right)$$
- $\mathsf{NextHS}(H, t, P \parallel Q) =$
$$\left(\begin{array}{l}\mathsf{let}\ H_P = H\ |_1^{\mathsf{NP}(P)}\ \mathsf{in}\\ \mathsf{let}\ H_Q = H\ \downarrow_{\mathsf{NP}(P)}|_1^{\mathsf{NP}(Q)}\ \mathsf{in}\\ \mathsf{let}\ H'_P = \mathsf{NextHS}(H_P, t, P)\ \mathsf{in}\\ \mathsf{let}\ H'_Q = \mathsf{NextHS}(H_Q, t, Q) \uparrow_{\mathsf{NP}(P)}\ \mathsf{in}\\ \quad \mathsf{if}\ \mathsf{InHS}(H_P, P) = \mathsf{InHS}(H_Q, Q)\ \mathsf{then}\ \mathsf{APPEND}\ H'_P\ H'_Q\\ \quad \mathsf{elseif}\ \mathsf{InHS}(H_P, P)\ \mathsf{then}\ H'_P\\ \quad \mathsf{else}\ H'_Q\end{array}\right)$$
- $\mathsf{NextHS}(H, t, \textbf{do}\ P\ \textbf{while}\ \sigma) =$
$$\left(\begin{array}{l}\mathsf{let}\ H_P = H\ |_1^{\mathsf{NP}(P)}\ \mathsf{in}\\ \mathsf{let}\ H'_P = \mathsf{NextHS}(H_P, t, P)\ \mathsf{in}\\ \quad \mathsf{if}\ \mathsf{InHS}(H_P, P) \wedge (H'_P = []) \wedge \sigma(t)\ \mathsf{then}\ \mathsf{NextHS}([], t, P)\\ \quad \mathsf{else}\ H'_P\end{array}\right)$$
- $\mathsf{NextHS}(H, t, \textbf{[weak] suspend}\ P\ \textbf{when}\ \sigma) =$
$$\left(\begin{array}{l}\mathsf{let}\ H_P = H\ |_1^{\mathsf{NP}(P)}\ \mathsf{in}\\ \mathsf{let}\ H'_P = \mathsf{NextHS}(H_P, t, P)\ \mathsf{in}\\ \quad \mathsf{if}\ \mathsf{InHS}(H_P, P) \wedge \sigma(t)\ \mathsf{then}\ H_P\\ \quad \mathsf{else}\ H'_P\end{array}\right)$$
- $\mathsf{NextHS}(H, t, \textbf{[weak] abort}\ P\ \textbf{when}\ \sigma) =$
$$\left(\begin{array}{l}\mathsf{let}\ H_P = H\ |_1^{\mathsf{NP}(P)}\ \mathsf{in}\\ \mathsf{let}\ H'_P = \mathsf{NextHS}(H_P, t, P)\ \mathsf{in}\\ \quad \mathsf{if}\ \mathsf{InHS}(H_P, P) \wedge \sigma(t)\ \mathsf{then}\ []\\ \quad \mathsf{else}\ H'_P\end{array}\right)$$
- $\mathsf{NextHS}(H, t, \textbf{now}\ \sigma) = []$
- $\mathsf{NextHS}(H, t, \textbf{during}\ P\ \textbf{holds}\ \sigma) = \mathsf{NextHS}(H, t, P)$

The above definition of $\mathsf{NextHS}(H, t, S)$ for all haltsets H and all statements S is directly one part of our HOL definition of the haltset encoded SOS rules. To be precise, it reflects the control flow part of the SOS semantics. Adding the data part, i.e. the set of actions $\mathcal{D}$ that have to be executed is rather simple. We omit that definition here due to lack of space (see however the appendix). Note furthermore that the next haltset $\mathsf{NextHS}(H,t,S)$ is deterministically computed from the current haltset and the variable's values, which shows the determinism of the control flow.

Using the above definition of $\mathsf{NextHS}(H,t,P)$, we have proved the desired relationship to the control flow predicate semantics. To this end, we have to translate haltsets to control flow conditions which is done with the following definition, where $\mathsf{labels}(P)$ is the set of labels that occur in P[6]:

$$\mathsf{HSLoc}(H,P,t) := \bigwedge_{\ell \in H \cap \mathsf{labels}(P)} \ell(t) \wedge \bigwedge_{\ell \in \mathsf{labels}(P) \setminus H} \neg \ell(t)$$

Note that only the indices between 1 and $\mathsf{NP}(P)$ are relevant for $\mathsf{HSLoc}(H,P,t)$, i.e. we have $\mathsf{HSLoc}(H,P,t) = \mathsf{HSLoc}(H\ |_1^{\mathsf{NP}(P)},P,t)$. This was in important property that was used in many lemmas. Analogously, $\mathsf{NextHS}(H,t,P)$ also depends only on the relevant indices of H, i.e. we have $\mathsf{NextHS}(H,t,P) = \mathsf{NextHS}(H\ |_1^{\mathsf{NP}(P)},t,P)$, that was also a helpful lemma.

We need further properties for establishing the desired equivalence between the semantics. In particular, we need the disjointness property $\mathsf{DisjointHS}(H,S)$ as an assumption. $\mathsf{DisjointHS}(H,S)$ means that the haltset H does not contain indices of both direct substatements of conditionals or sequences that occur as substatements of S, since the control flow can never be in both substatements of a conditional or a sequence. We have proved that this property is an invariant of the SOS rules. Finally, the predicate $\mathsf{NoInstantLoop}(S)$ states that S does not contain loops with instantaneous body. Using these definitions, the desired theorems that we have proved are the following ones:

Theorem 1 (Correctness of Control Flow Predicates). *Given a* Quartz *statement S, the following facts are valid and therefore prove the correctness of the control flow predicates defined in [29].*

1. $\mathsf{inst}\,(S)\,(t) \Leftrightarrow (\mathsf{NextHS}([],t,S) = [])$
2. $\neg\mathsf{inst}\,(S)\,(t) \vdash (\mathsf{enter}\,(S)\,(t) \Leftrightarrow \mathsf{HSLoc}(\mathsf{NextHS}([],t,S),S,t+1))$
3. $\left[\begin{array}{l}\mathsf{NoInstantLoop}(S),\\ \mathsf{DisjointHS}(H,S),\\ \mathsf{HSLoc}(H,S,t)\end{array}\right] \vdash \mathsf{term}\,(S)\,(t) \Leftrightarrow \mathsf{InHS}(H,S) \wedge (\mathsf{NextHS}(H,t,S) = [])$
4. $\left[\begin{array}{l}\mathsf{NoInstantLoop}(S),\\ \mathsf{DisjointHS}(H,S),\\ \mathsf{HSLoc}(H,S,t)\end{array}\right] \vdash \mathsf{move}\,(S)\,(t) \Leftrightarrow \left(\begin{array}{l}\mathsf{InHS}(H,S)\wedge\\ \mathsf{InHS}(\mathsf{NextHS}(H,t,S),S)\wedge\\ \mathsf{HSLoc}(\mathsf{NextHS}(H,t,S),S,t+1)\end{array}\right)$

The first fact says that a statement will be executed in zero time (it is instantaneous) at a point of time t iff the SOS rules started with the empty haltset yields the empty haltset as the next haltset. The second one says that the entering predicate covers those haltsets that are not empty and are reached from the empty haltset. The termination

[6] The HOL definition is rather different (it is again defined recursively along P), but the intuitive meaning is better understood with the listed 'definition'.

predicate describes all conditions where the SOS rules yield an empty haltset from a nonempty one, and the predicate for internal moves covers the conditions where the SOS rules yield an nonempty haltset from a nonempty one. Note that in the latter two facts, we used $\mathsf{InHS}(H, S)$ as assumption instead of $H \neq []$. This is a subtle difference concerning only the relevant indices, since $\mathsf{InHS}(H, S) \Leftrightarrow (H \mid_1^{\mathsf{NP}(S)} \neq [])$ holds.

The proofs of the above theorems were not too difficult, but produced large numbers of subgoals. Receiving about 150 subgoals after the application of the induction tactic for the programs and expansion of definitions was not unusual. It was therefore necessary to apply tactics in such a manner so that the number of subgoals was reduced as fast as possible. To this end, using tactics of the following form was convenient:

```
(tac THEN NO_TAC) ORELSE ALL_TAC
```

The meaning is that we apply tactic `tac` which is assumed to solve some of the subgoals. The predefined HOL tactic `NO_TAC` is therefore applied only to those subgoals that are not solved by `tac`, which has the effect that an exception is raised. Applying the predefined HOL tactic `ALL_TAC` will catch the exception so that all subgoals that are solved by `tac` disappear, and the other ones remain unchanged.

Beneath good strategies to handle large sets of subgoals, it was also a necessary to prove a set of lemmas to deal with technical problems like adding/ subtracting offsets or slicing haltsets.

5 Conclusions

In this paper, we have presented how the already existing deep embedding of our Esterel-variant Quartz has been extended by a definition of a process algebraic semantics. We have proved the equivalence between the three major classes of semantics, i.e. a control flow predicate semantics, a hardware circuit semantics, and the SOS semantics. Technical difficulties have been circumvented by using haltsets (or more precisely lists of indices that correspond to **pause** statements).

Beneath the satisfactory feeling that all given semantics are indeed equivalent to each other, and hence, define the same language, the newly added SOS semantics has its own advantages, since it is the only semantics that considers microsteps. Some applications like the investigation of causality analysis or schizophrenia detection are much better understood with a microstep semantics, although the semantics of synchronous programs can be solely explained with macrosteps.

References

1. L. Aceto, B. Bloom, and F. Vaandrager. Turning SOS rules into equations. *Information and Computation*, 111:1–52, 1994.
2. L. Aceto, W. Fokkink, and C. Verhoef. Structural operational semantics. In J. Bergstra, A. Ponse, and S. Smolka, editors, *Handbook of Process Algebra*, pages 197–292. Elsevier, Amsterdam, 2001.

3. C. Angelo, L. Claesen, and H. D. Man. Degrees of Formality in Shallow Embedding Hardware Description Languages in HOL. In J. Joyce and C.-J. Seger, editors, *Higher Order Logic Theorem Proving and Its Applications*, volume 780 of *LNCS*, pages 87–99, Vancouver, Canada, August 1993. University of British Columbia, Springer-Verlag, published 1994.
4. G. Berry. A hardware implementation of pure Esterel. In *ACM International Workshop on Formal Methods in VLSI Design*, Miami, Florida, January 1991.
5. G. Berry. The foundations of Esterel. In G. Plotkin, C. Stirling, and M. Tofte, editors, *Proof, Language and Interaction: Essays in Honour of Robin Milner*. MIT Press, 1998.
6. G. Berry. The constructive semantics of pure Esterel, July 1999.
7. G. Berry. The Esterel v5_91 language primer. `http://www.esterel.org`, June 2000.
8. G. Berry and G. Gonthier. The Esterel synchronous programming language: Design, semantics, implementation. *Science of Computer Programming*, 19(2):87–152, 1992.
9. R. Boulton. A HOL semantics for a subset of ELLA. technical report 254, University of Cambridge, Computer Laboratory, April 1992.
10. R. Boulton, A. Gordon, M. Gordon, J. Herbert, and J. van Tassel. Experience with embedding hardware description languages in HOL. In *International Conference on Theorem Provers in Circuit Design (TPCD)*, pages 129–156, Nijmegen, June 1992. IFIP TC10/WG 10.2, North-Holland.
11. F. Boussinot. SugarCubes implementation of causality. Research Report 3487, Institut National de Recherche en Informatique et en Automatique (INRIA), Sophia Antipolis Cedex (France), September 1998.
12. J. Brzozowski and C.-J. Seger. *Asynchronous Circuits*. Springer Verlag, 1995.
13. N. Day and J. Joyce. The semantics of statecharts in HOL. In J. Joyce and C.-J. Seger, editors, *Higher Order Logic Theorem Proving and its Applications*, volume 780 of *LNCS*, pages 338–352, Vancouver, Canada, August 1993. University of British Columbia, Springer-Verlag, published 1994.
14. Esterel-Technology. Website. `http://www.esterel-technologies.com`.
15. A. Girault and G. Berry. Circuit generation and verification of Esterel programs. Research report 3582, INRIA, December 1998.
16. M. Gordon and T. Melham. *Introduction to HOL: A Theorem Proving Environment for Higher Order Logic*. Cambridge University Press, 1993.
17. N. Halbwachs and F. Maraninchi. On the symbolic analysis of combinational loops in circuits and synchronous programs. In *Euromicro Conference*, Como, Italy, September 1995.
18. Jester Home Page. Website. `http://www.parades.rm.cnr.it/projects/jester/jester.html`.
19. L. Lavagno and E. Sentovich. ECL: A specification environment for system-level design. In *ACM/IEEE Design Automation Conference (DAC)*, 1999.
20. G. Logothetis and K. Schneider. Extending synchronous languages for generating abstract real-time models. In *European Conference on Design, Automation and Test in Europe (DATE)*, Paris, France, March 2002. IEEE Computer Society.
21. S. Malik. Analysis of cycle combinational circuits. *IEEE Transactions on Computer Aided Design*, 13(7):950–956, July 1994.
22. T. Melham. Automating recursive type definitions in higher order logic. Technical Report 146, University of Cambridge Computer Laboratory, Cambridge CB2 3QG, England, September 1988.
23. G. Plotkin. A Structural Approach to Operational Semantics. Technical Report FN–19, DAIMI, Aarhus University, 1981.
24. A. Poigné and L. Holenderski. Boolean automata for implementing pure Esterel. Arbeitspapiere 964, GMD, Sankt Augustin, 1995.
25. POLIS Homepage. Website. `http://www-cad.eecs.berkeley.edu/`.

26. R. Reetz. Deep Embedding VHDL. In E. Schubert, P. Windley, and J. Alves-Foss, editors, *Higher Order Logic Theorem Proving and its Applications*, volume 971 of *LNCS*, pages 277–292, Aspen Grove, Utah, USA, September 1995. Springer-Verlag.
27. F. Rocheteau and N. Halbwachs. Pollux, a Lustre-based hardware design environment. In P. Quinton and Y. Robert, editors, *Conference on Algorithms and Parallel VLSI Architectures II*, Chateau de Bonas, 1991.
28. K. Schneider. A verified hardware synthesis for Esterel. In F. Rammig, editor, *International IFIP Workshop on Distributed and Parallel Embedded Systems*, pages 205–214, Schloß Ehringerfeld, Germany, 2000. Kluwer Academic Publishers.
29. K. Schneider. Embedding imperative synchronous languages in interactive theorem provers. In *International Conference on Application of Concurrency to System Design (ICACSD 2001)*, pages 143–156, Newcastle upon Tyne, UK, June 2001. IEEE Computer Society Press.
30. K. Schneider and M. Wenz. A new method for compiling schizophrenic synchronous programs. In *International Conference on Compilers, Architecture, and Synthesis for Embedded Systems (CASES)*, pages 49–58, Atlanta, USA, November 2001. ACM.
31. T. Shiple, G. Berry, and H. Touati. Constructive analysis of cyclic circuits. In *European Design and Test Conference (EDTC)*, Paris, France, 1996. IEEE Computer Society Press.
32. S. Tini. *Structural Operational Semantics for Synchronous Languages*. PhD thesis, University of Pisa, 2000.
33. C. Zhang, R. Shaw, R. Olsson, K. Levitt, M. Archer, M. Heckman, and G. Benson. Mechanizing a programming logic for the concurrent programming language microSR in HOL. In J. Joyce and C.-J. Seger, editors, *Higher Order Logic Theorem Proving and its Applications*, volume 780 of *LNCS*, pages 29–43, Vancouver, Canada, August 1993. University of British Columbia, Springer-Verlag, published 1994.

A SOS Rules Based on a Haltset Encoding

As described in Sections 3 and 4, the rules below are of the form $(S, \Phi) \rhd H \xrightarrow{\mathcal{D}} H'$ with the meaning that statement S has a transition from haltset H to haltset H' that is enabled by condition Φ and that transition will execute the data actions contained in $\mathcal{D}$.

Atomic Statements:

$$(\textbf{nothing}, \mathsf{true}) \rhd H \xrightarrow{\{\}} \{\}$$

$$(\ell : \textbf{pause}, \mathsf{true}) \rhd H \xrightarrow{\{\}} (\ell \in H \Rightarrow \{\} | \{\ell\})$$

$$(\textbf{emit}\ x, \mathsf{true}) \rhd H \xrightarrow{\{\textbf{emit}\ x\}} \{\}$$

$$(\textbf{emit next}(x), \mathsf{true}) \rhd H \xrightarrow{\{\textbf{emit next}(x)\}} \{\}$$

$$(y := \tau, \mathsf{true}) \rhd H \xrightarrow{\{y := \tau\}} \{\}$$

$$(\textbf{next}(y) := \tau, \mathsf{true}) \rhd H \xrightarrow{\{\textbf{next}(y) := \tau\}} \{\}$$

Conditional:

$$\frac{(S_1,\Phi) \rhd H \xrightarrow{\mathcal{D}} H_1}{(\textbf{if } \sigma \textbf{ then } S_1 \textbf{ else } S_2 \textbf{ end}, \Phi \wedge \sigma \wedge \neg(\mathsf{InHS}(H,S_1) \vee \mathsf{InHS}(H,S_2))) \rhd H \xrightarrow{\mathcal{D}} H_1}$$

$$\frac{(S_2,\Phi) \rhd H \xrightarrow{\mathcal{D}} H_2}{(\textbf{if } \sigma \textbf{ then } S_1 \textbf{ else } S_2 \textbf{ end}, \Phi \wedge \neg\sigma \wedge \neg(\mathsf{InHS}(H,S_1) \vee \mathsf{InHS}(H,S_2))) \rhd H \xrightarrow{\mathcal{D}} H_2}$$

$$\frac{(S_1,\Phi) \rhd H \xrightarrow{\mathcal{D}} H_1}{(\textbf{if } \sigma \textbf{ then } S_1 \textbf{ else } S_2 \textbf{ end}, \Phi \wedge \mathsf{InHS}(H,S_1)) \rhd H \xrightarrow{\mathcal{D}} H_1}$$

$$\frac{(S_2,\Phi) \rhd H \xrightarrow{\mathcal{D}} H_2}{(\textbf{if } \sigma \textbf{ then } S_1 \textbf{ else } S_2 \textbf{ end}, \Phi \wedge \mathsf{InHS}(H,S_2)) \rhd H \xrightarrow{\mathcal{D}} H_2}$$

Sequence:

$$\frac{(S_1,\Phi_1) \rhd H \xrightarrow{\mathcal{D}_1} H_1}{(S_1; S_2, \Phi_1 \wedge \neg\mathsf{InHS}(H,S_2) \wedge \mathsf{InHS}(H_1,S_1)) \rhd H \xrightarrow{\mathcal{D}_1} H_1}$$

$$\frac{(S_1,\Phi_1) \rhd H \xrightarrow{\mathcal{D}_1} H_1 \quad (S_2,\Phi_2) \rhd \{\} \xrightarrow{\mathcal{D}_2} H_2}{(S_1; S_2, \Phi_1 \wedge \Phi_2 \wedge \neg\mathsf{InHS}(H,S_2) \wedge \neg\mathsf{InHS}(H_1,S_1)) \rhd H \xrightarrow{\mathcal{D}_1 \cup \mathcal{D}_2} H_2}$$

$$\frac{(S_2,\Phi_2) \rhd H \xrightarrow{\mathcal{D}_2} H_2}{(S_1; S_2, \Phi_2 \wedge \mathsf{InHS}(H,S_2)) \rhd H \xrightarrow{\mathcal{D}_2} H_2}$$

Synchronous Concurrency:

$$\frac{(S_1,\Phi_1) \rhd H \xrightarrow{\mathcal{D}_1} H_1}{(S_1 \parallel S_2, \Phi_1 \wedge \mathsf{InHS}(H,S_1) \wedge \neg\mathsf{InHS}(H,S_2)) \rhd H \xrightarrow{\mathcal{D}_1} H_1}$$

$$\frac{(S_2,\Phi_2) \rhd H \xrightarrow{\mathcal{D}_2} H_2}{(S_1 \parallel S_2, \Phi_2 \wedge \neg\mathsf{InHS}(H,S_1) \wedge \mathsf{InHS}(H,S_2)) \rhd H \xrightarrow{\mathcal{D}_2} H_2}$$

$$\frac{(S_1,\Phi_1) \rhd H \xrightarrow{\mathcal{D}_1} H_1 \quad (S_2,\Phi_2) \rhd H \xrightarrow{\mathcal{D}_2} H_2}{(S_1 \parallel S_2, \Phi_1 \wedge \Phi_2 \wedge (\mathsf{InHS}(H,S_1) = \mathsf{InHS}(H,S_2))) \rhd H \xrightarrow{\mathcal{D}_1 \cup \mathcal{D}_2} H_1 \cup H_2}$$

Loop:

$$\frac{(S,\Phi) \rhd H \xrightarrow{\mathcal{D}} H'}{(\textbf{do } S \textbf{ while } \sigma, \Phi \wedge (H' \neq \{\})) \rhd H \xrightarrow{\mathcal{D}} H'}$$

$$\frac{(S,\Phi_1) \rhd H \xrightarrow{\mathcal{D}} \{\} \qquad (S,\Phi_2) \rhd \{\} \xrightarrow{\mathcal{D}'} H'}{(\textbf{do } S \textbf{ while } \sigma, \Phi_1 \wedge \sigma \wedge \Phi_2) \rhd H \xrightarrow{\mathcal{D} \cup \mathcal{D}'} H'}$$

$$\frac{(S,\Phi) \rhd H \xrightarrow{\mathcal{D}} \{\}}{(\textbf{do } S \textbf{ while } \sigma, \Phi \wedge \neg\sigma) \rhd H \xrightarrow{\mathcal{D}} \{\}}$$

Suspension:

$$\frac{(S,\Phi) \rhd H \xrightarrow{\mathcal{D}} H'}{(\textbf{[weak] suspend } S \textbf{ when } \sigma, \Phi \wedge \neg \mathsf{InHS}(H,S)) \rhd H \xrightarrow{\mathcal{D}} H'}$$

$$\frac{(S,\Phi) \rhd H \xrightarrow{\mathcal{D}} H'}{(\textbf{[weak] suspend } S \textbf{ when } \sigma, \Phi \wedge \neg\sigma \wedge \mathsf{InHS}(H,S)) \rhd H \xrightarrow{\mathcal{D}} H'}$$

$$(\textbf{suspend } S \textbf{ when } \sigma, \sigma \wedge \mathsf{InHS}(H,S)) \rhd H \xrightarrow{\{\}} H$$

$$\frac{(S,\Phi) \rhd H \xrightarrow{\mathcal{D}} H'}{(\textbf{weak suspend } S \textbf{ when } \sigma, \Phi \wedge \sigma \wedge \mathsf{InHS}(H,S)) \rhd H \xrightarrow{\mathcal{D}} H}$$

Abortion:

$$\frac{(S,\Phi) \rhd H \xrightarrow{\mathcal{D}} H'}{(\textbf{[weak] abort } S \textbf{ when } \sigma, \Phi \wedge \neg \mathsf{InHS}(H,S)) \rhd H \xrightarrow{\mathcal{D}} H'}$$

$$\frac{(S,\Phi) \rhd H \xrightarrow{\mathcal{D}} H'}{(\textbf{[weak] abort } S \textbf{ when } \sigma, \Phi \wedge \neg\sigma \wedge \mathsf{InHS}(H,S)) \rhd H \xrightarrow{\mathcal{D}} H'}$$

$$(\textbf{abort } S \textbf{ when } \sigma, \sigma \wedge \mathsf{InHS}(H,S)) \rhd H \xrightarrow{\{\}} \{\}$$

$$\frac{(S,\Phi) \rhd H \xrightarrow{\mathcal{D}} H'}{(\textbf{weak abort } S \textbf{ when } \sigma, \Phi \wedge \sigma \wedge \mathsf{InHS}(H,S)) \rhd H \xrightarrow{\mathcal{D}} \{\}}$$

Choice:

$$\frac{(S_1, \Phi) \rhd H \xrightarrow{\mathcal{D}} H_1}{(\textbf{choose}\ S_1 \,[]\, S_2\ \textbf{end}, \Phi \wedge \neg(\mathsf{InHS}(H, S_1) \vee \mathsf{InHS}(H, S_2))) \rhd H \xrightarrow{\mathcal{D}} H_1}$$

$$\frac{(S_2, \Phi) \rhd H \xrightarrow{\mathcal{D}} H_2}{(\textbf{choose}\ S_1 \,[]\, S_2\ \textbf{end}, \Phi \wedge \neg(\mathsf{InHS}(H, S_1) \vee \mathsf{InHS}(H, S_2))) \rhd H \xrightarrow{\mathcal{D}} H_2}$$

$$\frac{(S_1, \Phi) \rhd H \xrightarrow{\mathcal{D}} H_1}{(\textbf{choose}\ S_1 \,[]\, S_2\ \textbf{end}, \Phi \wedge \mathsf{InHS}(H, S_1)) \rhd H \xrightarrow{\mathcal{D}} H_1}$$

$$\frac{(S_2, \Phi) \rhd H \xrightarrow{\mathcal{D}} H_2}{(\textbf{choose}\ S_1 \,[]\, S_2\ \textbf{end}, \Phi \wedge \mathsf{InHS}(H, S_2)) \rhd H \xrightarrow{\mathcal{D}} H_2}$$

Asynchronous Concurrency:

$$\frac{(S_1, \Phi_1) \rhd H \xrightarrow{\mathcal{D}_1} H_1}{(S_1 \,|||\, S_2, \Phi_1 \wedge \mathsf{InHS}(H, S_1) \wedge \neg \mathsf{InHS}(H, S_2)) \rhd H \xrightarrow{\mathcal{D}_1} H_1}$$

$$\frac{(S_2, \Phi_2) \rhd H \xrightarrow{\mathcal{D}_2} H_2}{(S_1 \,|||\, S_2, \Phi_2 \wedge \neg \mathsf{InHS}(H, S_1) \wedge \mathsf{InHS}(H, S_2)) \rhd H \xrightarrow{\mathcal{D}_2} H_2}$$

$$\frac{(S_1, \Phi_1) \rhd H \xrightarrow{\mathcal{D}_1} H_1 \quad (S_2, \Phi_2) \rhd H \xrightarrow{\mathcal{D}_2} H_2}{(S_1 \,|||\, S_2, \Phi_1 \wedge \Phi_2 \wedge \neg \mathsf{InHS}(H, S_1) \wedge \neg \mathsf{InHS}(H, S_2)) \rhd H \xrightarrow{\mathcal{D}_1 \cup \mathcal{D}_2} H_1 \cup H_2}$$

$$\frac{(S_1, \Phi_1) \rhd H \xrightarrow{\mathcal{D}_1} H_1 \quad (S_2, \Phi_2) \rhd H \xrightarrow{\mathcal{D}_2} H_2}{(S_1 \,|||\, S_2, \Phi_1 \wedge \Phi_2 \wedge \mathsf{InHS}(H, S_1) \wedge \mathsf{InHS}(H, S_2)) \rhd H \xrightarrow{\mathcal{D}_1 \cup \mathcal{D}_2} H_1 \cup H_2}$$

$$\frac{(S_1, \Phi_1) \rhd H \xrightarrow{\mathcal{D}_1} H_1}{(S_1 \,|||\, S_2, \Phi_1 \wedge \mathsf{InHS}(H, S_1) \wedge \mathsf{InHS}(H, S_2)) \rhd H \xrightarrow{\mathcal{D}_1} H_1 \cup (H \cap \mathsf{labels}\,(S_2))}$$

$$\frac{(S_2, \Phi_2) \rhd H \xrightarrow{\mathcal{D}_2} H_2}{(S_1 \,|||\, S_2, \Phi_2 \wedge \mathsf{InHS}(H, S_1) \wedge \mathsf{InHS}(H, S_2)) \rhd H \xrightarrow{\mathcal{D}_2} (H \cap \mathsf{labels}\,(S_1)) \cup H_2}$$

Weakest Precondition for General Recursive Programs Formalized in Coq*

Xingyuan Zhang[1], Malcolm Munro[1], Mark Harman[2], and Lin Hu[2]

[1] Department of Computer Science
University of Durham, Science Laboratories
South Road, Durhram, DH1 3LE, UK
{Xingyuan.Zhang,Malcolm.Munro}@durham.ac.uk
[2] Department of Information Systems and Computing
Brunel University, Uxbridge, Middlesex, UB8 3PH, UK
{Mark.Harman,Lin.Hu}@brunel.ac.uk

Abstract. This paper describes a formalization of the weakest precondition, wp, for general recursive programs using the type-theoretical proof assistant Coq. The formalization is a deep embedding using the computational power intrinsic to type theory. Since Coq accepts only structural recursive functions, the computational embedding of general recursive programs is non-trivial. To justify the embedding, an operational semantics is defined and the equivalence between wp and the operational semantics is proved. Three major healthiness conditions, namely: Strictness, Monotonicity and Conjunctivity are proved as well.
Keywords: Weakest Precondition, Operational Semantics, Formal Verification, Coq.

1 Introduction

The weakest precondition, wp, proposed by E. W. Dijkstra [5] proved to be useful in various areas of software development and has been investigated extensively [1, 13, 12]. There have been a number of attempts to support wp and refinement calculus with computer assisted reasoning systems such as HOL [16, 19, 9], Isabelle [14, 17, 18], Ergo [3], PVS [8] and Alf [10]. Unfortunately, among these works, only Laibinis and Wright [9] deals with general recursion.

In this paper, an embedding of wp is given for general recursive programs using the intentional type theory supported by Coq [2]. Since the computational mechanism peculiar to type theory is used, we name such a style of embedding 'computational embedding'. The importance of computational embedding is that it can be seen as a bridge between deep and shallow embedding. Since wp is defined as a function from statement terms to predicate transformers, before the definition of wp in wp(c) is expanded, by accessing the syntax structure of c, we enjoy the benefit of deep embedding, so that meta-level operations such as program transformations (usually expressed as functions on program terms)

* The work in this paper is sponsored by the EPSRC project GUSTT.

V.A. Carreño, C. Muñoz, S. Tahar (Eds.): TPHOLs 2002, LNCS 2410, pp. 332–347, 2002.

can be verified. On the other hand, after expanding wp, $\mathsf{wp}\,(c)$ becomes the semantics of c. With the computational mechanism taking care of the expanding process, we can focus on the semantics without worrying about syntax details, a benefit traditionally enjoyed by shallow embedding. Therefore, computational embedding enables us to switch between deep and shallow embedding with ease.

The language investigated in this paper is general recursive in the sense that it contains a statement which takes the form of a parameter-less recursive procedure : $\mathsf{proc}\ p \equiv c$, the execution of which starts with the execution of the procedure body c. When a recursive call $\mathsf{pcall}\ p$ is encountered during the execution, the $\mathsf{pcall}\ p$ is replaced by the procedure body c and the execution continues from the beginning of c.

A naïve definition of $\mathsf{wp}\,(\mathsf{proc}\ p \equiv c)$ could be:

$$\mathsf{wp}\,(\mathsf{proc}\ p \equiv c) \overset{\mathsf{def}}{\Longrightarrow} \bigvee_{n<\omega} \mathsf{wp}\left(\boxed{\mathsf{proc}\ p \equiv c}^{\,n}\right) \tag{1}$$

where the expansion operation $\boxed{\mathsf{proc}\ p \equiv c}^{\,n}$ is defined as:

$$\begin{cases} \boxed{\mathsf{proc}\ p \equiv c}^{\,0} \overset{\mathsf{def}}{\Longrightarrow} \mathsf{assert}(\overline{\mathsf{false}}) \\ \boxed{\mathsf{proc}\ p \equiv c}^{\,n+1} \overset{\mathsf{def}}{\Longrightarrow} c\left\{p/\boxed{\mathsf{proc}\ p \equiv c}^{\,n}\right\} \end{cases} \tag{2}$$

where in each round of expansion, p in c is substituted by $\boxed{\mathsf{proc}\ p \equiv c}^{\,n}$.

The problem with this definition is that: in (1), the recursive call of wp:

$$\mathsf{wp}\left(\boxed{\mathsf{proc}\ p \equiv c}^{\,n}\right)$$

is not structural recursive, since the argument $\boxed{\mathsf{proc}\ p \equiv c}^{\,n}$ is structurally larger than the argument $\mathsf{proc}\ p \equiv c$ on the left. Therefore, (1) is rejected by the function definition mechanism of Coq, which only accepts structural recursive functions.

This paper proposes a formulation of wp which solves this problem. To justify such a formulation, an operational semantics is defined and the equivalence of wp and the operational semantics is proved. Three major healthiness conditions are proved for wp as well, namely: Strictness, Monotonicity and Conjunctivity. For brevity, in this paper, only the equivalence proof is discussed in full detail.

The rest of the paper is arranged as follows: Section 2 defines the notions of predicate, predicate transformer, transformer of predicate transformers together with various partial orders and monotonicity predicates. Section 3 gives the definition wp. Section 4 presents the operational semantics. Section 5 relates operational semantics to wp. Section 6 relates wp to operational semantics. Section 7 concludes the whole paper. Additional technical detail can be found in Appendices. The technical development of this paper has been fully formalized and checked by Coq. The Coq scripts are available from

`http://www.dur.ac.uk/xingyuan.zhang/tphol`

Conventions. Because type theory [4, 15, 11] was first proposed to formalize constructive mathematics, we are able to present the work in standard mathematical notation. For brevity, we use the name of a variable to suggest its type. For example, $s, s', \overline{s}, \ldots$ in this paper always represent program stores.

Type theory has a notion of computation, which is used as a definition mechanism, where the equation $a \stackrel{\mathsf{def}}{\Longrightarrow} b$ represents a 'computational rule' used to expand the definition of a to b.

Free variables in formulæ are assumed to be universally quantified, for example, $n_1 + n_2 = n_2 + n_1$ is an abbreviation of $\forall\, n_1, n_2.\; n_1 + n_2 = n_2 + n_1$.

For $A : \texttt{Set}$, the exceptional set $\mathcal{M}(A)$ is defined as:

$$
\begin{array}{c}
\dfrac{A : \texttt{Set}}{\mathcal{M}(A) : \texttt{Set}}\ \textbf{M\_formation} \\[2ex]
\dfrac{A : \texttt{Set}}{\bot : \mathcal{M}(A)}\ \textbf{bottom\_value} \quad \dfrac{A : \texttt{Set} \quad a : A}{\mathsf{unit}\,(a) : \mathcal{M}(A)}\ \textbf{normal\_value}
\end{array}
\tag{3}
$$

Intuitively, $\mathcal{M}(A)$ represents the type obtained from A by adding a special element $\bot$ to represent undefined value. A normal element a of the original type A is represented as $\mathsf{unit}\,(a)$ in exceptional set $\mathcal{M}(A)$. However, the unit is usually omitted, unless there is possibility of confusion.

2 Predicates and Predicate Transformers

We take the view that the predicates transformed by wp are predicates on program stores. Therefore, the type of predicates $\mathcal{PD}$ is defined as:

$$\mathcal{PD} \stackrel{\mathsf{def}}{\Longrightarrow} \mathcal{PS} \rightarrow \texttt{Prop} \tag{4}$$

where $\texttt{Prop}$ is the type of propositions, $\mathcal{PS}$ is the type of program stores. For simplicity, this paper is abstract about program stores and denotable values. However, there is a comprehensive treatment of program stores, denotable values and expression evaluation in the Coq scripts.

The type of predicate transformers $\mathcal{PT}$ is defined as:

$$\mathcal{PT} \stackrel{\mathsf{def}}{\Longrightarrow} \mathcal{PD} \rightarrow \mathcal{PD} \tag{5}$$

The type of transformers of predicate transformers $\mathcal{PTT}$ is defined as:

$$\mathcal{PTT} \stackrel{\mathsf{def}}{\Longrightarrow} \mathcal{PT} \rightarrow \mathcal{PT} \tag{6}$$

The partial order $\preccurlyeq_P$ between predicates is defined as:

$$P_1 \preccurlyeq_P P_2 \stackrel{\mathsf{def}}{\Longrightarrow} \forall\, s\,.\, P_1\,(s) \Rightarrow P_2\,(s) \tag{7}$$

The partial order $\preccurlyeq_{pt}$ between predicate transformers is defined as:

$$pt_1 \preccurlyeq_{pt} pt_2 \stackrel{\mathsf{def}}{\Longrightarrow} \forall\, P\,.\, pt_1\,(P) \preccurlyeq_P pt_2\,(P) \tag{8}$$

The monotonicity predicate on $\mathcal{PT}$ is defined as:

$$\text{mono}(pt) \stackrel{\mathsf{def}}{\Longrightarrow} \forall P_1, P_2 .\, P_1 \preccurlyeq_P P_2 \Rightarrow pt\,(P_1) \preccurlyeq_P pt\,(P_2) \tag{9}$$

The partial order on $\preccurlyeq_\sigma$ between environments is defined as:

$$\sigma_1 \preccurlyeq_\sigma \sigma_2 \stackrel{\mathsf{def}}{\Longrightarrow} \forall p \,.\, \sigma_1\,(p) \preccurlyeq_{pt} \sigma_2\,(p) \tag{10}$$

The partial order $\preccurlyeq_{ptt}$ between transformers of predicate transformers is defined as:

$$ptt_1 \preccurlyeq_{ptt} ptt_2 \stackrel{\mathsf{def}}{\Longrightarrow} \forall pt \,.\, \text{mono}(pt) \Rightarrow ptt_1\,(pt) \preccurlyeq_{pt} ptt_2\,(pt) \tag{11}$$

The corresponding derived equivalence relations are defined as:

$$\begin{cases} P_1 \approx_P P_2 \stackrel{\mathsf{def}}{\Longrightarrow} P_1 \preccurlyeq_P P_2 \wedge P_2 \preccurlyeq_P P_1 \\ pt_1 \approx_{\mathsf{pt}} pt_2 \stackrel{\mathsf{def}}{\Longrightarrow} pt_1 \preccurlyeq_{pt} pt_2 \wedge pt_2 \preccurlyeq_{pt} pt_1 \\ ptt_1 \approx_{\mathsf{ptt}} ptt_2 \stackrel{\mathsf{def}}{\Longrightarrow} ptt_1 \preccurlyeq_{ptt} ptt_2 \wedge ptt_2 \preccurlyeq_{ptt} ptt_1 \end{cases} \tag{12}$$

The monotonicity predicate on environments is defined as:

$$\text{mono}(\sigma) \stackrel{\mathsf{def}}{\Longrightarrow} \forall p \,.\, \text{mono}(\sigma\,(p)) \tag{13}$$

3 Formalization of wp

In order to overcome the problem mentioned in the Introduction, a notion of environment is introduced, which is represented by the type Σ:

$$\Sigma \stackrel{\mathsf{def}}{\Longrightarrow} \mathcal{ID} \rightarrow \mathcal{PT} \tag{14}$$

which is a function from identifiers to predicate transformers. In this paper, procedure names are represented as identifiers.

The empty environment $\varepsilon : \Sigma$, which maps all procedure names to the false predicate transformer $\lambda P.\,\mathsf{F}$, is defined as:

$$\varepsilon \stackrel{\mathsf{def}}{\Longrightarrow} \lambda p.\, \lambda P.\, \mathsf{F} \tag{15}$$

where p is procedure name, P is predicate and F is the false predicate, which is defined as:

$$\mathsf{F} \stackrel{\mathsf{def}}{\Longrightarrow} \lambda s.\, \mathsf{False} \tag{16}$$

where s is program store, False is the false proposition in Coq. Therefore, F does not hold on any program store.

The operation $\sigma\,[p \mapsto pt]$ is defined, to add the mapping $p \mapsto pt$ to environment σ:

$$\begin{cases} \sigma\,[p \mapsto pt]\,(\overline{p}) \stackrel{\mathsf{def}}{\Longrightarrow} pt & \text{if } \overline{p} = p \\ \sigma\,[p \mapsto pt]\,(\overline{p}) \stackrel{\mathsf{def}}{\Longrightarrow} \sigma\,(\overline{p}) & \text{if } \overline{p} \neq p \end{cases} \tag{17}$$

$$\frac{}{\mathcal{C} : \mathsf{Set}}\ \textbf{C\_fmt}$$

$$\frac{i : \mathcal{ID} \quad e : \mathcal{E}}{i := e : \mathcal{C}}\ \textbf{s\_asgn} \qquad \frac{e : \mathcal{E}}{\mathsf{assert}(e) : \mathcal{C}}\ \textbf{s\_assert} \qquad \frac{c_1 : \mathcal{C} \quad c_2 : \mathcal{C}}{c_1;\ c_2 : \mathcal{C}}\ \textbf{s\_seq}$$

$$\frac{e : \mathcal{E} \quad c_1 : \mathcal{C} \quad c_2 : \mathcal{C}}{\mathsf{if}\ e\ \mathsf{then}\ c_1\ \mathsf{else}\ c_2 : \mathcal{C}}\ \textbf{s\_ifs} \qquad \frac{p : \mathcal{ID} \quad c : \mathcal{C}}{\mathsf{proc}\ p\ \equiv\ c : \mathcal{C}}\ \textbf{s\_proc} \qquad \frac{p : \mathcal{ID}}{\mathsf{pcall}\ p : \mathcal{C}}\ \textbf{s\_pcall}$$

Fig. 1. The Definition of $\mathcal{C}$

$$\begin{cases} \mathsf{wpc}\,(\sigma,\ i := e) \stackrel{\mathsf{def}}{\Longrightarrow} \lambda\, P, s.\, \exists\, v\, .\, [\![e]\!]_s = v\ \wedge\ P\,(s\,[v^{\,i}]) & \text{(18a)} \\ \mathsf{wpc}\,(\sigma,\ \mathsf{assert}(e)) \stackrel{\mathsf{def}}{\Longrightarrow} \lambda\, P, s.\, [\![e]\!]_s = \mathsf{true}\ \wedge\ P\,(s) & \text{(18b)} \\ \mathsf{wpc}\,(\sigma,\ c_1;\ c_2) \stackrel{\mathsf{def}}{\Longrightarrow} \lambda\, P, s.\, \mathsf{wpc}\,(\sigma,\ c_1)\,(\mathsf{wpc}\,(\sigma,\ c_2)\,(P))\,(s) & \text{(18c)} \\ \mathsf{wpc}\,(\sigma,\ \mathsf{if}\ e\ \mathsf{then}\ c_1\ \mathsf{else}\ c_2) \stackrel{\mathsf{def}}{\Longrightarrow} \lambda\, P, s\, .([\![e]\!]_s = \mathsf{true}\ \wedge\ \mathsf{wpc}\,(\sigma,\ c_1)\,(P)\,(s)) \vee \\ \qquad ([\![e]\!]_s = \mathsf{false}\ \wedge\ \mathsf{wpc}\,(\sigma,\ c_2)\,(P)\,(s)) & \text{(18d)} \\ \mathsf{wpc}\,(\sigma,\ \mathsf{proc}\ p\ \equiv\ c) \stackrel{\mathsf{def}}{\Longrightarrow} \lambda\, P, s.\, \exists\, n\, .\, \boxed{\lambda\, \overline{pt}.\, \mathsf{wpc}\,(\sigma\,[p \mapsto \overline{pt}]\,,\ c)}^{\,n}\,(P)\,(s) & \text{(18e)} \\ \mathsf{wpc}\,(\sigma,\ \mathsf{pcall}\ p) \stackrel{\mathsf{def}}{\Longrightarrow} \lambda\, P, s.\, \sigma\,(p)\,(P)\,(s) & \text{(18f)} \end{cases}$$

Fig. 2. The Definition of wpc

Instead of defining wp directly, an operation $\mathsf{wpc}\,(\sigma,\ c)$ is defined to compute a predicate transformer for command c under the environment σ. By using the environment σ, $\mathsf{wpc}\,(\sigma,\ c)$ can be defined using structural recursion. With wpc, the normal wp is now defined as: $\mathsf{wp}\,(c) \stackrel{\mathsf{def}}{\Longrightarrow} \mathsf{wpc}\,(\varepsilon,\ c)$.

The syntax of the programming language formalized in this paper is given in Figure 1 as the inductive type $\mathcal{C}$. The type of expressions is formalized as an abstract type $\mathcal{E}$. The evaluation of expressions is formalized as the operation $[\![e]\!]_s$, where the expression $[\![e]\!]_s = v$ means that expression e evaluate to value v under program store s and the expression $[\![e]\!]_s = \bot$ means there is no valuation of expression e under program store s. The definition of $\mathsf{wpc}\,(\sigma,\ c)$ is given in Figure 2, where each command corresponds to an equation. The right-hand-side of each equation is a lambda abstraction $\lambda\, P, s.\,(\ldots)$, where P is the predicate required to hold after execution of the corresponding command, and s is the program store before execution. The operation $s\,[v^{\,i}]$ is the program store obtained from program store s by setting the value of variable i to v.

Equations (18a) – (18d) are quite standard. Only (18e) and (18f) need more explanation. Equation (18e) is for $\mathsf{wpc}\,(\sigma,\ \mathsf{proc}\ p\ \equiv\ c)$, the originally problematic case. The recursive call on the right-hand-side is $\mathsf{wpc}\,(\sigma\,[p \mapsto \overline{pt}]\,,\ c)$, which is structural recursive with respect to the second argument. The key idea is that: a mapping $p \mapsto \overline{pt}$ is added to the environment σ, which maps p to the formal parameter $\overline{pt}$. By abstracting on $\overline{pt}$, a 'transformer of predicate transformers':

$$\lambda \overline{pt}.\,\mathsf{wpc}\left(\sigma\left[p \mapsto \overline{pt}\right],\ c\right) : \mathcal{PT} \rightarrow \mathcal{PT}$$

is obtained. Notice that the term $\boxed{\lambda \overline{pt}.\,\mathsf{wpc}\left(\sigma\left[p \mapsto \overline{pt}\right],\ c\right)}^{n}$ in (18e) is no longer the expansion operation defined in (2), but a folding operation defined as:

$$\begin{cases} \boxed{ptt}^{0} \stackrel{\mathsf{def}}{\Longrightarrow} \lambda P.\,\mathsf{F} \\ \boxed{ptt}^{n+1} \stackrel{\mathsf{def}}{\Longrightarrow} ptt\left(\boxed{ptt}^{n}\right) \end{cases} \tag{19}$$

In place of $\bigvee_{n<\omega} \mathsf{wp}\left(\boxed{\mathsf{proc}\ p\ \equiv\ c}^{n}\right)$, we write

$$\exists n.\,\boxed{\lambda \overline{pt}.\,\mathsf{wpc}\left(\sigma\left[p \mapsto \overline{pt}\right],\ c\right)}^{n} (P)\,(s)$$

which is semantically identical, but expressible in Coq.

The equation (18f) defines wpc for pcall p, which is the predicate transformer assigned to p by σ.

As a sanity checking, three healthiness conditions, namely: Strictness, Monotonicity and Conjunctivity, are proved. For brevity, only the one, which is used in this paper, is listed here:

Lemma 1 (Generalized Monotonicity Lemma).

$$\begin{aligned} &\forall c. \\ &\quad (\sigma_1 \preccurlyeq_\sigma \sigma_2 \Rightarrow \mathrm{mono}(\sigma_1) \Rightarrow \mathrm{mono}(\sigma_2) \Rightarrow \\ &\qquad\qquad \mathsf{wpc}\,(\sigma_1,\ c) \preccurlyeq_{pt} \mathsf{wpc}\,(\sigma_2,\ c)) \wedge & \text{(20a)} \\ &\quad (\mathrm{mono}(\sigma) \Rightarrow \mathrm{mono}(\mathsf{wpc}\,(\sigma,\ c))) & \text{(20b)} \end{aligned}$$

The proof for other healthiness conditions can be found in the Coq scripts.

4 The Operational Semantics

To strengthen the justification of wp, an operational semantics is defined and a formal relationship between the operational semantics and wp is established. The operational semantics is given in Figure 3 as an inductively defined relation $s \xrightarrow[cs]{c} s'$, which means that: the execution of command c transforms program store from s to s'. The cs is the 'call stack' under which c is executed. The type of call stack $\mathcal{CS}$ is defined as:

$$\frac{}{\mathcal{CS} : \mathtt{Set}}\ \mathcal{CS}\mathbf{\_form} \qquad \frac{}{\vartheta : \mathcal{CS}}\ \mathbf{nil\_cs} \qquad \frac{p : \mathcal{ID} \quad c : \mathcal{C} \quad cs, \overline{cs} : \mathcal{CS}}{cs[p \rightsquigarrow (c, \overline{cs})] : \mathcal{CS}}\ \mathbf{cons\_cs} \tag{21}$$

where ϑ is the empty call stack, and $cs[p \rightsquigarrow (c, \overline{cs})]$ is the call stack obtained from cs by pushing $(p, c, \overline{cs})$.

In the rule **e_proc**, when a recursive procedure $\mathsf{proc}\ p\ \equiv\ c$ is executed, the operation $cs[p \leadsto (c, cs)]$ pushes procedure body c together with the call stack cs, under which the c is going to be executed onto the call stack, and then the procedure body c is executed.

In the rule **e_pcall**, when a recursive call $\mathsf{pcall}\ p$ is executed, the operation $\mathsf{lookup}(cs, p)$ is used to look up the procedure body being called and the call stack under which it is going to be executed. Suppose $(c, \overline{cs})$ is found, then c is executed under the call stack $\overline{cs}$.

The definition of lookup is:

$$\begin{cases} \mathsf{lookup}(cs[\overline{p} \leadsto (c, \overline{cs})], p) \stackrel{\mathsf{def}}{\Longrightarrow} (c, \overline{cs}) & \text{if } p = \overline{p} \\ \mathsf{lookup}(cs[\overline{p} \leadsto (c, \overline{cs})], p) \stackrel{\mathsf{def}}{\Longrightarrow} \mathsf{lookup}(cs, p) & \text{if } p \neq \overline{p} \\ \mathsf{lookup}(\vartheta, p) \stackrel{\mathsf{def}}{\Longrightarrow} \perp \end{cases} \tag{22}$$

$$\frac{[\![e]\!]_s = \mathsf{true}}{s \xrightarrow[cs]{\mathsf{assert}(e)} s}\ \mathbf{e\_asrt} \qquad \frac{[\![e]\!]_s = v}{s \xrightarrow[cs]{i:=e} s[v^i]}\ \mathbf{e\_asgn}$$

$$\frac{[\![e]\!]_s = \mathsf{true} \quad s \xrightarrow[cs]{c_1} s'}{s \xrightarrow[cs]{\mathsf{if}\ e\ \mathsf{then}\ c_1\ \mathsf{else}\ c_2} s'}\ \mathbf{e\_ifs\_t} \qquad \frac{[\![e]\!]_s = \mathsf{false} \quad s \xrightarrow[cs]{c_2} s'}{s \xrightarrow[cs]{\mathsf{if}\ e\ \mathsf{then}\ c_1\ \mathsf{else}\ c_2} s'}\ \mathbf{e\_ifs\_f}$$

$$\frac{s \xrightarrow[cs]{c_1} s' \quad s' \xrightarrow[cs]{c_2} s''}{s \xrightarrow[cs]{c_1;\ c_2} s''}\ \mathbf{e\_seq} \qquad \frac{s \xrightarrow[cs[p \leadsto (c, cs)]]{c} s'}{s \xrightarrow[cs]{\mathsf{proc}\ p\ \equiv\ c} s'}\ \mathbf{e\_proc}$$

$$\frac{\mathsf{lookup}(cs, p) = (c, \overline{cs}) \quad s \xrightarrow[\overline{cs}[p \leadsto (c, \overline{cs})]]{c} s'}{s \xrightarrow[cs]{\mathsf{pcall}\ p} s'}\ \mathbf{e\_pcall}$$

Fig. 3. Definition of the operational semantics

5 Relating Operational Semantics to wp

The operational semantics can be related to wp by the following lemma:

Lemma 2 (Operational Demantics to wp).

$$s \xrightarrow[\vartheta]{c} s' \Rightarrow P(s') \Rightarrow \mathsf{wp}(c)(P)(s)$$

which says: if the execution of c under the empty call stack yields program store s', then for any predicate P, if P holds on s', then the predicate $\mathsf{wp}(c)(P)$ (the predicate P transformed by $\mathsf{wp}(c)$) holds on the initial program store s.

Instead of proving Lemma 2 directly, the following generalized lemma is proved first, and Lemma 2 is treated as a corollary of Lemma 3.

Lemma 3 ('Operational Semantics to wp' Generalized).

$$s \xrightarrow[cs]{c} s' \Rightarrow \tag{23a}$$

$$\mathsf{can}(cs) \Rightarrow \tag{23b}$$

$$\exists n.$$

$$(\forall P . P(s') \Rightarrow \mathsf{wpc}(\{\!|cs|\!\}^n,\ c)(P)(s)) \tag{23c}$$

where the predicate can is used to constrain the form of cs, so that it can be guaranteed that the execution $s \xrightarrow[cs]{c} s'$ is a sub-execution of some top level execution $\overline{s} \xrightarrow[\vartheta]{\overline{c}} \overline{s'}$. Therefore, can is formalized as the following inductively defined predicate:

$$\frac{}{\mathsf{can}(\vartheta)}\ \textbf{can\_nil} \qquad \frac{\mathsf{can}(cs)}{\mathsf{can}(cs[p \leadsto (c, cs)])}\ \textbf{can\_cons} \tag{24}$$

The operation $\{\!|cs|\!\}^n$ is used to transform a call stack to environment. It is defined as:

$$\begin{cases} \{\!|\vartheta|\!\}^n \stackrel{\mathsf{def}}{\Longrightarrow} \varepsilon \\ \{\!|cs[p \leadsto (c, \overline{cs})]|\!\}^n \stackrel{\mathsf{def}}{\Longrightarrow} \\ \quad \{\!|cs|\!\}^n \left[p \mapsto \boxed{\lambda \overline{pt} . \mathsf{wpc}\left(\{\!|\overline{cs}|\!\}^n\left[p \mapsto \overline{pt}\right],\ c\right)}^{\,n}\right] \end{cases} \tag{25}$$

The idea behind can and $\{\!|cs|\!\}^n$ is explained in detail in Section 5.1. The proof of Lemma 3 is given in Appendix A. The preliminary lemmas used in the proof are given in Appendix C.1.

Since $\mathsf{can}(\vartheta)$ is trivially true, by instantiating cs to ϑ, Lemma 2 follows directly from Lemma 3.

5.1 Informal Explanation of Lemma 3

In (18e), the n is determined by the number of recursive calls to p during the execution of c. n can be any natural number larger than this number. By expanding wpc in $\mathsf{wpc}(\sigma,\ \mathsf{proc}\ p\ \equiv\ c)(P)(s)$, we have:

$$\exists\, n . \boxed{\lambda \overline{pt} . \mathsf{wpc}\left(\sigma\left[p \mapsto \overline{pt}\right],\ c\right)}^{\,n} (P)(s)$$

If $n = \overline{n} + 1$, then by expanding the definition of $\boxed{\cdots}^{\,\overline{n}+1}$, we have:

$$\mathsf{wpc}\left(\sigma\left[p \mapsto \boxed{\lambda \overline{pt} . \mathsf{wpc}\left(\sigma\left[p \mapsto \overline{pt}\right],\ c\right)}^{\,\overline{n}}\right],\ c\right)(P)(s) \tag{26}$$

The natural number $\overline{n}$ is the number of recursive calls of p during the execution of c starting from program store s.

An analysis of the operational semantics in Figure 3 may reveal that: if the execution $s \xrightarrow[cs]{c} s'$ is a sub-execution of some 'top level execution' $\overline{s} \xrightarrow[\vartheta]{\overline{c}} \overline{s'}$, then cs must be of the form:

$$\vartheta[p_0 \rightsquigarrow (c_0, cs_0)][p_1 \rightsquigarrow (c_1, cs_1)] \ldots [p_m \rightsquigarrow (c_m, cs_m)] \tag{27}$$

where the $[p_i \rightsquigarrow (c_i, cs_i)]$ $(i \in \{0, \ldots, m\})$ are pushed onto cs through the execution of $\mathsf{proc}\ p_i \equiv c_i$ $(i \in \{0, \ldots, m\})$. During the execution of c, there must be a number of recursive calls on each of the procedures p_i $(i \in \{0, \ldots, m\})$. Let these numbers be $n_{p_0}, n_{p_1}, \ldots, n_{p_m}$.

In the design of $\{\!|cs|\!\}^n$, inspired by the analysis in (26), we first intended to transform cs into:

$$\varepsilon\,[p_0 \mapsto \mathsf{fd}(\{\!|cs_0|\!\}, n_{p_0}, p_0, c_0)]\,[p_1 \mapsto \mathsf{fd}(\{\!|cs_1|\!\}, n_{p_1}, p_1, c_1)] \ldots [p_m \mapsto \mathsf{fd}(\{\!|cs_m|\!\}, n_{p_m}, p_m, c_m)] \tag{28}$$

where $\mathsf{fd}(\sigma, n, p, c)$ is the abbreviation defined as:

$$\mathsf{fd}(\sigma, n, p, c) \stackrel{\mathsf{def}}{\Longrightarrow} \boxed{\lambda\,\overline{pt}.\,\mathsf{wpc}\left(\sigma\left[p \mapsto \overline{pt}\right],\ c\right)}^{\,n} \tag{29}$$

However, it is quite inconvenient to find the values for all the natural numbers $n_{p_0}, n_{p_1}, \ldots, n_{p_m}$. Fortunately, since $\mathsf{fd}(\sigma, n, p, c)$ is monotonous with respect to n, it is sufficient to deal only with the upper bound of them. It is usually easier to deal with one natural number than a group of natural numbers. Therefore, the cs is finally transformed into:

$$\varepsilon\,[p_0 \mapsto \mathsf{fd}(\{\!|cs_0|\!\}^{n_{max}}, n_{max}, p_0, c_0)]\,[p_1 \mapsto \mathsf{fd}(\{\!|cs_1|\!\}^{n_{max}}, n_{max}, p_1, c_1)] \ldots [p_m \mapsto \mathsf{fd}(\{\!|cs_m|\!\}^{n_{max}}, n_{max}, p_m, c_m)] \tag{30}$$

where n_{max} is a natural number larger than any n_{p_i} $(i \in \{0, \ldots, m\})$. The n in Lemma 3 is actually the n_{max} in (30). The equation (30) also explains the definition of $\{\!|cs|\!\}^n$ in (25).

6 Relating wp to Operational Semantics

wp can be related to the operational semantics by the following lemma:

Lemma 4 (wp to Operational Semantics).

$$\mathsf{wp}\,(c)\,(P)\,(s) \Rightarrow \exists\, s'.\left(s \xrightarrow[\vartheta]{c} s' \wedge P\,(s')\right)$$

which says: for any predicate P, if the transformation of any predicate P by $\mathsf{wp}(c)$ holds on s, then the execution of c under the empty call stack terminates and transforms program store from s to s', and P holds on s'.

Instead of proving Lemma 4 directly, the following generalized lemma is proved first and Lemma 4 follows as a corollary of Lemma 5.

Lemma 5 ('wp to Operational Semantics' Generalized).

$$
\begin{aligned}
&\mathsf{wpc}\,(\mathsf{envof}(ecs),\ c)\,(P)\,(s) \Rightarrow && \text{(31a)}\\
&\mathsf{ecan}(ecs) \Rightarrow && \text{(31b)}\\
&\exists\, s'\,.\\
&\quad (s \xrightarrow[\mathsf{csof}(ecs)]{c} s' \land && \text{(31c)}\\
&\quad P\,(s')) && \text{(31d)}
\end{aligned}
$$

where, ecs is an 'extended call stack', the type of which – $\mathcal{ECS}$ is defined as:

$$
\frac{}{\mathcal{ECS} : \mathsf{Type}}\ \textbf{ECS\_form} \qquad \frac{}{\theta : \mathcal{ECS}}\ \textbf{nil\_ecs}
$$
$$
\frac{p : \mathcal{ID} \quad c : \mathcal{C} \quad \overline{ecs} : \mathcal{ECS} \quad pt : \mathcal{PT} \quad ecs : \mathcal{ECS}}{ecs[p \hookrightarrow (c, \overline{ecs}, pt)] : \mathcal{ECS}}\ \textbf{cons\_ecs} \tag{32}
$$

where θ is the 'empty extended call stack' and $ecs[p \hookrightarrow (c, \overline{ecs}, pt)]$ is the extended call stack obtained by adding $(p, c, \overline{ecs}, pt)$ to the head of ecs. It is obvious from the definition of $\mathcal{ECS}$ that an extended call stack is a combination of call stack and environment, with each procedure name p being mapped to a triple $(c, \overline{ecs}, pt)$. Therefore, by forgetting pt in the triple, a normal call stack is obtained. This is implemented by the operation $\mathsf{csof}(ecs)$:

$$
\begin{cases}
\mathsf{csof}(\theta) \stackrel{\mathsf{def}}{\Longrightarrow} \vartheta \\
\mathsf{csof}(ecs[p \hookrightarrow (c, \overline{ecs}, pt)]) \stackrel{\mathsf{def}}{\Longrightarrow} \mathsf{csof}(ecs)[p \rightsquigarrow (c, \mathsf{csof}(\overline{ecs}))]
\end{cases} \tag{33}
$$

By forgetting $c, \overline{ecs}$ in the triple, a normal environment is obtained. This is implemented by the operation $\mathsf{envof}(ecs)$:

$$
\begin{cases}
\mathsf{envof}(\theta) \stackrel{\mathsf{def}}{\Longrightarrow} \varepsilon \\
\mathsf{envof}(ecs[p \hookrightarrow (c, \overline{ecs}, pt)]) \stackrel{\mathsf{def}}{\Longrightarrow} \mathsf{envof}(ecs)\,[p \mapsto pt]
\end{cases} \tag{34}
$$

The operation $\mathsf{lookup\_ecs}(p, ecs)$ is defined to lookup (in ecs) the triple $(c, \overline{ecs}, pt)$ mapped to p:

$$
\begin{cases}
\mathsf{lookup\_ecs}(p, ecs[\overline{p} \hookrightarrow (c, \overline{ecs}, pt)]) \stackrel{\mathsf{def}}{\Longrightarrow} (\overline{p}, c, \overline{ecs}, pt) & \text{if } p = \overline{p} \\
\mathsf{lookup\_ecs}(p, ecs[\overline{p} \hookrightarrow (c, \overline{ecs}, pt)]) \stackrel{\mathsf{def}}{\Longrightarrow} \mathsf{lookup\_ecs}(p, ecs) & \text{if } p \neq \overline{p} \\
\mathsf{lookup\_ecs}(p, \theta) \stackrel{\mathsf{def}}{\Longrightarrow} \bot
\end{cases} \tag{35}
$$

The notation $\perp$ is overloaded here, it is an undefined value in exceptional type, instead of exceptional set.

The predicate $\mathsf{ecan}(ecs)$ is defined to constrain the form of ecs, so that in each triple $(c, \overline{ecs}, pt)$, pt can be related to the $(c, \overline{ecs})$ in the following sense:

$$\mathsf{ecan}(ecs) \stackrel{\mathsf{def}}{\Longrightarrow} \mathsf{lookup\_ecs}(p, ecs) = (c, \overline{ecs}, pt) \Rightarrow \\ pt\,(P)\,(s) \Rightarrow \exists\, s' .\, \left(s \xrightarrow[\mathsf{csof}(\overline{ecs})]{\mathsf{proc}\ p\ \equiv\ c} s' \wedge P\,(s') \right) \quad (36)$$

Since $\mathsf{lookup\_ecs}(p, \theta) = \perp$, it clear that $\mathsf{ecan}(\theta)$ holds. Therefore, by instantiating ecs to θ, Lemma 4 can be proved as a corollary of Lemma 5.

The proof of Lemma 5 is given in Appendix B. The preliminary lemmas used in the proof are given in Appendix C.2.

7 Conclusion

We have given a computational embedding of wp in Coq. The definition is verified by relating it to an operational semantics. Since such a style of embedding has the benefits of both deep and shallow embedding, it can be used to verify both program transformations and concrete programs.

Laibinis and Wright [9] treats general recursion in HOL. But that is a shallow embedding and there is no relation between wp and operational semantics.

There have been some efforts to verification imperative programs using type theory. Fillitre [6] implemented an extension of Coq to generate proof obligations from annotated imperative programs. The proof of these proof obligations in Coq will guarantee the correctness of the annotated imperative programs. Since it uses shallow embedding, meta programming (such as program transformation) can not be verified in Fillitre's setting.

Kleymann [7] derived Hoare logic directly from operational semantics. Since Kleymann's treatment is a deep embedding, program transformations can be verified. However, because the operational semantics is formalized as an inductive relation (rather than using computation), verifying concrete programs in Kleymann's setting is not very convenient. This paper can be seen as an effort to overcome this problem through a computational treatment of wp. In our setting, computation mechanism can be used to simplify proof obligations when verifying concrete programs. Hoare triple can be defined as:

$$\{P_1\}\ c\ \{P_2\} \stackrel{\mathsf{def}}{\Longrightarrow} P_1\,(s) \Rightarrow \mathsf{wp}\,(c)\,(P_2)\,(s) \quad (37)$$

From this, Hoare logic rules for structure statements can be derived. For example, the rule for $\mathsf{if}\ e\ \mathsf{then}\ c_1\ \mathsf{else}\ c_2$ is:

Lemma 6 (The Proof Rule for $\mathsf{if}\ e\ \mathsf{then}\ c_1\ \mathsf{else}\ c_2$**).**

$$\{\lambda\, s.\, ([\![e]\!]_s = \mathsf{true} \wedge P_1\,(s))\}\ c_1\ \{P_2\} \Rightarrow \\ \{\lambda\, s.\, ([\![e]\!]_s = \mathsf{false} \wedge P_1\,(s))\}\ c_2\ \{P_2\} \Rightarrow \\ \{P_1\}\ \mathsf{if}\ e\ \mathsf{then}\ c_1\ \mathsf{else}\ c_2\ \{P_2\}$$

Hoare logic rules can be used to propagate proof obligations from parent statements to its sub-statements. When the propagation process reaches atomic statements (such as $i := e$), by expanding the definition of wp, the proof obligation can be simplified by the computation mechanism. Since users have access to the definition of wp all the time, they can choose whatever convenient for their purpose, either Hoare logic rules (such as Lemma 6) or the direct definition of wp.

We have gone as far as the verification of insertion sorting program. Admittedly, verification of concrete program in our setting is slightly complex than in Fillitre's. However, the ability to verify both program transformations and concrete programs makes our approach unique. The treatment of program verification in our setting will be detailed in a separate paper.

Acknowledgement

I must first thank Prof. Zhaohui Luo for a lot of technical directions into type theory. Dr. James McKinna provided many good comments on an earlier version of this paper. The anonymous referees are thanked for the helpful suggestions of improvement.

A Proof of Lemma 3

The proof is by induction on the structure of the $s \xrightarrow[cs]{c} s'$ in (23a). Some preliminary lemmas used in the proof are listed in Appendix C.1. There is one case corresponding to each execution rule. For brevity, only two of the more interesting cases are discussed here:

1. When the execution is constructed using **e_proc**, we have $c = (\mathsf{proc}\ p\ \equiv\ \overline{c})$ and $s \xrightarrow[cs[p \rightsquigarrow (\overline{c}, cs)]]{\overline{c}} s'$. From the induction hypothesis for this execution and (23b), it can be derived that:

$$\exists \overline{n} . \forall P . P(s') \Rightarrow \mathsf{wpc}\left(\{\!|cs[p \rightsquigarrow (\overline{c}, cs)]|\!\}^{\overline{n}},\ \overline{c}\right)(P)(s) \tag{38}$$

   By assigning $\overline{n}$ to n and expanding the definition of wpc, the goal (23c) becomes:

$$\forall P . P(s') \Rightarrow \exists n . \boxed{\lambda \overline{pt} . \mathsf{wpc}\left(\{\!|cs|\!\}^{\overline{n}}\left[p \mapsto \overline{pt}\right],\ \overline{c}\right)}^{\,n}(P)(s) \tag{39}$$

   By assigning $\overline{n}+1$ to n and expanding the definition of $\boxed{\cdots}^{\overline{n}+1}$, it becomes:

$$\forall P . P(s') \Rightarrow \mathsf{wpc}\left(\{\!|cs|\!\}^{\overline{n}}\left[p \mapsto \boxed{\lambda \overline{pt} . \mathsf{wpc}\left(\{\!|cs|\!\}^{\overline{n}}\left[p \mapsto \overline{pt}\right],\ \overline{c}\right)}^{\,\overline{n}}\right],\ \overline{c}\right)(P)(s) \tag{40}$$

   which is exactly (38) with the definition of $\{\!|\cdots|\!\}^{\overline{n}}$ expanded.

2. When the execution is constructed using **e_proc**, we have $c = (\mathsf{pcall}\ p)$ and

$$\mathsf{lookup}(cs, p) = (\overline{c}, \overline{cs}) \tag{41a}$$

$$s \xrightarrow[\overline{cs}[p \rightsquigarrow (\overline{c}, \overline{cs})]]{\overline{c}} s' \tag{41b}$$

By applying Lemma 7 to (23b) and (41a), it can be deduced that $\mathsf{can}(\overline{cs})$, from which, $\mathsf{can}(\overline{cs}[p \rightsquigarrow (\overline{c}, \overline{cs})])$ can be deduced. By applying induction hypothesis for (41b) to this, it can be deduced that:

$$\exists \overline{n} \,.\, \forall P \,.\, P\,(s') \Rightarrow \mathsf{wpc}\left(\{\!|\overline{cs}|\!\}^{\overline{n}} \left[p \mapsto \boxed{\lambda \overline{pt}.\, \mathsf{wpc}\,(\{\!|\overline{cs}|\!\}^{\overline{n}}\,[p \mapsto \overline{pt}]\,,\,\overline{c})}^{\,\overline{n}}\right],\,\overline{c}\right)(P)\,(s) \tag{42}$$

By assigning $\overline{n} + 1$ to n, the goal (23c) is specialized to:

$$\overline{P}\,(s') \Rightarrow \{\!|cs|\!\}^{\overline{n}+1}\,(p)\,(\overline{P})\,(s) \tag{43}$$

By applying Lemma 8 to (41a) and expanding the definition of $\boxed{\cdots}^{\overline{n}+1}$, it can be deduced that:

$$\{\!|cs|\!\}^{\overline{n}+1}\,(p) = \mathsf{wpc}\left(\{\!|\overline{cs}|\!\}^{\overline{n}+1} \left[p \mapsto \boxed{\lambda \overline{pt}.\, \mathsf{wpc}\,(\{\!|\overline{cs}|\!\}^{\overline{n}+1}\,[p \mapsto \overline{pt}]\,,\,\overline{c})}^{\,\overline{n}}\right],\,\overline{c}\right) \tag{44}$$

After rewritten using (44), the goal (43) becomes:

$$\overline{P}\,(s') \Rightarrow \mathsf{wpc}\left(\{\!|\overline{cs}|\!\}^{\overline{n}+1} \left[p \mapsto \boxed{\lambda \overline{pt}.\, \mathsf{wpc}\,(\{\!|\overline{cs}|\!\}^{\overline{n}+1}\,[p \mapsto \overline{pt}]\,,\,\overline{c})}^{\,\overline{n}}\right],\,\overline{c}\right)(\overline{P})\,(s) \tag{45}$$

By applying (42) to the $\overline{P}\,(s')$ in (45), we have:

$$\mathsf{wpc}\left(\{\!|\overline{cs}|\!\}^{\overline{n}} \left[p \mapsto \boxed{\lambda \overline{pt}.\, \mathsf{wpc}\,(\{\!|\overline{cs}|\!\}^{\overline{n}}\,[p \mapsto \overline{pt}]\,,\,\overline{c})}^{\,\overline{n}}\right],\,\overline{c}\right)(\overline{P})\,(s) \tag{46}$$

By applying Lemma 9 to the fact that $\overline{n} \le \overline{n}+1$ and (23b), it can be deduced that $\{\!|\overline{cs}|\!\}^{\overline{n}} \preccurlyeq_{\sigma} \{\!|\overline{cs}|\!\}^{\overline{n}+1}$. By applying Lemma 10 to this, it can be deduced that:

$$\boxed{\lambda \overline{pt}.\, \mathsf{wpc}\,(\{\!|\overline{cs}|\!\}^{\overline{n}}\,[p \mapsto \overline{pt}]\,,\,\overline{c})}^{\,\overline{n}} \preccurlyeq_{pt} \boxed{\lambda \overline{pt}.\, \mathsf{wpc}\,(\{\!|\overline{cs}|\!\}^{\overline{n}+1}\,[p \mapsto \overline{pt}]\,,\,\overline{c})}^{\,\overline{n}} \tag{47}$$

By combining $\{\!|\overline{cs}|\!\}^{\overline{n}} \preccurlyeq_{\sigma} \{\!|\overline{cs}|\!\}^{\overline{n}+1}$ and (47), it can be deduced that:

$$\begin{aligned} &\{\!|\overline{cs}|\!\}^{\overline{n}} \left[p \mapsto \boxed{\lambda \overline{pt}.\, \mathsf{wpc}\,(\{\!|\overline{cs}|\!\}^{\overline{n}}\,[p \mapsto \overline{pt}]\,,\,\overline{c})}^{\,\overline{n}}\right] \preccurlyeq_{\sigma} \\ &\qquad \{\!|\overline{cs}|\!\}^{\overline{n}+1} \left[p \mapsto \boxed{\lambda \overline{pt}.\, \mathsf{wpc}\,(\{\!|\overline{cs}|\!\}^{\overline{n}+1}\,[p \mapsto \overline{pt}]\,,\,\overline{c})}^{\,\overline{n}}\right] \end{aligned} \tag{48}$$

By applying (20a) to this, it can be deduced that:

$$\begin{aligned} &\mathsf{wpc}\left(\{\!|\overline{cs}|\!\}^{\overline{n}} \left[p \mapsto \boxed{\lambda \overline{pt}.\, \mathsf{wpc}\,(\{\!|\overline{cs}|\!\}^{\overline{n}}\,[p \mapsto \overline{pt}]\,,\,\overline{c})}^{\,\overline{n}}\right],\,\overline{c}\right) \preccurlyeq_{pt} \\ &\qquad \mathsf{wpc}\left(\{\!|\overline{cs}|\!\}^{\overline{n}+1} \left[p \mapsto \boxed{\lambda \overline{pt}.\, \mathsf{wpc}\,(\{\!|\overline{cs}|\!\}^{\overline{n}+1}\,[p \mapsto \overline{pt}]\,,\,\overline{c})}^{\,\overline{n}}\right],\,\overline{c}\right) \end{aligned} \tag{49}$$

From this and (46), the goal (45) can be proved.

B Proof of Lemma 5

The proof is by induction on the structure of c. Some preliminary lemmas used in the proof are listed in Appendix C.2. There is one case for each type of command. For brevity, only two of the more interesting cases are discussed here:

1. When $c = (\mathsf{proc}\ p\ \equiv\ \overline{c})$, after expanding the definition of wpc, the premise (31a) becomes:

$$\exists\,\overline{n}\,.\,\forall P\,.\,\boxed{\lambda\,\overline{pt}\,.\,\mathsf{wpc}\left(\mathsf{envof}(ecs)\left[p\mapsto\overline{pt}\right],\ \overline{c}\right)}^{\overline{n}}(P)\,(s) \tag{50}$$

A nested induction on $\overline{n}$ is used to prove the goal, which gives rise to two cases:

(a) When $\overline{n} = 0$, this case can be refuted. Since

$$\boxed{\lambda\,\overline{pt}\,.\,\mathsf{wpc}\left(\mathsf{envof}(ecs)\left[p\mapsto\overline{pt}\right],\ \overline{c}\right)}^{\overline{n}}$$

reduces to $\lambda\,P\,.\,\mathsf{F}$, it can not hold on P and s. And this is in contradiction with (50).

(b) When $\overline{n} = \underline{n}{+}1$, after expanding the definition of $\boxed{\cdots}^{\underline{n}+1}$, (50) becomes:

$$\mathsf{wpc}\left(\mathsf{envof}(ecs)\left[p\mapsto\boxed{\lambda\,\overline{pt}\,.\,\mathsf{wpc}\left(\mathsf{envof}(ecs)\left[p\mapsto\overline{pt}\right],\ c\right)}^{\underline{n}}\right],\ \overline{c}\right)(P)\,(s) \tag{51}$$

which is exactly

$$\mathsf{wpc}\left(\mathsf{envof}(ecs[p\hookrightarrow(\overline{c},ecs,\boxed{\lambda\,\overline{pt}\,.\,\mathsf{wpc}\left(\mathsf{envof}(ecs)\left[p\mapsto\overline{pt}\right],\ \overline{c}\right)}^{\underline{n}})]),\ \overline{c}\right)(P)\,(s) \tag{52}$$

with the definition of $\mathsf{envof}(\cdots)$ expanded. From the nested induction hypothesis for $\underline{n}$, it can be proved that:

$$\mathsf{ecan}(ecs[p\hookrightarrow(\overline{c},ecs,\boxed{\lambda\,\overline{pt}\,.\,\mathsf{wpc}\left(\mathsf{envof}(ecs)\left[p\mapsto\overline{pt}\right],\ \overline{c}\right)}^{\underline{n}})])$$

By applying the main induction hypothesis to this and (52), after expanding the definition of csof, it can be deduced that:

$$\exists\,\underline{s}\,.\,s\xrightarrow[\mathsf{csof}(ecs)[p\rightsquigarrow(\overline{c},\mathsf{csof}(ecs))]]{\overline{c}}\underline{s}\ \wedge P\,(\underline{s}) \tag{53}$$

By assigning $\underline{s}$ to s', the goal (31d) can be proved directly from the $P\,(\underline{s})$ in (53). Also the goal (31c) can be proved by applying **e_proc** to the $s\xrightarrow[\mathsf{csof}(ecs)[p\rightsquigarrow(\overline{c},\mathsf{csof}(ecs))]]{\overline{c}}\underline{s}$ in (53).

2. When $c = (\mathsf{pcall}\ p)$, after expanding the definition of wpc, the premise (31a) becomes $\mathsf{envof}(ecs)\,(p)\,(P)\,(s)$. By applying Lemma 11 to this, we have:

$$\exists\,\overline{c},\overline{ecs}\,.\,\mathsf{lookup\_ecs}(p,ecs) = (\overline{c},\overline{ecs},\mathsf{envof}(ecs)\,(p)) \tag{54}$$

After expanding the definition of ecan in (31b), it can be applied to (54) and the $\mathsf{envof}(ecs)\,(p)\,(P)\,(s)$ at the beginning to yield:

$$\exists \underline{s}\,.\, s \xrightarrow[\mathsf{csof}(ecs)]{\mathsf{proc}\ p\ \equiv\ \overline{c}} \underline{s} \ \wedge\ P\,(\underline{s}) \tag{55}$$

By assigning $\underline{s}$ to s', the goal (31d) can be proved directly from the $P\,(\underline{s})$ in (55). By inversion on the $s \xrightarrow[\mathsf{csof}(ecs)]{\mathsf{proc}\ p\ \equiv\ \overline{c}} \underline{s}$ in (55), we have: $s \xrightarrow[\mathsf{csof}(ecs)[p\rightsquigarrow(\overline{c},\mathsf{csof}(ecs))]]{\overline{c}} \underline{s}$. By applying Lemma 12 to (54), we have: $\mathsf{lookup}(p, \mathsf{csof}(ecs)) = (\overline{c}, \mathsf{csof}(\overline{ecs}))$. Therefore, the goal (31c) can be proved by applying **e_pcall** to these two results.

C Preliminary Lemmas

C.1 Lemmas Used in the Proof of Lemma 3

Lemma 7. $\mathsf{can}(cs) \Rightarrow \mathsf{lookup}(p, cs) = (\overline{c}, \overline{cs}) \Rightarrow \mathsf{can}(\overline{cs})$

Lemma 8. $\mathsf{lookup}(p, cs) = (\overline{c}, \overline{cs}) \Rightarrow \{\!|cs|\!\}^n\,(p) = \boxed{\lambda\,\overline{pt}.\,\mathsf{wpc}\left(\{\!|\overline{cs}|\!\}^n\left[p \mapsto \overline{pt}\right],\ \overline{c}\right)}^{\,n}$

Lemma 9. $n_1 \le n_2 \Rightarrow \mathsf{can}(cs) \Rightarrow \{\!|cs|\!\}^{n_1} \preccurlyeq_\sigma \{\!|cs|\!\}^{n_2}$

Lemma 10.

$$\sigma_1 \preccurlyeq_\sigma \sigma_2 \Rightarrow \mathrm{mono}(\sigma_1) \Rightarrow \mathrm{mono}(\sigma_2) \Rightarrow \\ \boxed{\lambda\,\overline{pt}.\,\mathsf{wpc}\left(\sigma_1\left[p \mapsto \overline{pt}\right],\ c\right)}^{\,n} \preccurlyeq_{pt} \boxed{\lambda\,\overline{pt}.\,\mathsf{wpc}\left(\sigma_2\left[p \mapsto \overline{pt}\right],\ c\right)}^{\,n}$$

C.2 Lemmas Used in the Proof of Lemma 5

Lemma 11.

$$\mathsf{envof}(ecs)\,(p)\,(P)\,(s) \Rightarrow \exists\,\overline{c}, \overline{ecs}\,. \\ \mathsf{lookup\_ecs}(p, ecs) = (\overline{c}, \overline{ecs}, \mathsf{envof}(ecs)\,(p))$$

Lemma 12. $\mathsf{lookup\_ecs}(p, ecs) = (\overline{c}, \overline{ecs}, \overline{pt}) \Rightarrow \mathsf{lookup}(p, \mathsf{csof}(ecs)) = (\overline{c}, \mathsf{csof}(\overline{ecs}))$

References

1. R. J. R. Back. A calculus of refinements for program derivations. *Acta Informatica*, 25(6):593–624, August 1988.
2. B. Barras, S. Boutin, C. Cornes, J. Courant, J.C. Filliatre, E. Giménez, H. Herbelin, G. Huet, C. Mu noz, C. Murthy, C. Parent, C. Paulin, A. Saïbi, and B. Werner. The Coq Proof Assistant Reference Manual – Version V6.1. Technical Report 0203, INRIA, August 1997.

3. D. Carrington, I. Hayes, R. Nickson, G. Watson, and J. Welsh. Refinement in Ergo. Technical report 94-44, Software Verification Research Centre, School of Information Technology, The University of Queensland, Brisbane 4072. Australia, November 1994.
4. T. Coquand and G. Huet. The Calculus of Constructions. *Information and Computation*, 76:96–120, 1988.
5. E. W. Dijkstra. *A Discipline of Programming.* Prentice-Hall, 1976.
6. J.-C. Filliâtre. Proof of Imperative Programs in Type Theory. In *International Workshop, TYPES '98, Kloster Irsee, Germany*, volume 1657 of *Lecture Notes in Computer Science.* Springer-Verlag, March 1998.
7. T. Kleymann. *Hoare Logic and VDM: Machine-Checked Soundness and Completeness Proofs.* Ph.D. thesis, University of Edinburgh, 1998.
8. J. Knappmann. *A PVS based tool for developing programs in the refinement calculus.* Marster's Thesis, Christian-Albrechts-University, 1996.
9. L. Laibinis and J. von Wright. Functional procedures in higher-order logic. Technical Report TUCS-TR-252, Turku Centre for Computer Science, Finland, March 15, 1999.
10. L. Lindqvist. *A formalization of Dijkstra's predicate transformer* wp *in Martin-Lof type theory.* Master's Thesis, Linkopin University, Sweden, 1997.
11. Z. Luo. *Computation and Reasoning: A Type Theory for Computer Science.* Number 11 in International Series of Monographs on Computer Science. Oxford University Press, 1994.
12. C. Morgan. The specification statement. *ACM Transactions on Programming Languages and Systems*, 10(3):403–419, July 1988.
13. J. M. Morris. A theoretical basis for stepwise refinement and the programming calculus. *Science of Computer Programming*, 9(3):287–306, December 1987.
14. T. Nipkow. Winskel is (almost) right: Towards a mechanized semantics textbook. In V. Chandru and V. Vinay, editors, *Proceedings of the Conference on Foundations of Software Technology and Theoretical Computer Science*, pages 180–192. Springer-Verlag LNCS 1180, 1996.
15. B. Nordström, K. Peterson, and J. M. Smith. *Programming in Martin-Lof's Type Theory*, volume 7 of *International Series of Monographs on Computer Science.* Oxford University Press, New York, NY, 1990.
16. R. J. R. Back and J. von Wright. Refinement concepts formalized in higher-order logic. Reports on Computer Science & Mathematics Series A—85, Institutionen för Informationsbehandling & Mathematiska Institutet, Åbo Akademi, Lemminkäinengatan 14, SF-20520 Turku, Finland, September 1989.
17. M. Staples. *A Mechanised Theory of Refinement.* Ph.D. Dissertation, Computer Laboratory, University of Cambridge, 1998.
18. M. Staples. Program transformations and refinements in HOL. In Y. Bertot G. Dowek, C. Paulin, editor, *TPHOLs: The 12th International Conference on Theorem Proving in Higher-Order Logics. LNCS, Springer-Verlag.*, 1999.
19. J. von Wright and K. Sere. Program transformations and refinements in HOL. In Myla Archer, Jennifer J. Joyce, Karl N. Levitt, and Phillip J. Windley, editors, *Proceedigns of the International Workshop on the HOL Theorem Proving System and its Applications*, pages 231–241, Los Alamitos, CA, USA, August 1992. IEEE Computer Society Press.

Author Index

Lecture Notes in Computer Science

For information about Vols. 1–2331
please contact your bookseller or Springer-Verlag

Vol. 2332: L. Knudsen (Ed.), Advances in Cryptology – EUROCRYPT 2002. Proceedings, 2002. XII, 547 pages. 2002.

Vol. 2333: J.-J.Ch. Meyer, M. Tambe (Eds.), Intelligent Agents VIII. Revised Papers, 2001. XI, 461 pages. 2001. (Subseries LNAI).

Vol. 2334: G. Carle, M. Zitterbart (Eds.), Protocols for High Speed Networks. Proceedings, 2002. X, 267 pages. 2002.

Vol. 2335: M. Butler, L. Petre, K. Sere (Eds.), Integrated Formal Methods. Proceedings, 2002. X, 401 pages. 2002.

Vol. 2336: M.-S. Chen, P.S. Yu, B. Liu (Eds.), Advances in Knowledge Discovery and Data Mining. Proceedings, 2002. XIII, 568 pages. 2002. (Subseries LNAI).

Vol. 2337: W.J. Cook, A.S. Schulz (Eds.), Integer Programming and Combinatorial Optimization. Proceedings, 2002. XI, 487 pages. 2002.

Vol. 2338: R. Cohen, B. Spencer (Eds.), Advances in Artificial Intelligence. Proceedings, 2002. XII, 373 pages. 2002. (Subseries LNAI).

Vol. 2340: N. Jonoska, N.C. Seeman (Eds.), DNA Computing. Proceedings, 2001. XI, 392 pages. 2002.

Vol. 2342: I. Horrocks, J. Hendler (Eds.), The Semantic Web – ISCW 2002. Proceedings, 2002. XVI, 476 pages. 2002.

Vol. 2345: E. Gregori, M. Conti, A.T. Campbell, G. Omidyar, M. Zukerman (Eds.), NETWORKING 2002. Proceedings, 2002. XXVI, 1256 pages. 2002.

Vol. 2346: H. Unger, T. Böhme, A. Mikler (Eds.), Innovative Internet Computing Systems. Proceedings, 2002. VIII, 251 pages. 2002.

Vol. 2347: P. De Bra, P. Brusilovsky, R. Conejo (Eds.), Adaptive Hypermedia and Adaptive Web-Based Systems. Proceedings, 2002. XV, 615 pages. 2002.

Vol. 2348: A. Banks Pidduck, J. Mylopoulos, C.C. Woo, M. Tamer Ozsu (Eds.), Advanced Information Systems Engineering. Proceedings, 2002. XIV, 799 pages. 2002.

Vol. 2349: J. Kontio, R. Conradi (Eds.), Software Quality – ECSQ 2002. Proceedings, 2002. XIV, 363 pages. 2002.

Vol. 2350: A. Heyden, G. Sparr, M. Nielsen, P. Johansen (Eds.), Computer Vision – ECCV 2002. Proceedings, Part I. XXVIII, 817 pages. 2002.

Vol. 2351: A. Heyden, G. Sparr, M. Nielsen, P. Johansen (Eds.), Computer Vision – ECCV 2002. Proceedings, Part II. XXVIII, 903 pages. 2002.

Vol. 2352: A. Heyden, G. Sparr, M. Nielsen, P. Johansen (Eds.), Computer Vision – ECCV 2002. Proceedings, Part III. XXVIII, 919 pages. 2002.

Vol. 2353: A. Heyden, G. Sparr, M. Nielsen, P. Johansen (Eds.), Computer Vision – ECCV 2002. Proceedings, Part IV. XXVIII, 841 pages. 2002.

Vol. 2355: M. Matsui (Ed.), Fast Software Encryption. Proceedings, 2001. VIII, 169 pages. 2001.

Vol. 2356: R. Kohavi, B.M. Masand, M. Spiliopoulou, J. Srivastava (Eds.), WEBKDD 2002 – Mining Log Data Across All Customers Touch Points. Proceedings, 2001. XI, 167 pages. 2002. (Subseries LNAI).

Vol. 2358: T. Hendtlass, M. Ali (Eds.), Developments in Applied Artificial Intelligence. Proceedings, 2002 XIII, 833 pages. 2002. (Subseries LNAI).

Vol. 2359: M. Tistarelli, J. Bigun, A.K. Jain (Eds.), Biometric Authentication. Proceedings, 2002. X, 197 pages. 2002.

Vol. 2360: J. Esparza, C. Lakos (Eds.), Application and Theory of Petri Nets 2002. Proceedings, 2002. X, 445 pages. 2002.

Vol. 2361: J. Blieberger, A. Strohmeier (Eds.), Reliable Software Technologies – Ada-Europe 2002. Proceedings, 2002 XIII, 367 pages. 2002.

Vol. 2362: M. Tanabe, P. van den Besselaar, T. Ishida (Eds.), Digital Cities II. Proceedings, 2001. XI, 399 pages. 2002.

Vol. 2363: S.A. Cerri, G. Gouardères, F. Paraguaçu (Eds.), Intelligent Tutoring Systems. Proceedings, 2002. XXVIII, 1016 pages. 2002.

Vol. 2364: F. Roli, J. Kittler (Eds.), Multiple Classifier Systems. Proceedings, 2002. XI, 337 pages. 2002.

Vol. 2365: J. Daemen, V. Rijmen (Eds.), Fast Software Encryption. Proceedings, 2002. XI, 277 pages. 2002.

Vol. 2366: M.-S. Hacid, Z.W. Raś, D.A. Zighed, Y. Kodratoff (Eds.), Foundations of Intelligent Systems. Proceedings, 2002. XII, 614 pages. 2002. (Subseries LNAI).

Vol. 2367: J. Fagerholm, J. Haataja, J. Järvinen, M. Lyly. P. Råback, V. Savolainen (Eds.), Applied Parallel Computing. Proceedings, 2002. XIV, 612 pages. 2002.

Vol. 2368: M. Penttonen, E. Meineche Schmidt (Eds.), Algorithm Theory – SWAT 2002. Proceedings, 2002. XIV, 450 pages. 2002.

Vol. 2369: C. Fieker, D.R. Kohel (Eds.), Algorithmic Number Theory. Proceedings, 2002. IX, 517 pages. 2002.

Vol. 2370: J. Bishop (Ed.), Component Deployment. Proceedings, 2002. XII, 269 pages. 2002.

Vol. 2371: S. Koenig, R.C. Holte (Eds.), Abstraction, Reformulation, and Approximation. Proceedings, 2002. XI, 349 pages. 2002. (Subseries LNAI).

Vol. 2372: A. Pettorossi (Ed.), Logic Based Program Synthesis and Transformation. Proceedings, 2001. VIII, 267 pages. 2002.

Vol. 2373: A. Apostolico, M. Takeda (Eds.), Combinatorial Pattern Matching. Proceedings, 2002. VIII, 289 pages. 2002.

Vol. 2374: B. Magnusson (Ed.), ECOOP 2002 – Object-Oriented Programming. XI, 637 pages. 2002.

Vol. 2375: J. Kivinen, R.H. Sloan (Eds.), Computational Learning Theory. Proceedings, 2002. XI, 397 pages. 2002. (Subseries LNAI).

Vol. 2377: A. Birk, S. Coradeschi, T. Satoshi (Eds.), RoboCup 2001: Robot Soccer World Cup V. XIX, 763 pages. 2002. (Subseries LNAI).

Vol. 2378: S. Tison (Ed.), Rewriting Techniques and Applications. Proceedings, 2002. XI, 387 pages. 2002.

Vol. 2379: G.J. Chastek (Ed.), Software Product Lines. Proceedings, 2002. X, 399 pages. 2002.

Vol. 2380: P. Widmayer, F. Triguero, R. Morales, M. Hennessy, S. Eidenbenz, R. Conejo (Eds.), Automata, Languages and Programming. Proceedings, 2002. XXI, 1069 pages. 2002.

Vol. 2381: U. Egly, C.G. Fermüller (Eds.), Automated Reasoning with Analytic Tableaux and Related Methods. Proceedings, 2002. X, 341 pages. 2002 .(Subseries LNAI).

Vol. 2382: A. Halevy, A. Gal (Eds.), Next Generation Information Technologies and Systems. Proceedings, 2002. VIII, 169 pages. 2002.

Vol. 2383: M.S. Lew, N. Sebe, J.P. Eakins (Eds.), Image and Video Retrieval. Proceedings, 2002. XII, 388 pages. 2002.

Vol. 2384: L. Batten, J. Seberry (Eds.), Information Security and Privacy. Proceedings, 2002. XII, 514 pages. 2002.

Vol. 2385: J. Calmet, B. Benhamou, O. Caprotti, L. Henocque, V. Sorge (Eds.), Artificial Intelligence, Automated Reasoning, and Symbolic Computation. Proceedings, 2002. XI, 343 pages. 2002. (Subseries LNAI).

Vol. 2386: E.A. Boiten, B. Möller (Eds.), Mathematics of Program Construction. Proceedings, 2002. X, 263 pages. 2002.

Vol. 2387: O.H. Ibarra, L. Zhang (Eds.), Computing and Combinatorics. Proceedings, 2002. XIII, 606 pages. 2002.

Vol. 2388: S.-W. Lee, A. Verri (Eds.), Pattern Recognition with Support Vector Machines. Proceedings, 2002. XI, 420 pages. 2002.

Vol. 2389: E. Ranchhod, N.J. Mamede (Eds.), Advances in Natural Language Processing. Proceedings, 2002. XII, 275 pages. 2002. (Subseries LNAI).

Vol. 2391: L.-H. Eriksson, P.A. Lindsay (Eds.), FME 2002: Formal Methods – Getting IT Right. Proceedings, 2002. XI, 625 pages. 2002.

Vol. 2392: A. Voronkov (Ed.), Automated Deduction – CADE-18. Proceedings, 2002. XII, 534 pages. 2002. (Subseries LNAI).

Vol. 2393: U. Priss, D. Corbett, G. Angelova (Eds.), Conceptual Structures: Integration and Interfaces. Proceedings, 2002. XI, 397 pages. 2002. (Subseries LNAI).

Vol. 2396: T. Caelli, A. Amin, R.P.W. Duin, M. Kamel, D. de Ridder (Eds.), Structural, Syntactic, and Statistical Pattern Recognition. Proceedings, 2002. XVI, 863 pages. 2002.

Vol. 2398: K. Miesenberger, J. Klaus, W. Zagler (Eds.), Computers Helping People with Special Needs. Proceedings, 2002. XXII, 794 pages. 2002.

Vol. 2399: H. Hermanns, R. Segala (Eds.), Process Algebra and Probabilistic Methods. Proceedings, 2002. X, 215 pages. 2002.

Vol. 2401: P.J. Stuckey (Ed.), Logic Programming. Proceedings, 2002. XI, 486 pages. 2002.

Vol. 2402: W. Chang (Ed.), Advanced Internet Services and Applications. Proceedings, 2002. XI, 307 pages. 2002.

Vol. 2403: Mark d'Inverno, M. Luck, M. Fisher, C. Preist (Eds.), Foundations and Applications of Multi-Agent Systems. Proceedings, 1996-2000. X, 261 pages. 2002. (Subseries LNAI).

Vol. 2404: E. Brinksma, K.G. Larsen (Eds.), Computer Aided Verification. Proceedings, 2002. XIII, 626 pages. 2002.

Vol. 2405: B. Eaglestone, S. North, A. Poulovassilis (Eds.), Advances in Databases. Proceedings, 2002. XII, 199 pages. 2002.

Vol. 2406: C. Peters, M. Braschler, J. Gonzalo, M. Kluck (Eds.), Evaluation of Cross-Language Information Retrieval Systems. Proceedings, 2001. X, 601 pages. 2002.

Vol. 2407: A.C. Kakas, F. Sadri (Eds.), Computational Logic: Logic Programming and Beyond. Part I. XII, 678 pages. 2002. (Subseries LNAI).

Vol. 2408: A.C. Kakas, F. Sadri (Eds.), Computational Logic: Logic Programming and Beyond. Part II. XII, 628 pages. 2002. (Subseries LNAI).

Vol. 2409: D.M. Mount, C. Stein (Eds.), Algorithm Engineering and Experiments. Proceedings, 2002. VIII, 207 pages. 2002.

Vol. 2410: V.A. Carreño, C. Muñoz, S. Tahar (Eds.), Theorem Proving in Higher Order Logics. Proceedings, 2002. X, 349 pages. 2002.

Vol. 2412: H. Yin, N. Allinson, R. Freeman, J. Keane, S. Hubbard (Eds.), Intelligent Data Engineering and Automated Learning – IDEAL 2002. Proceedings, 2002. XV, 597 pages. 2002.

Vol. 2413: K. Kuwabara, J. Lee (Eds.), Intelligent Agents and Multi-Agent Systems. Proceedings, 2002. X, 221 pages. 2002. (Subseries LNAI).

Vol. 2418: D. Wells, L. Williams (Eds.), Extreme Programming and Agile Methods – XP/Agile Universe 2002. Proceedings, 2002. XII, 292 pages. 2002.

Vol. 2419: X. Meng, J. Su, Y. Wang (Eds.), Advances in Web-Age Information Management. Proceedings, 2002. XV, 446 pages. 2002.

Vol. 2421: L. Brim, P. Jancar, M. Krstinsky, A. Kucera (Eds.), CONCUR 2002 – Concurrency Theory. Proceedings, 2002. XII, 611 pages. 2002.y

Vol. 2430: T. Elomaa, H. Mannila, H. Toivonen (Eds.), Machine Learning: ECML 2002. Proceedings, 2002. XIII, 532 pages. 2002. (Subseries LNAI).

Vol. 2431: T. Elomaa, H. Mannila, H. Toivonen (Eds.), Principles of Data Mining and Knowledge Discovery. Proceedings, 2002. XIV, 514 pages. 2002. (Subseries LNAI).

Vol. 2442: M. Yung (Ed.), Advances in Cryptology – CRYPTO 2002. Proceedings, 2002. XIV, 627 pages. 2002.